Semiconductor Devices and Technologies for Future Ultra Low Power Electronics

Semiconductor Devices and Technologies for Future Ultra Low Power Electronics

Edited by

D. Nirmal
J. Ajayan
Patrick J. Fay

CRC Press
Taylor & Francis Group
Boca Raton London New York

CRC Press is an imprint of the
Taylor & Francis Group, an **informa** business

First edition published 2022
by CRC Press
6000 Broken Sound Parkway NW, Suite 300, Boca Raton, FL 33487–2742

and by CRC Press
2 Park Square, Milton Park, Abingdon, Oxon, OX14 4RN

CRC Press is an imprint of Taylor & Francis Group, LLC

ISBN: 978-1-032-06161-0 (hbk)
ISBN: 978-1-032-06162-7 (pbk)
ISBN: 978-1-003-20098-7 (ebk)

DOI: 10.1201/9781003200987

Typeset in Times
by Apex CoVantage, LLC

Contents

Preface

What's the future of the semiconductor industry? In an era in which everything is becoming increasingly computerized, computer circuits power practically everything. Modern electronics depends on semiconductor chips and transistors on silicon-based integrated circuits, which have the power to switch electronic signals on and off. This is because silicon is a very cheap material compared with other elemental and compound semiconductors and is widely available in nature in the form of sand. The modern digital age that semiconductors helped create has led to the advent of the Internet of Things (IoT), autonomous vehicles, artificial intelligence (AI), and 5G communication systems. Moore's law has been the driving force behind the success of the semiconductor industry in the past decades. In 1956, Gordon Moore, the cofounder of Intel, predicted that the number of transistors on a semiconductor chip would double each year, while costs would halve. With the progress of nano-electronic circuit miniaturization, lowering power consumption has become the important roadblock due to the fundamental thermal limit of the subthreshold swing for complementary metal oxide semiconductor (CMOS) technologies. "Silicon is reaching the fundamental limit of its performance in a growing number of applications that require increased speed, reduced latency, and light detection". However, it is still too early to think of a successor to the traditional silicon material. "That suggests silicon will be completely replaced, which is unlikely to happen anytime soon, and may well never happen". Researchers are looking for new ways of achieving enhanced performance in computers and also in communication systems while using less power. It is high time to think beyond traditional bulk silicon CMOS-based integrated circuit chips for high-speed and low power applications. Innovations are essential for driving Moore's law over the next decade. Current Intel processors are based on FinFET technology. FinFETs still have plenty of life; at some point in the near future, the semiconductor industry will switch over to a new type of transistor technology like gate-all-around (GAA) FETs, carbon nanotube-based FETs, III-V compound semiconductor-based QWFETs, graphene transistors, atomic transistors, light controlled transistors, and so on. Ultra low power integrated circuits are important for IoT applications and energy harvesting systems. Devices with a higher I_{ON}/I_{OFF} ratio and steep slope switching are essential to achieve the requirement of ultra low power systems. Negative capacitance field-effect transistors (NCFETs) and tunnel field-effect transistors (TFETs) have emerged as promising semiconductor transistors for future ultra low power applications owing to their better subthreshold swing and scalability. This book deals with the design considerations and performance analysis of NCFETs, FinFETs, TFETs, and flexible transistors. This book also highlights the significance of 2-D materials and 2-D material-based transistors for future ultra low power applications.

Editors

D. Nirmal obtained his BE degree in electrical and electronics engineering from Anna University, and MTech in VLSI design from Karunya Institute of Technology and Sciences. He completed his PhD in information and communication engineering from Anna University in 2013. He is an associate professor and head in the Department of Electronics and Communication Engineering at Karunya. His research interests include nanoelectronics, 1-D/2-D materials, carbon nanotubes, GaN technology, device and circuit simulation—GSL, sensors, and nanoscale device design and modeling. He is a founding chair of the IEEE EDS Coimbatore chapter. He has completed a project funded by DRDO, Department of Space and Ministry of Electronics and Information Technology, Government of India and is currently handling a funded project from ISRO, India. He has published more than 100 research papers. He is also a senior IEEE member. He has received several awards to his credit and has delivered many lectures in national and international conferences and programs.

J. Ajayan received his BTech degree in electronics and communication engineering from Kerala University in 2009; and MTech and PhD degrees in electronics and communication engineering from Karunya University, Coimbatore, India, in 2012 and 2017, respectively. He is an associate professor in the Department of Electronics and Communication Engineering at SR University, Telangana, India. He has published more than 100 research articles in various journals (Elsevier, Taylor & Francis, Springer, IOP Science, Wiley) and international conferences. He has published one book, more than ten book chapters, and two patents. He is a reviewer of more than 30 journals from various publishers, including IEEE, IET, Elsevier, and Springer journals. He was a guest editor for the Special Issue on P2P Computing for Beyond 5G Network (B5G) and Internet-of-Everything (IoE) by *Peer-to-Peer Networking and Applications*, Springer and the Special Issue on Energy Harvesting Devices, Circuits and Systems for Internet of Things by *Microelectronics Journal*, Elsevier. He has served as a member of the technical advisory/reviewer committee on more than ten conferences. His areas of interest are microelectronics, semiconductor devices, nanotechnology, RF integrated circuits, and photovoltaics.

Patrick J. Fay (Fellow, IEEE), received a doctorate in electrical engineering from the University of Illinois at Urbana—Champaign, Illinois, USA, in 1996. He is a professor with the Department of Electrical Engineering, University of Notre Dame, Notre Dame, Indiana, USA. He established the High Speed Circuits and Devices Laboratory at Notre Dame and oversaw the design, construction, and commissioning of the 9,000-ft^2 class 100 cleanroom housed in Stinson-Remick Hall at Notre Dame. He has served as the director of this facility since 2003. He has published 11 book chapters and more than 300 articles in scientific journals and conference proceedings. His research focuses on the design, fabrication, and characterization of microwave and millimeter-wave electronic devices and circuits, the use of micromachining techniques for the fabrication of RF through submillimeter-wave packaging, and power devices. Dr. Fay is a distinguished lecturer of the IEEE Electron Devices Society.

Contributors

J. Ajayan
Department of Electronics and Communication Engineering
SR University
Warangal, India

Ganesan Anushya
Department of Physics
S.A.V. Sahaya Thai Arts and Science (Women) College
Tirunelveli, India

Rock-Hyun Baek
Department of Electrical Engineering
Pohang University of Science and Technology (POSTECH)
Korea

Arighna Basak
Department of Electronics and Communication Engineering
Brainware University
West Bengal, India

Michael Benjamin
Department of Science and Humanities
Stella Mary's College of Engineering
Kanyakumari District, India

Ankur Beohar
School of Electrical and Electronics Engineering (SEEE)
VIT Bhopal University
Bhopal, India

Sandip Bhattacharya
Department of ECE
SR University
Warangal, India

Kalyan Biswas
Department of Electronics and Communication Engineering
MCKV Institute of Engineering
West Bengal, India

Arpan Deyasi
Department of Electronics and Communication Engineering
RCC Institute of Information Technology
West Bengal, India

Nilesh Kumar Jaiswal
Department of Micro and Nanoelectronics School of Electronics
Vellore Institute of Technology
Vellore, India

Harsupreet Kaur
Department of Electronic Science
University of Delhi South Campus
New Delhi, India

Ribu Mathew
School of Electrical and Electronics Engineering (SEEE)
VIT Bhopal University
Bhopal, India

D. Nirmal
Department of Electronics and Communication Engineering
Karunya Institute of Technology and Science
Coimbatore, India

Kumar Prasannajit Pradhan
Indian Institute of Information Technology, Design and Manufacturing
Kancheepuram, Chennai, India

J. Charles Pravin
Centre for VLSI Design, Department of Electronics and Communication Engineering
Kalasalingam Academy of Research and Education
Krishnankoil, India

Shiromani Balmukund Rahi
Department of Electrical Engineering
Indian Institute of Technology
Kanpur, India

Rasu Ramachandran
Department of Chemistry
The Madura College
Madurai, India

V. N. Ramakrishnan
Department of Micro and Nanoelectronics
School of Electronics
VIT
Vellore, India

K. Ramkumar
Department of Micro and Nanoelectronics
VIT
Vellore, India

Raj Sarika
Department of Chemistry
Fatima College (Autonomous)
Madurai, India

Angsuman Sarkar
Department of Electronics and Communication Engineering
Kalyani Govt. Engineering College
West Bengal, India

Singh Rohitkumar Shailendra
Research Institute of Electronics
Shizuoka University
Hamamatsu, Japan

R. Sridevi
Department of ECE
M. Kumarasamy College of Engineering
Karur, India

Jay Prakash Srivastava
Department of Mechanical Engineering
SR University
Warangal, India

Shubham Tayal
Department of ECE
SR University
Warangal, India

Abhishek Kumar Upadhyay
Institute for Fundamental of Electrical Engineering and Electronics
Technische Universitate
Dresden, Germany

Jun-Sik Yoon
Department of Electrical Engineering
Pohang University of Science and Technology (POSTECH)
Korea

1 An Introduction to Nanoscale CMOS Technology Transistors

A Future Perspective

Kumar Prasannajit Pradhan

CONTENTS

1.1 INTRODUCTION: BACKGROUND AND DRIVING FORCES

There is a great demand for electronic gadgets as space, consumer, and defense industries, and the most recent evolution of artificial intelligence and Internet of

DOI: 10.1201/9781003200987-1

Things (IoT) era have created the opportunity for further research in this field. The main driving power to make everything possible is the innovation of semiconductor transistors such as metal oxide semiconductor field-effect transistors (MOSFETs) and integrated circuits (ICs). The worldwide popularity of the transistors has helped create multimillion semiconductor industries with continued progress. To assimilate more functions and fulfill significant demands, the IC industries continuously focused on achieving high performance with low-cost devices through large-volume production. As the MOSFETs are the unit component of ICs, the main focus is on establishing new transistors that are faster in performance and smaller in size than their predecessors. And the ultimate concern regarding the novel transistors is that they should support a high grade of reliability with low cost. Hence, accomplishing these critical goals is the primary driving force in further exploring a comprehensive transistor research in academia and industry.

The journey started with the invention of first point-contact field-effect transistor by John Bardeen, William Shockley, and Walter Brattain in 1947 [1]. Hereafter, there has been irrepressible advancement after the announcement of Kahng and Atalla in 1960 regarding the very first invention of MOSFET [2].

Further, in 1965, Gordon E. Moore, cofounder of the Intel Corporation, formulated a law popularly known as Moore's law. The law affirms that "The number of transistors incorporated in a chip will approximately double in every 24 months" [3] and after that, it is widely accepted by the semiconductor industries. According to the predictions, the miniaturization in transistor size with increasing package density will be continued, which enables the present-day innovations in the era of electronics. A basic understanding of Moore's law is shown in Figure 1.1.

In the continued effort to miniaturize or make the transistors small, the dimensions have become the primary bottleneck for further improvement in the device performance. Along the endless innovations in the fabrication processes, the semiconductor foundries (as ST-Microelectronics, TSMC, Intel, SMIC, CEA-Leti, Global Foundries, etc.) have entered from bulk to nano electronics era. Figure 1.2 (a) demonstrates the primary requirement of technology improvement to accomplish the market demand. The constant progress in semiconductor processing has made it possible to move beyond "Moore's Law," that is, "More Moore" and "More-Than-Moore (MtM)" with state-of-the-art IC chips (beyond CMOS technology) containing more than a billion transistors.

The International Technology Roadmap for Semiconductors (ITRS) [4] is giving the 15-year assessment predictions of the semiconductor industry's future technological requirements. The ITRS predictions have assisted in creating opportunities for research and development (R&D) and academia to come up with strategic outcomes for fulfilling the requirements of future demands. The Executive Summary of the ITRS furnish a taxonomy of scaling in the classical "More Moore" and "MtM" sense as shown in Figure 1.2 (b).

Moreover, the "More Moore" describes the three types of scaling approaches:

- Geometrical (constant field scaling): shrinking of horizontal and vertical physical feature sizes of the devices to improve density, performance, and reliability.

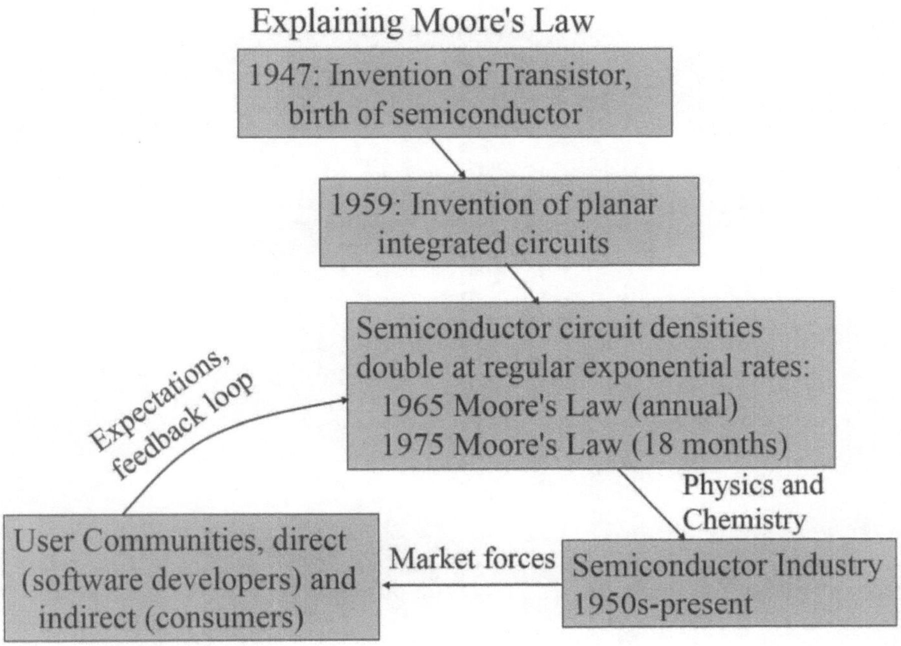

FIGURE 1.1 Basic Understanding of Moore's Law

- 3-D devices, modification of nongeometrical process techniques and new materials that affect the electrical performance of the chip.
- It enables HP, LP, reliability, and low-cost productivity. Specific design technologies that address the power and performance trade-offs associated with functionality needs.

And finally, the functional diversification of the MtM approach typically admits for the non-digital functionalities (e.g., power control, analog and RF communication, passive components, actuators, sensors, etc.) to migrate from the system board-level to an exact package-level like system-in-package (SiP) or system-on-a-chip (SoC) for potential solutions. The aspiration to integrate additional functionality into the ICs results in an endless race of shrinking the device dimensions. The benefits of scaling are higher packaging density with a high switching speed and low power dissipation.

In the history of electronics era, all the emerging innovations have been feasible due to the strong association among the devices and materials research. Primarily, the market demand for three different kinds of application-oriented ICs as low power (LP), high performance (HP), and low standby power (LSTP) became a major challenge to fulfill the demand for the CMOS technology of sub 50 nm gate length [5]. To overcome the challenges that appeared in miniaturizing, the ITRS offers diverse

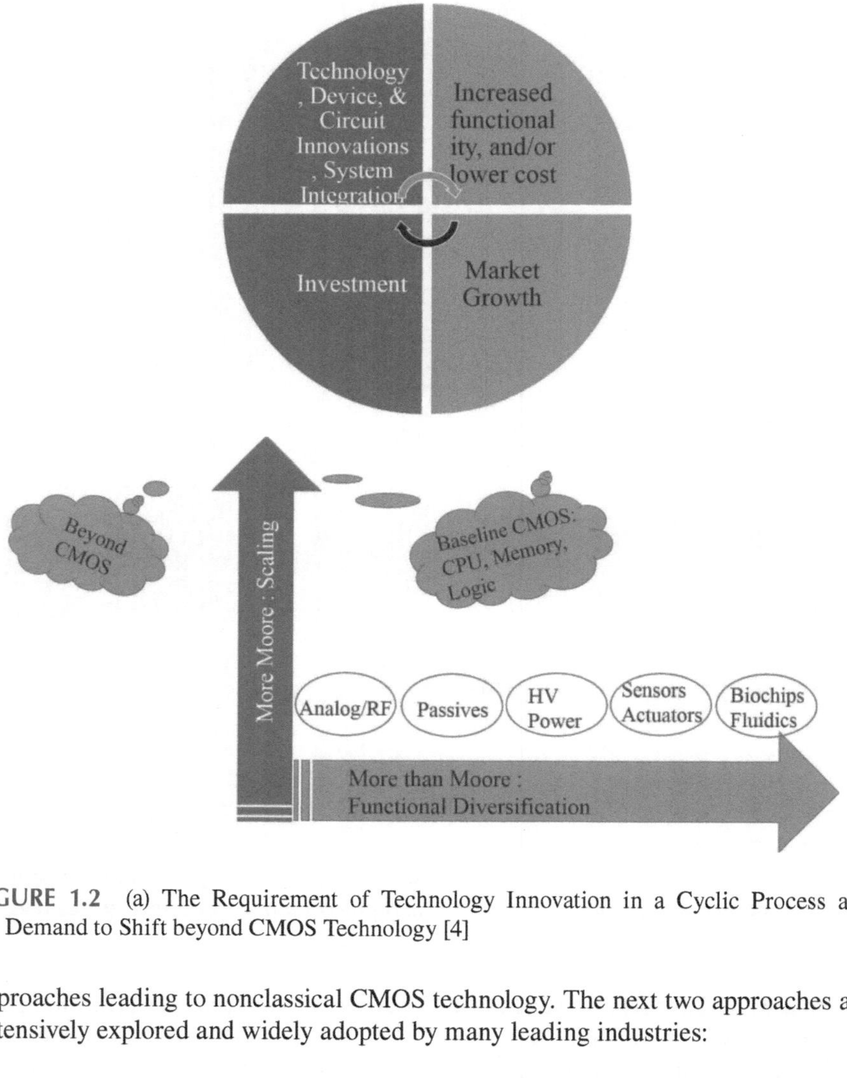

FIGURE 1.2 (a) The Requirement of Technology Innovation in a Cyclic Process and (b) Demand to Shift beyond CMOS Technology [4]

approaches leading to nonclassical CMOS technology. The next two approaches are extensively explored and widely adopted by many leading industries:

1. The structural modifications leading to novel transistor architectures with enhanced electrostatic performance.
2. The evolution of new materials with better carrier transport efficiency.

As semiconductor designs become progressively complex with the devices getting smaller and smaller, the optimization of the structures becomes a hurdle for the designers. Also, the semiconductor manufacturers face the challenge of establishing process technologies within the cost constraints and stipulated time frame. The number of wafers required to implement the new process is one key factor for deciding the cost and time. Hence, modeling and Technology CAD (Computer Aided Design, or TCAD) simulation become the computing paradigm prior to the high-cost fabrication processes. TCAD simulations reduce the number of engineering wafers and consequently save money and time. Moreover, the researchers can analyze important electrical insights of the semiconductor devices that lead to novel device concepts.

1.2 THE ERA OF NANOELECTRONICS

The fundamental revolution in the field of science and technology that covers the designed model with at least one characteristic dimension measured in nanometers (in the range of about 10^{-9} m to 10^{-7} m (1 to 100 nm). The evolution of nanoscale in electronics has the ability to understand, control, and manipulate matter at the level of <100 nm in order to create materials, devices, and systems with fundamentally new properties, functions, and performance.

The unprecedented market growth with fulfilling thousands of demands like cost-effective devices through large-volume production and achieving application-based performance are possible due to the evolution of MOSFETs and henceforth Moore's law. There are several perspectives on the concept of nanoelectronics as per the technology domains described in Figure 1.3.

- Initially, every two years the number of components has been approximately doubled per chip. This has been the trend since 1970, widely known as "Moore's Law".
- The increased performance can be traded with power depending upon the application. This opens the scope for integration of new materials into devices that further progresses toward "More Moore".
- Another trend is described by functional diversity of semiconductor devices. These non-digital functionalities added a fundamental advantage with device scaling. This is designated as "More-than-Moore" (MtM) [4].

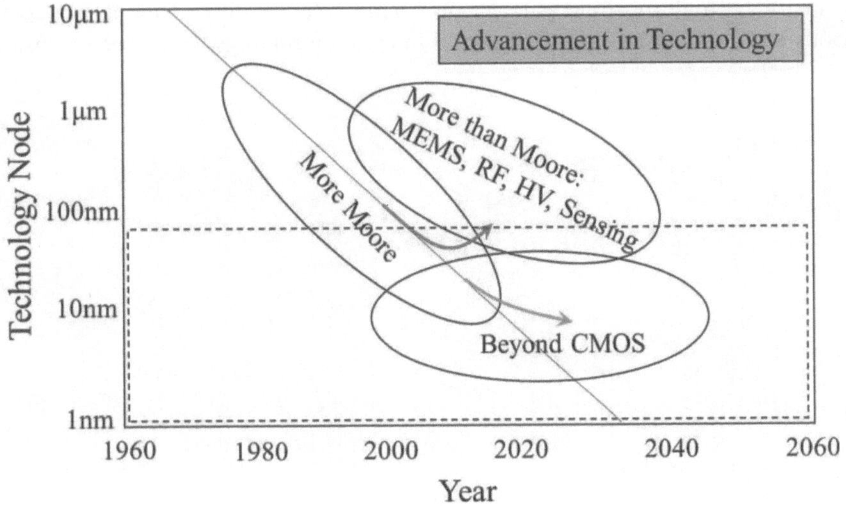

FIGURE 1.3 Advancement in Technology [6]

- Finally, the traditional scaling has reached the limits of standard CMOS technology. Hence, searching for new nanoscale electronic devices has become the primary obligation. This evolution in the era of nanoelectronics is "Beyond CMOS".

As transistors are the most important and basic element of any integrated circuits (ICs), the MOSFET has proven its fundamental advantage over other semiconductor devices due to the following notable properties:

- Unipolar device
- High switching speed
- Higher package density
- High operating frequency
- High linearity

The typical structural representation of conventional and silicon-on-insulator (SOI) MOSFETs is shown in Figure 1.4 (a) and (b), respectively. It is the essential component of high frequency and switching applications across the electronics industry.

The progress of semiconductor electronics devices makes a rapid improvement of performance and low-cost applications. The scaling, that is, continuous reduction in both horizontal and vertical dimensions of the devices in electronic circuits, has led to an increase in the performance in electronic systems with more features and higher processing speed, while performance-to-cost ratio is also rising.

To satisfy the market demand, the density of transistors in a chip and the performance in terms of speed and power consumption need to increase [7–8]. However, the major issue of concern in miniaturizing the Si-based bulk MOSFETs is the outburst of short-channel effects (SCEs). The SCEs can be defined in terms of losing control on electrostatic integrity, threshold (V_{th}) variation, and band-to-band tunneling that facilitates high substrate leakage, etc. To mitigate the aforementioned outcomes, various architectural levels and materials integration findings have been extensively

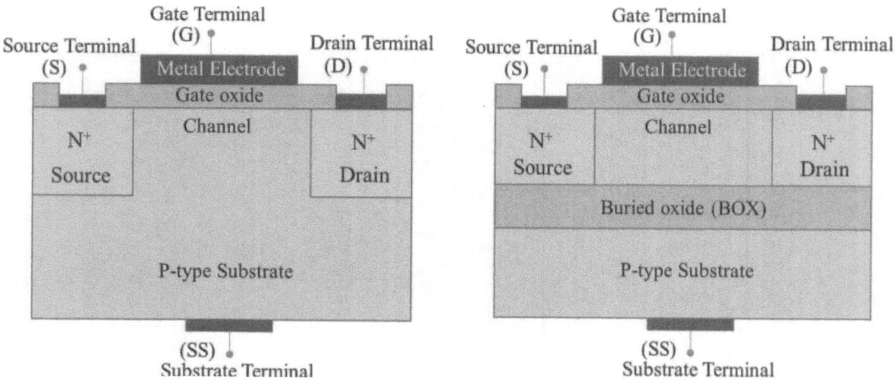

FIGURE 1.4 Typical Structural Representation of (a) Conventional MOSFET and (b) Silicon-on-Insulator (SOI) MOSFET

investigated that led to the invention of silicon on insulator (SOI), ultra-thin-body (UTB), III-V MOSFETs, and lastly, multiple-gate MOSFETs. The need for more performance and package density has accelerated the miniaturization trends in almost every aspects of the device, such as channel length, gate dielectric thickness, source/drain and channel doping levels, supply voltage, etc. that are approaching their fundamental limits. Hence, emergency alternatives to the existing material and structures may need to be identified in order to continue further growth to fulfill the requirements.

1.3 ITRS TECHNOLOGY TARGETS

The International Technology Roadmap for Semiconductors (ITRS) efforts have always estimated the validity of Moore's law and the continuance of the principled cycle. The roadmap has provided the executive summaries for future technology nodes, which helped to preserve the principled cycle by describing the gaps for the trend to progress and assisting the R&D endeavor [4]. Each new technology node remarks a new generation of chip manufacturing. The progression of node names (180 nm, 90 nm, 45 nm, 32 nm, 22 nm, etc.) over the years reflects the steady momentum in both logic and memory chips. The smaller the node number, the smaller is the feature size that increases the package density, producing less costly chips.

In a new technology node, all features of the circuit has reduced by 70% of the previous node making the area half. Normally, this practice of periodic size reduction is referred to as scaling. The ITRS predictions are pretty much beneficial to design certain application-oriented (high performance, low power, low standby power) novel devices for a specific technology node.

1.3.1 High Performance (HP)

The present-day market is more cautious regarding high-performance (HP) consumer applications, and this mainly aims to achieve high current drivability as well as packing density instead of thinking about the off-state leakage current. The devices with high V_{th} have less leakage but consequently greater delay, whereas low V_{th} devices have less delay and high leakage. Primarily, the V_{th} controls the transistor switching speed, that is, the less the V_{th}, the higher the switching speed of transistor for an active state. It is required to design the device for HP application in such a manner that it should provide low $V_{th,}$ which further predicts high drive current [9–11].

1.3.2 Low Power (LP)

The power consumption has became an important aspect of nanoscale device applications in the advent of the mobile era. Many key features for future applications like low power consumption, high speed, and high density are all equally significant but one or two become dominantly important depending on the applications [12–13]. Low power circuit designs have three different aspects involving area, performance, and power trade-offs [14–17]. The need for low power synthesis is driven by battery life and efficiency, device reliability, and demand for portable systems. In low power designs, the reduction of peak power, peak power differential, cycle difference power, average power, and energy are essential [18]. These forms of power dissipation affect different attributes of a CMOS circuit and needs to be optimized for meaningful low power design.

1.3.3 Low Standby Power (LSTP)

Because of the supply voltage (V_{DD}) scaling, the power consumption remains under control. Hence, V_{th} of the transistor has to be commensurately scaled to maintain a high drive current and to achieve performance improvement. However, the threshold voltage scaling results in the substantial increase of the subthreshold leakage current [15, 19]. The two statements given by C. Hu, "Small is beautiful" and "OFF is not totally OFF", always exist in a trade-off manner [20]. In a typical example—suppose I_{off} is 100 nA per transistor and a cell phone with 100 million transistors would consume 10 A even in standby. Then the battery will drain in a minute without receiving and transmitting any call. So, for low standby operating power, the desirable device need to be of high V_{th}.

1.4 ISSUES DUE TO THE MINIATURIZATION OF DEVICES

This section enumerates some of these challenges given due to conventional scaling of MOSFET toward nanoscale as shown in Figure 1.5.

- I_1: reverse bias pn junction leakage
- I_2: the subthreshold leakage (includes drain-induced barrier lowering, DIBL)
- I_3: the gate oxide leakage
- I_4: gate current due to hot carrier injection
- I_5: gate-induced drain leakage (GIDL)
- I_6: channel punch through current

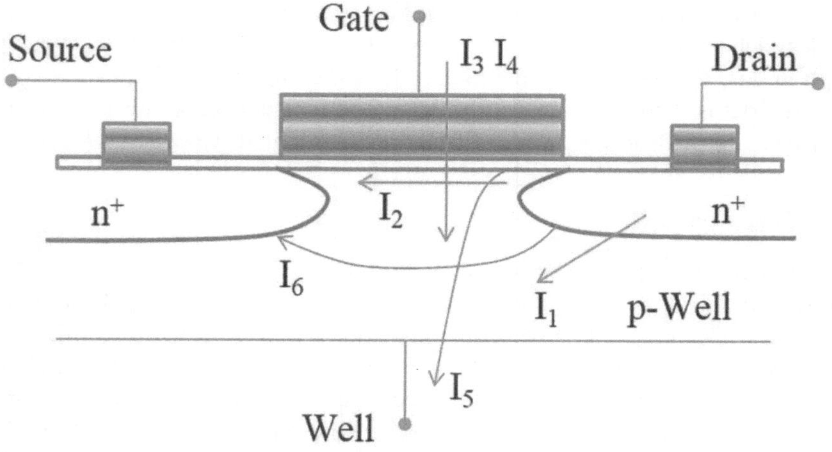

FIGURE 1.5 Leakage Currents Due to Miniaturization of the Device [21]

The previously listed leakage currents are primarily due to the prominent short-channel effects (SCEs) while scaling the device. The fundamental roots for these static leakage currents are enumerated:

- *High electric fields*: This is the root of "avalanche breakdown", which consequently generates current surges that progressively damage the nanodevices. The effect can be negotiable with applied bias voltages in nanoelectronic devices.
- *Heat dissipation*: This causes transistors/other switching devices to malfunction in high package density circuits. The failure is due to the limited thermodynamic efficiency of nanodevices that restricts their density.
- *Vanishing bulk properties*: This occurs because of the nonuniformity in doping profile on small scale semiconductors. This can be overcome either by not doping at all, that is, accumulating the electrons entirely using gates as in the case of GaAs heterostructure or by accomplishing the dopant atoms to form a regular array. The latter can be achieved through molecular nanoelectronics.
- *Shrinking of depletion regions*: They should be too small for a large effective gate length (L_{eff}), which consequently prevents the quantum mechanical tunneling of carriers from source to drain when the device is supposed to be turned off. As such, direct tunneling of electrons are always through a barrier, which ensures an acceptable functionality of nanoelectronic devices.
- Depreciation in the thin gate oxide layer beneath the metal electrode, which prevents the leaking of electrons from gate to drain. The same leakage through thin gate oxide also involves in direct electron tunneling to channel.

The crucial design and technological issues, including the critical barriers and their feasible solutions, are reviewed in terms of scalability, power consumption, and performance matrices to understand the predictions of future technologies of semiconductor devices.

To suppress the SCEs, a novel 3-dimensional nonclassical CMOS technology termed as FinFET [22–24] has been proposed and thereafter became the driving force for further development for both industry and academic researchers.

1.5 HISTORICAL SURVEY

Young [25] in 1989 clearly demonstrated that the major issues commonly faced by MOSFETs at nanoscale were SCEs. Thereafter, people are searching for novel device architectures to mitigate the SCEs at nanoscale. Figure 1.6 explains the common features of initial development of 3-D device, that is, the DELTA transistor proposed by Hisamoto et al. [22] in 1989 and traditional FinFET architecture adopted from Samsung Foundry. The structural configurations and working principles of both DELTA transistor and FinFETs are nearly the same. The body of the device, that is, Fin, is a thin structure having some height (H_{Fin}) and width (W_{Fin}) and forms the source, channel, and drain region. The gate oxide and metal gate electrode are wrapped from the three sides of the Fin. The flow of electron and current conduction is in the Fin, which connects source and drain. Further, the drive current of the device can be easily enhanced by adding parallel Fins as per the dimension.

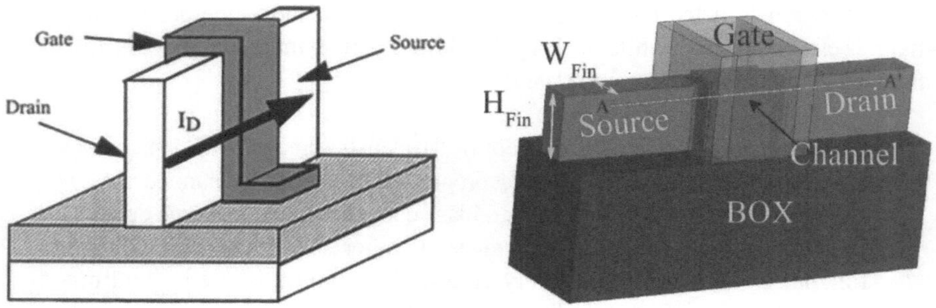

FIGURE 1.6 (a) Depleted Lean-Channel Transistor (DELTA) [5] and (b) Silicon-on-Insulator (SOI) FinFET from Samsung Foundry

Although it is familiar from over a decade, still these devices have earned a lot of attention in later days. There are several groups working in the FinFETs and mostly from IBM, UC Berkeley, Global Foundries, and Intel. From the literature, it is worth noting that for FinFETs, channel width is generally described as height of the Fin. But while compared to classical MOSFETs, the channel width is actually twice that of the Fin height. The important part of the FinFETs is Fin creation, and there are several approaches for this involving various techniques like electron beam lithography or ultraviolet photolithography or a combination of both [26].

1.6 FinFETs: PROS, CONS, AND POSSIBLE ALTERNATIVES

The numerous research in both industry and academia under various architectures of 3-D FinFET domain has produced remarkable results. However, it is not that the FinFETs are the only solution to the obstacles of miniaturization. The primary most obvious option is always to go with the conventional planar CMOS technology until the fundamental barrier is achieved, that is, the atomic size of silicon. While transiting from an older technology to a completely newer one, the cost factor must be considered. Hence, the industries are always focusing on alternative approaches considering the investment and time required for manufacturing. The "multiple gate" technologies, to which FinFETs belong, accommodate a number of solutions to address the miniaturization problem. Some of them are double-gate [27], cylindrical gate-all-around (CGAA) [28], and vertical replacement-gate [29] MOSFETs. All the previously listed configurations also offer benefits that are analogous to FinFETs in terms of performance like improved subthreshold slope (SS), and short-channel effects (SCEs), with a higher drive current. However, the basic advantage in FinFETs is that the effective width can be controlled by the number of Fins, whereas the DG MOSFETs are limited to a width of less than one micron, and the CGAA MOSFETs are very complex in design and manufacturing. Also, issues like mismatch in top- and bottom-gate alignment in case of DG MOSFET make the fabrication process difficult. Similarly, another important device is vertical MOSFETs (V-MOSFETs), which are crucial in providing similar benefits as CGAAA MOSFETs with lithography independent gate material lengths. But the issue of V-MOSFET lies in adjustment of threshold voltage and defining source/drain and channel junctions that limit the use.

In addition to devices that are on a field-effect basis, studies are going beyond to find more ambitious solutions like tunnel FETs (TFETs) [30] and junctionless (JL) MOSFETs [31]. TFETs are promising devices that can be used for low-power applications due to its low subthreshold swing (SS), which is less than 60 mV/decade that consequently lowers the leakage current with a better immunity toward short-channel effects (SCEs). JL MOSFETs avoid the doping complexity as they have uniform doping profile throughout the source, channel, and drain. However, the disadvantage of TFETs and JL MOSFETs is that they are low on state drive current compared with MOSFETs. Hence, these devices are considered to replace the conventional CMOS technology in the near future for LP and LSTP applications.

From all the previously listed proposals, the FinFETs show as a potential candidate to enhance the lifetime of the FETs to further continue Moore's law as a near-term solution to the problems faced by the IC industries. The manufacturing is also feasible with implementation in circuit design. The FinFETs are structures that have already been in featured research for more than a decade and are the most readily available form of the "non-classical CMOS" technology.

1.6.1 FinFETs from an Industry Aspect

Previously, the manufacturing of FinFETs has been placed in the far future list of ITRS prediction. But the reality comes sooner after some important announcements from the industries. First of this kind is from Intel, declaring its version of FinFET, that is, tri-gate MOSFET has transferred from research to manufacturing. "Our latest research indicates that the scalability, performance and excellent manufacturability of our tri-gate transistor makes it a strong contender for production as early as 2007 on our 45-nm process technology" [32].

This statement was given by Sunlin Chou, senior vice president and general manager of Intel's Technology and Manufacturing Group at Kyoto, Japan, June 12, 2003 in the Symposia of VLSI Technology and Circuits. Other leading industries such as IBM, TSMC, and AMD are also in line to produce the 3-D FinFETs. Another statement from Chenming Hu, CTO of TSMC Ltd., follows:

> I think the FinFET is going to be used, but not wholesale. There won't be a chip where you look in and every transistor is a FinFET. No, it will be used selectively where the performance requires it. I think it could come as early as the 65-nm node [33].

From all these announcements, it can be identified that FinFETs are of immense interest for the IC industries. Thus, a good understanding of FinFETs with advanced domain is apparent. The current work presented here gives a detailed investigation of the hybrid nature of FinFETs and provides a platform for future research on hybrid FinFETs.

1.6.2 FinFETs: A Journey

Here some previous works are demonstrated to establish the journey of technology advancements toward FinFETs. The research to mitigate the problems that occurred in nanoscale devices has taken two different paths. The first is the introduction of novel materials into the classical MOSFETs to establish strain [34] and III-V material in the channel, which improves the carrier mobility and, consequently, the drive current. The second is the development of nonclassical multi-gate MOSFETs (MuGFETs), which is an exceptional concept for further diminishing of the device dimensions [35]. The following sections summarize some of the featured technologies that help in enhancing device performance.

1.6.3 Ultra-Thin Body Devices

The continuous downscaling of physical device dimensions to meet Moore's law has become the primary agent for failure of standard manufacturing processes in the deep sub-micron regime. Various SCEs and floating-body effects are notably difficult concerns, which hinder further development. Some of the important issues are hot carrier effect (HCE), scattering effects, CMOS latch-up, degradation of gate dielectric, increase of parasitic capacitances, mobility degradation, body effects, and a large subthreshold leakage current. These issues hamper the device performance as well as the reliability. While few innovative approaches for scaling have mitigated some of these issues, one such technological breakthrough is silicon-on-insulator (SOI) that was proposed in the late 1980s and has gained momentous attraction. SOI technology predicts a low capacitance that enables high-speed operation. Further, the supply voltage (V_{DD}) can be reduced to diminish the power consumption while providing adequate speed. SOI technology has the ability to resist high temperatures with favorable radiation hardness, which enables the fabrication of micro-electro-mechanical system (MEMS) devices for control systems, and the capability to handle high voltages. SOI has moved to the mainstream, where it can contribute important enhancements in circuit performance [36].

It should be noted that in the case of PD-SOI, the $t_{Si}>X_d$ and in FD-SOI, the $t_{Si}<X_d$. Figure 1.7 examines the structures of bulk-Si and SOI substrates. The primary feature of the SOI structure is the layer of oxide just below the silicon channel like a "sandwich" (silicon/oxide/silicon). The in-between oxide region is commonly referred to as buried oxide (BOX) and is formed by the oxidation of Si or oxygen ion implantation into Si. The top silicon layer is termed as thin film or silicon body thickness (t_{Si}). If the t_{Si} is single crystalline, the MOSFETs made in it are called SOI devices; but if the t_{Si} is polycrystalline, then it leads to the formation of thin-film transistors (TFTs). The Si substrate beneath the BOX is called Si substrate or base wafer or supporting substrate, or handle wafer. The terminologies "Si body" and "SOI body" attribute to the part of SOI layer that constitutes the body of a MOSFET. The SOI technology adequately suppresses many of the terminal origin of capacitances by diminishing the depletion capacitance in the channel and substrate regions, consequently reducing the delay of the device and enhancing the performance. The parasitic capacitances like gate to substrate are also decreased, which reduces the power consumption, escalating the technology attractive.

Cristoloveanu and Hu in 2001 [37–38] demonstrated that the devices fabricated on SOI substrates offer superior characteristics over bulk MOS devices. SOI CMOS can provide 20–50% higher switching speeds or two to three times lower power consumption compared to the conventional CMOS technology. Advantages of SOI platform are reduced junction capacitance, increased channel mobility, suppressed short-channel effect, and reduced second-order effects [39]. Further, Colinge in 2004 explained that

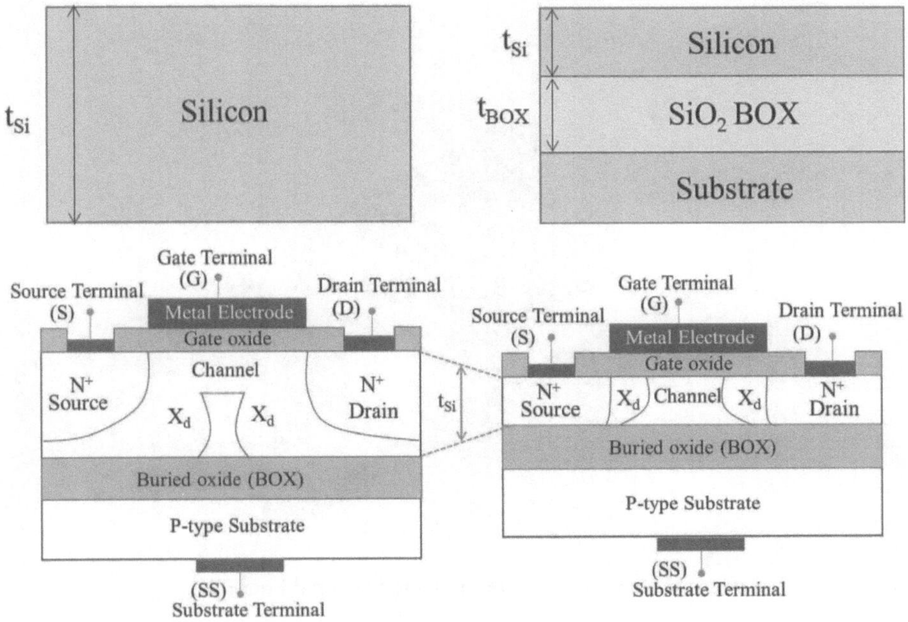

FIGURE 1.7 Schematic Diagram of (a) Bulk-Si and SOI Substrates and (b) PD SOI and FD SOI

the SOI technologies have the flexibility to enable the fabrication of radiation-hardened nonplanar (3-D) integrated circuits [40–41]. More investigation on SOI gives rise to two flavors like partially depleted (PD) and fully depleted (FD) as shown in Figure 1.7 (b). The advantages of SOI over bulk are heightened when the devices are fabricated in FD SOI (thin-film) platform. The improvements with thin films lead to near ideal subthreshold slope (~60 mV/dec), increased saturation current and transconductance, reduction in hot electron effect, and elimination of the kink effect.

1.6.4 Multi-Gate MOSFETs

Further structural innovations lead to multi-gate technology. Figure 1.8 represents the journey from SG to DG to tri-gate FinFET. DG-SOI MOSFETs have two symmetrical interconnected gates. DG MOSFETs can be illustrated with three possible orientations as planar, vertical, or mixed-mode [7]. The fabrication of such devices has been achieved with the delta process [22], Fin process [24], epitaxial lateral overgrowth [42], wafer bonding [43–44], and tunnel epitaxy [45].

The concept of DG was initially demonstrated by Balestra et al. [46] back in 1987 on conventional SOI MOSFETs with concurrently biasing the front and back gates of the FD transistor. The Si film of DG-MOSFETs has to be thicker than the cumulative sum of depletion regions induced by both the gates. There is no interaction produced among the two inversion layers, and the operation of this device is very much similar to that of two conventional MOSFETs linked in parallel. But if the thickness of

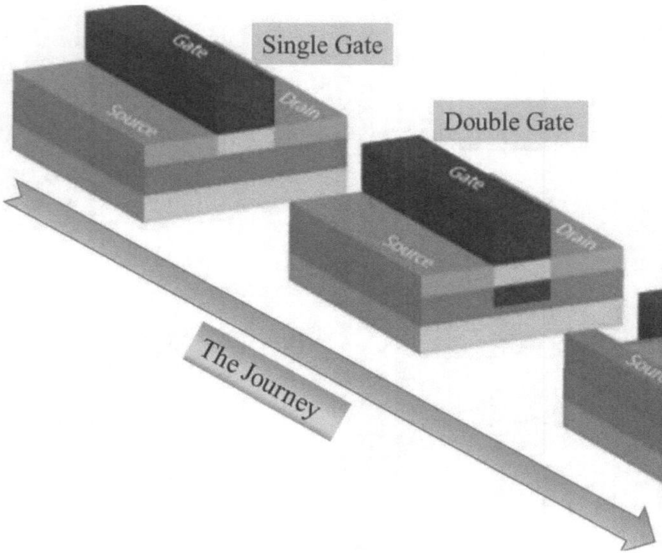

FIGURE 1.8 Schematic Representation of the Journey toward FinFETs

silicon is sufficiently lower, the whole Si film is depleted, and a substantial amount of interaction takes place between two potential wells. In such cases, the inversion layer is formed not only at the bottom and top of the silicon film, that is, near the two silicon-oxide interfaces, but also throughout the entire silicon body thickness. Then the device is said to operate in volume inversion (VI), that is, the carriers are no longer confined to the interfaces but are distributed throughout the entire silicon volume. The development of the front and back inversion layers are caused by VI and continuity in thin SOI film. The minority carriers flow in the middle of the film and experience less surface scattering and hence improving the drive current, mobility, and transconductance of the device. The two gates exercise ideal control on the inversion charge and potential, so that SCEs (DIBL, charge sharing, fringing field, and punch through) are eminently minimized. This is due to the reduction of natural length λ_{DG} in DG MOSFETs compared to that of SG transistors.

In addition, like other dual-gated devices, DG-MOSFETs are claimed to be more immune to short-channel effects (SCEs) than bulk silicon MOSFETs and even than single-gate fully depleted SOI MOSFETs [47]. Wind et al. in 1996 demonstrated that the two-gate electrodes jointly control the carriers, thus screening the drain field away from the channel [48]. And Wong et al. further established in 1998 that this characteristic would support a much greater endeavor for scaling than ever imagined for conventional MOSFETs [45]. Along with all these advantages of DG MOSFETs, it also suffers from considerable SCEs in the deep sub-nanometer regime. Hence, the novel device structure tri-gate FinFET is developed to improve the device immunity against the SCEs consequently improving the reliability in circuit applications. The research on 3-D FinFET devices has became the hot topic after the formal announcement from leading industries (Intel, Samsung, TSMC, and Global Foundries). Sun et al. [49] have studied and compared the performance of three analogous architectures on 3-D platform as planar-bulk, tri-gate bulk, and SOI FinFET. Their findings are notable for further research in the 3-D architectural level. The performance of FinFETs by considering III-V (InGaAs) and strained-Si as the channel material are also explored by Park et al. [50]. They found that tri-gate FinFETs are the best possible architecture for gate contacts of sub-15 nm technology.

Similarly, researchers like Manoj et al. [51] and Agrawal et al. [52] have continued the investigation with alternative gate oxide materials. The high-k dielectric materials such as Si_3N_4, HfO_2, TiO_2, La_2O_3, $LaAlO_3$, and $SrTiO_3$ with high dielectric constants can be directly used in place of SiO_2 as gate oxide or as gate stack engineering, that is, a thick layer of high-k dielectric is epitaxially grown over a thin layer of low-k dielectric. The high-k materials will help to prevent the direct gate tunneling that occurred through thin SiO_2 gate oxide. While considering the high dielectric materials over the SiO_2 layer, the effective oxide thickness (EOT) will be calculated as:

$$\text{Equivalent high-k} = T_{k*}(k_{SiO2}/k_{high\text{-}k})$$

$$\text{EOT} = \text{Equivalent } SiO_2 + \text{Equivalent high-k}$$

where T_k is the physical thickness of the high-k material and k_{SiO2} and $k_{high\text{-}k}$ are permittivity of SiO_2 and high-k materials, respectively. From the relation, films with high-k value are thicker than SiO_2 film and have prominent control over the SCEs. Hence, it is not wrong to state that the marriage among conventional semiconductors, and oxides clearly manifests to combine semiconductor electronics with corresponding electronic systems. Again, the same high-k materials are also explored by incorporating it over the underlap region of the FinFETs as symmetric (both sides of the channel) and asymmetric (any one side of the channel) way. Pal et al. [53–54] have systematically evaluated the impact of dual-k spacers (inner high-k and outer low-k) on various performance of 3-D FinFETs in a symmetric and asymmetric (only in the source-to-channel underlap region) way.

1.6.5 Hybrid FinFETs

Furthermore, to achieve the market demand regarding HP consumer applications, a novel hybrid transistor, that is, inverted-T FET (ITFET) [55–57] was first proposed by Mathew et al. and Zhang et al.

Figure 1.9 has demonstrated the process flow of inverted-T FET architecture. They found that the vertical devices, which merge both FinFETs and UTB devices, exhibit a good SCEs control and proposal for future scaling. The study was further extended by Fahad et al. [58–59] to optimize the UTB layer thickness, and the device was renamed Wavy FinFET. Figure 1.10 differentiates the hybrid FinFETs from the conventional FinFETs. The extra UTB layer and the height of the Fin is calculated by including the UTB thickness in case of hybrid device. Hence, the hybrid FinFETs can be outperformed compared to the conventional one without increasing the area of the chip [59]. As the foundries, electronic design automation (EDA) and integrated device manufacturer (IDM) companies show more emphasis and investments on most promising 3-D technologies. The hybrid FinFETs will be a potential candidate for future technology nodes.

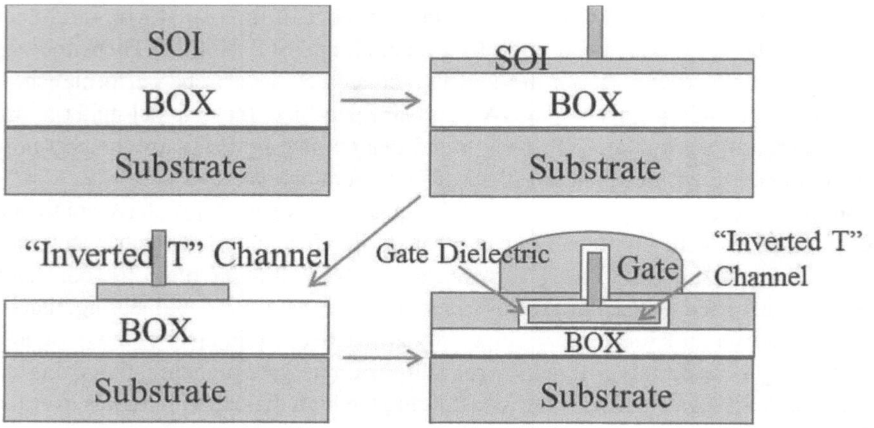

FIGURE 1.9 Process Flow of Inverted-T FET (ITFET)

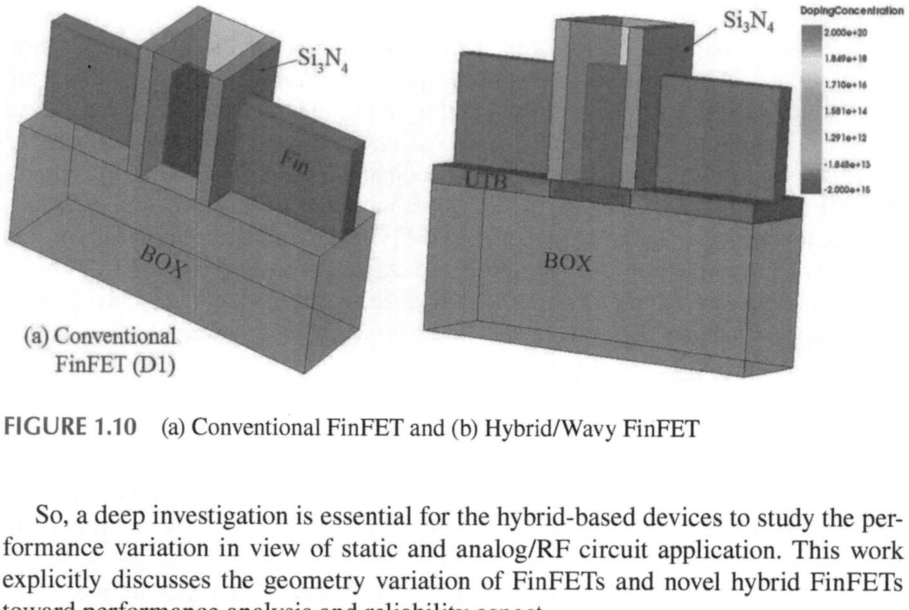

FIGURE 1.10 (a) Conventional FinFET and (b) Hybrid/Wavy FinFET

So, a deep investigation is essential for the hybrid-based devices to study the performance variation in view of static and analog/RF circuit application. This work explicitly discusses the geometry variation of FinFETs and novel hybrid FinFETs toward performance analysis and reliability aspect.

1.7 DEVICE PERFORMANCE AND RELIABILITY METRICS

The important device performance with reliability analysis are studied extensively. From previously mentioned literature, the investigation moves toward the SCEs by using nonclassical MOSFET structures as ultra-thin-body (UTB) SOI, DG-MOSFET, 3-D FinFETs, and hybrid FinFETs. The threshold-voltage adjustment of the device for HP applications is a challenging requirement that can be achieved by tuning the work function of gate electrode. Reducing the parasitic capacitances as well as obtaining the highest attainable gate overdrive at an appropriate I_{off} is another question. Nonplanar structures such as hybrid FinFETs with diverse device engineering to optimize the structural configurations in view of better performance and more reliability will be the primary requirement for future technology nodes.

1.7.1 Static Performance

Moreover, the device performance is categorized as static and analog and RF performance. Starting with the static performance, subthreshold swing (SS) and drain-induced barrier lowering (DIBL) are two primary short-channel parameters, and it is essential to analyze these parameters in nanoscale device designs. In deep sub-micron technology, the drain-to-source biasing (V_{DS}) has a strong impact on the band bending or over the inversion layer of the device. Thus, the V_{th} and the subthreshold leakage current of short-channel devices are sensitive toward V_{DS} variation. This is referred to as DIBL and can be evaluated as Equation 1.1. The DIBL is the cause for punch-through effect, that is, the merging of depletion regions of drain and source in the channel surface to lower the potential barrier of the source.

$$DIBL = \frac{\Delta V_{th}}{\Delta V_{DS}} = \frac{V_{th1} - V_{th2}}{V_{DS2} - V_{DS1}} \quad (1.1)$$

The CMOS standby power is derived from the subthreshold current. The MOSFETs that are designed for high-speed and low-power consumption digital applications are primarily operated in the subthreshold region. The subthreshold current has a considerable influence on the circuit performance. Hence, a steeper SS is the basic requirement for a good switching operation. The SS can be defined as Equation 1.2 [60].

$$SS = \left[\frac{\partial \log_{10}(I_D)}{\partial V_{GS}}\right]^{-1} \quad (1.2)$$

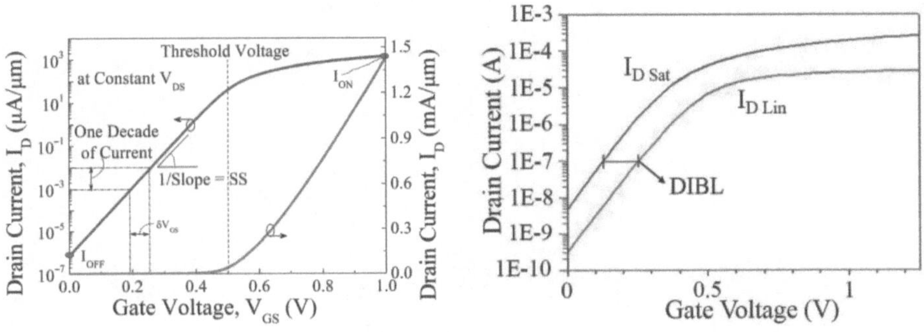

FIGURE 1.11 (a) Pictorial Representation to Define Subthreshold Slope (SS), On Current (I_{on}), Leakage Current (I_{off}), and Threshold Voltage (V_{th}) from I_D-V_{GS} Characteristics and (b) DIBL

1.7.2 Switching Parameters

The fundamental components to attain high performance are the reduction of time delay, power loss, and supply voltage to a minimal extent. Hence, an ideal device demands low voltage drop, fast on-off switching ratio and low parasitic capacitances. Amplifier efficiency is derived from total power losses by the MOSFETs. The total power loss of MOSFETs is the resultant of switching loss, conduction loss, and gate charge. Again, the conduction loss is directly associated to output resistance (R_o) and is expressed as:

$$P_{CONDUCTION} = \left(I_D^2\right) \times R_0 = I_D \times V_{DD} \tag{1.3}$$

Therefore, lower R_o results in lower MOSFETs conduction loss and consequently gets better amplifier efficiency. This can be illustrated as:

Lower R_o ⟹ Lower $P_{CONDUCTION}$ ⟹ Better efficiency

For ultra-low power systems, the V_{DD} must be lowered without limiting the operating speed. This can be achievable by reducing the V_{th} of the MOSFETs. There are mainly two types of power dissipation: (i) static power dissipation that is because of the subthreshold leakage current and (ii) dynamic power dissipation involved in total capacitance. Figure 1.12 explains the relationship between the two power dissipations. Devices being low-V_{th} with small leakage current are beneficial for fabricating suitable low voltage analog switches. The dynamic power dissipation is becoming dominant at higher supply voltages (V_{DD}) that increases total power in proportion to

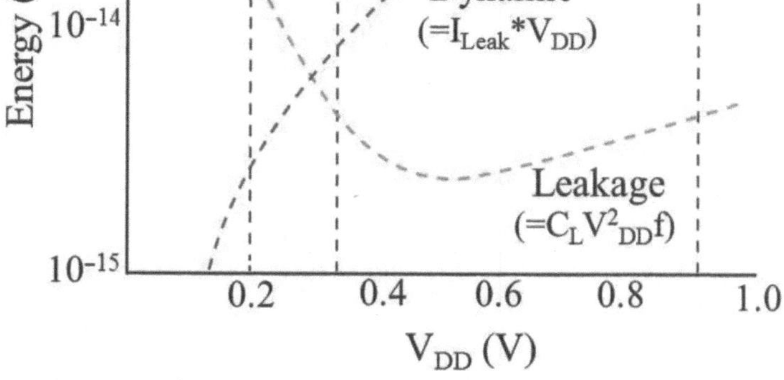

FIGURE 1.12 Power Dissipation versus Supply Voltage

the square of the V_{DD}. However, for lower V_{DD}, the V_{th} of the MOSFETs becomes low and consequently facilitates the OFF-state leakage current. It causes the increase in static power dissipation further enhancing the total power dissipation. Thus, the suppression in the leakage current is the primary issue in nanoscale devices.

1.7.3 Intrinsic Delay

At the time of proposing any new devices, the researchers or industries always rely on ITRS projections with an aim to meet the required I_{on} (for the specified V_{DD}), I_{off}, parasitic components, and electrostatic integrity [4]. The projected I_{on} in ITRS is to fulfill the requirement of the intrinsic delay of the transistor based on the CV/I metric that is an optimistic figure of merit (FOM) for the circuit performance.

The transistor delay can be approximated by $C_{tot}V_{DD}/I_D$, where V_{DD} is the supply voltage, I_D is the drain current at $V_{GS} = V_{DS} = V_{DD}$, and C_{tot} represents the total capacitance of the transistor. For a good transistor delay, the I_D must be high with a lower C_{tot}. In other words, the improved carrier mobility and velocity is the main driving force for reducing the delay.

1.7.4 Analog and RF Performance

Most of the time, the MOS technology is optimized for digital designs, and the devices are characterized by the drive current and intrinsic delay. And the CMOS technology meant for analog and RF circuits suffers from many challenges. Similarly, for system-in-package (SiP) and system-on-a-chip (SoC) applications, the optimization of devices becomes a hindrance [4, 61–62]. The SoC indicates a device that includes a system of differently functioning circuit blocks into a single silicon chip, while SiP refers to a semiconductor device that incorporates multiple readily available chips into a single package. Hence, it is required to improve the device performance for both digital and analog/RF circuit applications [63–70]. Dominant semiconductor foundries such as IBM, OKI, RFMD, Honeywell, etc. have already manufactured several products for the telecommunication market by implementing SOI RF technologies.

An additional number of electronic systems is integrated on a single chip parallel to the advancement in technology. The interfaces are increasing with the growing demand of analog applications along with the advancements in networking; as a consequence, the importance of analog and RF performance can be crucial.

1.7.5 Analog Performance

Most of the analog circuits handle a consolidation of two fundamental analog functions [71] as:

- Transformation from gate voltage (V_{GS})-to-drain current (I_D) (i.e., V-to-I) and the coefficient is transconductance (g_m).
- Transformation from drain current (I_D)-to-drain voltage (V_{DS}) (i.e., I-to-V) and the coefficient is drain resistance (r_d).

The ratio of transconductance-to-drain current (TGF = g_m/I_D) is one of the crucial figures of merit (FoMs) as far as analog circuits are concerned. The TGF (transconductance generation factor) demonstrates the efficient use of current to achieve a desired value of transconductance. High TGF value is advantageous for realization of analog circuits, which are meant to operate at low supply voltage. The analytical relation of TGF gives rise to an interesting fact expressed as Equation 1.4.

$$\frac{\Delta V_{out}}{\Delta V_{in}} = \frac{\Delta I_D}{g_d}\frac{1}{\Delta V_{in}} = \frac{g_m}{g_d} = \frac{g_m}{I_D}V_{EA} \tag{1.4}$$

where g_d is the output drain conductance and $V_{EA} = I_D/g_d$ is the early voltage [72]. From this equation, larger g_m/I_D ratio will be better for the analog devices (more gain, g_m/g_d).

The SCEs [73] commonly prominent in shorter-channel devices tend to degrade analog FoMs such as transconductance (g_m), g_m/I_D ratio, early voltage (V_{EA}), and the intrinsic dc gain (g_m/g_d). Moreover, TGF and V_{EA} are the key analog FoMs and exhibit the efficiency of the device to convert DC power into AC frequency, and the gain.

1.7.6 RF Performance

- Cutoff frequency (f_T), that is, the frequency at which current gain is unity
- Maximum operating frequency ($f_{\max}$), that is, the frequency at which the power gain is unity.

$$f_T = \frac{g_m}{2\pi\left(C_{gs} + C_{gd}\right)}$$

$$f_{\max} = \frac{f_T}{2\sqrt{\left(R_g + R_i\right)\left(g_d + 2f_T C_{gd}\right)}} \tag{1.5}$$

where g_m: transconductance, C_{gs}: gate-to-source capacitance, C_{gd}: gate-to-drain capacitance, R_g: gate resistance, R_i: real part of the input impedance because of non-quasi-static effects, g_d: drain conductance. The continuous miniaturization of CMOS devices has resulted in significant enhancement in the RF FOMs of bulk as well as SOI MOSFETs [74–76]. It is also evident from Equation 1.5 that an increase in f_T is possible by reducing the gate length. Similarly, $f_{\max}$ has a strong dependency not only on g_m and g_d but also on parasitic components.

1.7.7 Reliability Aspects

Primarily, the reliability aspect of any device can be measured from two analyses:

(i) The impact of temperature variation on device characteristics toward evaluation of the zero temperature coefficient/temperature compensation point (ZTC/TCP) [77–78]. It is the point where the I-V characteristics show a little or no variation with respect to temperature that is required for biasing the digital and analog circuits meant for high-temperature applications.

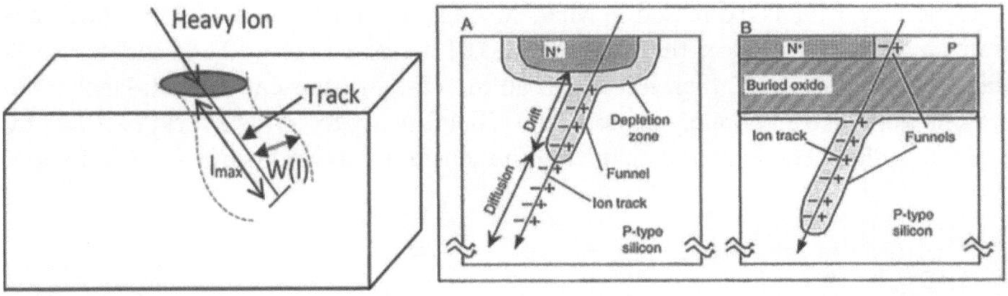

FIGURE 1.13 Ion Strike on a Conventional MOSFET and SOI MOSFET (Adopted from [41].)

(ii) To measure the device sensitivity toward heavy ion irradiation. The electronic devices, which are designed for space and atmospheric applications, are affected from radiation due to the continuous shrinking of devices [79]. In nanoelectronics, the natural radiation is one of the major failures in terms of reliability of a device [4, 80]. Silicon-on-insulator (SOI) circuits and devices are widely used in aerospace/military applications due to its exposure toward high hardness against transient radiation effects [40]. Single event effects (SEE) or single event upset (SEU) is induced due to the interaction of ionizing particles like heavy ion, neutrons, and protons with the electronic device [81]. If such an ionizing particle penetrates the reverse biased junction, depletion region, and the bulk silicon of the device, then a plasma track is produced along its path, resulting in the generation of electron-hole pairs. During this ion track, the depletion layer of the device is distorted in its vicinity and creates a transient current. The distortion of the depletion layer is known as "funnel", which extends the depletion layer along the ion track in such a way that the electrons move toward the junction and holes move toward the grounded substrate, resulting in the creation of substrate current. The electrons are collected by the drain contact, which leads to current transient that changes the logic state of the node. If a heavy ion penetrates into an integrated circuit, then it is called a multiple bit upset (MBU) [40]. However, SOI devices are less susceptible to radiation due to the presence of a buried oxide (BOX) layer between the silicon film and the substrate. Hence, the charges generated in the substrate cannot be collected at the junction of the device, resulting in less transient current compared to bulk devices. The formation of funnel at bulk P-N junction and at SOI P-N junction is illustrated in Figure 1.13 [40].

The energy lost by an ionizing particle along its track is defined as "linear energy transfer" (*LET*) and it can be expressed as in Equation 1.6 [40]. *LET* is one of the crucial parameters for a heavy ion, which perturbs the logic state of the device.

$$LET = \frac{1}{m_v}\frac{dW}{dx} \tag{1.6}$$

where m_v is volumic mass of the silicon, dW is the energy lost by the ionizing particle and absorbed by silicon, and dx is the linear distance along the ion track.

Generally, LET is expressed in MeV.cm²/mg and an LET of 1 MeV.cm²/mg generates a very low charge, that is, around 0.01pC for silicon [40]. If the device is subjected to more LET, then it leads to an increase of transient current. Hence, the present work also elaborately explores the reliability aspect of the hybrid FinFETs by evaluating the ZTC point and analyzing the sensitivity toward heavy ion irradiation.

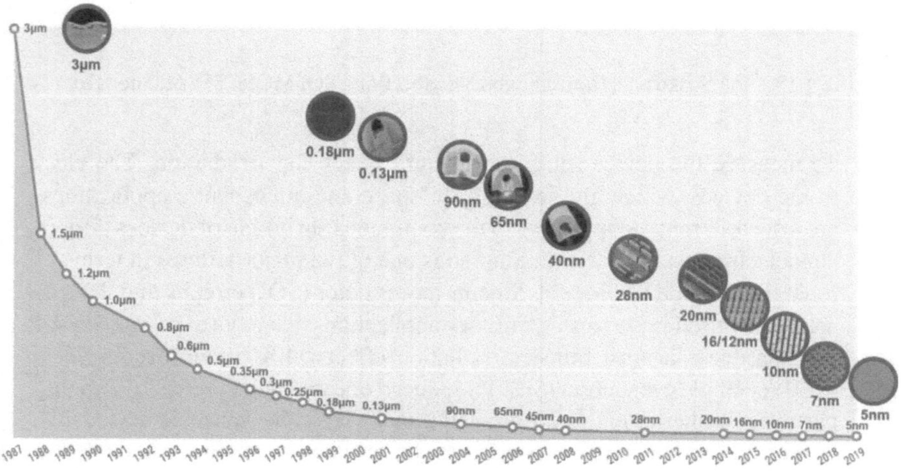

FIGURE 1.14 TSMC Technology Roadmap [82]

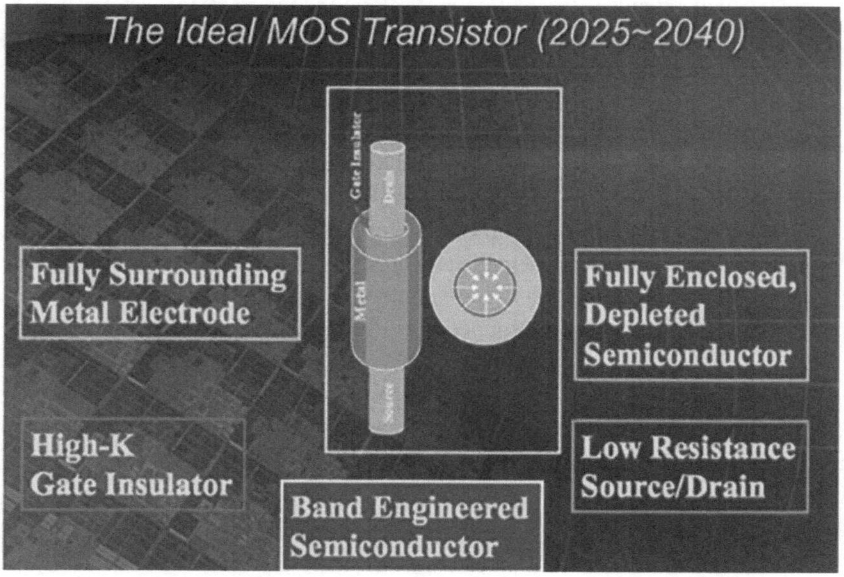

FIGURE 1.15 Ideal MOS Transistor under Vertical Gate-All-Around (GAA) Technology Released by IEEE Electron Devices Society for Next-Generation Technology [83]

1.8 DEVICES IN THE RACE FOR FUTURE TECHNOLOGY

To follow Moore's law, different foundries have adopted technologies for further miniaturization without much compromise in the device performance. One such example is depicted under Figure 1.14 (adopted from TSMC technology roadmap) even to support "More Moore" and "More-than-Moore". The figure shows FinFETs are taking care till 5 nm technology with inclusion of ample advancements.

Thanks to the FinFET technology for saving the semiconductor and electronic industry by taking care of 22 nm and beyond till 5 nm [82].

Now, the race has begun for future 3 nm technology node. As described under Figure 1.15, the foundries can utilize the vertical dimensions to further increase the packing density. And one of the emerging solutions is to go with the vertical gate-all-around (GAA) MOSFET [83] as predicted by the ITRS roadmap. Other possibilities for the 3 nm technology are the critical breakthrough in GAA design released by Samsung, that is, multi-bridge channel (MBC) FET or nanosheet transistor [84] (Figure 1.16). The design is unique and will follow the larger current per stack that will result in simpler device integration. As predicted, the MBC FETs will be the next and may be the last step to follow Moore's law.

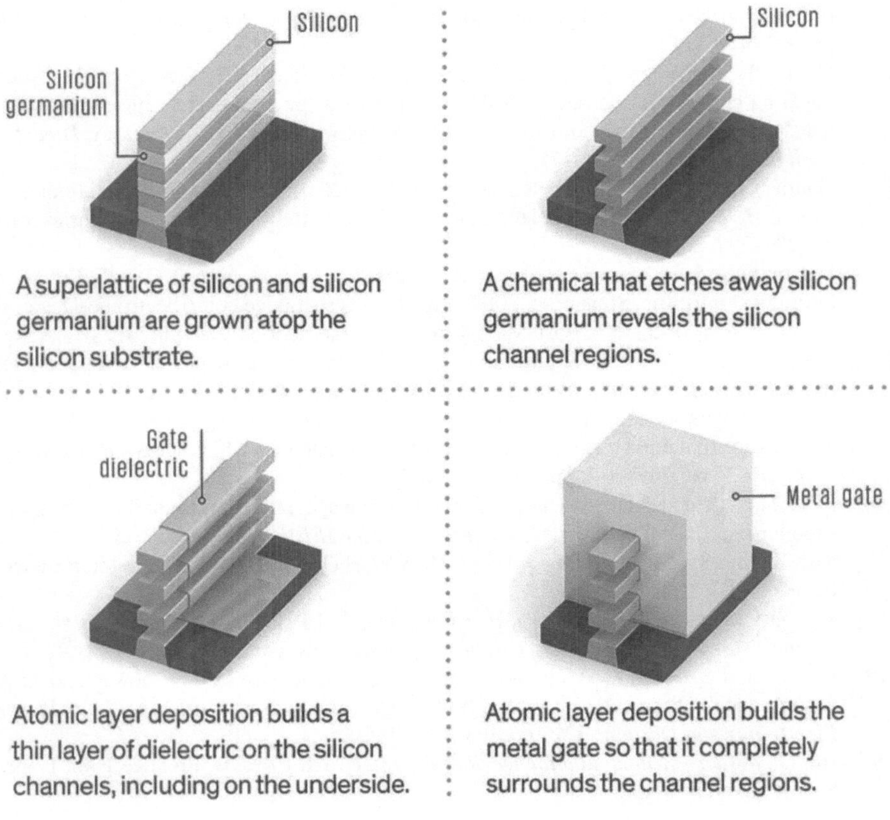

FIGURE 1.16 Nanosheet Transistor Scheduled for 3 nm Technology [84]

REFERENCES

[1] Shockley, W., "Shockley", Sept. 25 1951, US Patent 2,569,347.
[2] Dawon, K., "Electric field controlled semiconductor device", Aug. 27 1963, US Patent 3,102,230.
[3] Moore, G. E., et al., "Cramming more components onto integrated circuits", *Electronics*, vol. 38, no. 8, pp. 114 ff, 1965.
[4] "The international technology roadmap for semiconductors", *Technical Reports*, 2011. [Online]. Available: http://public.itrs.net
[5] Horowitz, M., T. Indermaur, and R. Gonzalez, "Low-power digital design", *IEEE Symposium in Low Power Electronics*, pp. 8–11, 1994.
[6] Ionescu, A. M., "Nanoelectronics roadmap: Evading Moore's law", 2009. http://citeseerx.ist.psu.edu/viewdoc/download?doi=10.1.1.581.3089&rep=rep1&type=pdf
[7] Taur, Y., et al., "CMOS scaling into the nanometer regime", *Proceedings of the IEEE*, vol. 85, no. 4, pp. 486–504, 1997.
[8] Wong, H.-S. P., D. J. Frank, P. M. Solomon, C. H. J. Wann, and J. J. Welser, "Nanoscale CMOS", *Proceedings of the IEEE*, vol. 87, no. 4, pp. 537–570, 1999.
[9] Wann, C., F. Assaderaghi, R. Dennard, C. Hu, G. Shahidi, and Y. Taur, "Channel profile optimization and device design for low-power high-performance dynamic-threshold MOSFET", *IEEE International Electron Devices Meeting*, pp. 113–116, 1996.
[10] Bernstein, K., D. J. Frank, A. E. Gattiker, W. Haensch, B. L. Ji, S. R. Nassif, E. J. Nowak, D. J. Pearson, and N. J. Rohrer, "High-performance CMOS variability in the 65-nm regime and beyond", *IBM Journal of Research and Development*, vol. 50, no. 4.5, pp. 433–449, 2006.
[11] Ghani, T., K. Mistry, P. Packan, S. Thompson, M. Stettler, S. Tyagi, and M. Bohr, "Scaling challenges and device design requirements for high performance sub-50 nm gate length planar CMOS transistors", *IEEE Symposium on VLSI Technology, Digest of Technical Papers*, pp. 174–175, 2000.
[12] Gautier, J., X. Jehl, and M. Sanquer, "Single electron devices and applications", in *NanoCMOS Era* (Simon Deleonibus, ed), Pan Standford Publishing, Singapore, pp. 279–297, 2009.
[13] Deleonibus, S., *Electronic Devices Architectures for the NANO-CMOS Era*, Pan Stanford Publishing, Singapore, 2009. https://www.routledge.com/Electronic-Devices-Architectures-for-the-NANO-CMOS-Era/Deleonibus/p/book/9789814241281
[14] Chandrakasan, A. P., S. Sheng, and R. W. Brodersen, "Low-power CMOS digital design", *IEICE Transactions on Electronics*, vol. 75, no. 4, pp. 371–382, 1992.
[15] Roy, K., S. Mukhopadhyay, and H. Mahmoodi-Meimand, "Leakage current mechanisms and leakage reduction techniques in deep-submicrometer CMOS circuits", *Proceedings of the IEEE*, vol. 91, no. 2, pp. 305–327, 2003.
[16] Falk, H., "Leakage current mechanisms and leakage reduction techniques in deep-submicrometer CMOS circuits", *Proceedings of the IEEE*, vol. 91, no. 2, 2003.
[17] Roy, K., and S. C. Prasad, *Low-power CMOS VLSI Circuit Design*, Wiley-Interscience, Hoboken, NJ, 2009.
[18] Mohanty, S. P., N. Ranganathan, E. Kougianos, and P. Patra, *Low-power High-level Synthesis for Nanoscale CMOS Circuits*, Springer, New York, NY, 2008.
[19] De, V., and S. Borkar, "Technology and design challenges for low power and high performance", *Proceedings of the 1999 International Symposium on Low Power Electronics and Design. ACM*, pp. 163–168, 1999.
[20] Hu, C., *Modern Semiconductor Devices for Integrated Circuits*, Prentice Hall, Upper Saddle River, NJ, 2010.

[21] Yeo, K.-S., and K. Roy, *Low Voltage, Low Power VLSI Subsystems*, McGraw-Hill, Inc., New York, 2004.
[22] Hisamoto, D., T. Kaga, Y. Kawamoto, and E. Takeda, "A fully depleted lean-channel transistor (delta)-a novel vertical ultra thin soi MOSFET", *IEEE International Electron Devices Meeting*, pp. 833–836, 1989.
[23] Hisamoto, D., W.-C. Lee, J. Kedzierski, H. Takeuchi, K. Asano, C. Kuo, E. Anderson, T.-J. King, J. Bokor, and C. Hu, "FinFET-a self-aligned double-gate MOSFET scalable to 20 nm", *IEEE Transactions on Electron Devices*, vol. 47, no. 12, pp. 2320–2325, Dec. 2000.
[24] Hisamoto, D., W.-C. Lee, J. Kedzierski, E. Anderson, H. Takeuchi, K. Asano, T.-J. King, J. Bokor, and C. Hu, "A folded-channel MOSFET for deep-sub-tenth micron era", *IEEE International Electron Devices Meeting*, pp. 1032–1034, 1998.
[25] Young, K., "Short-channel effect in fully depleted SOI MOSFETs", *IEEE Transactions on Electron Devices*, vol. 36, no. 2, pp. 399–402, 1989.
[26] Curanovic, B., "Development of a fully-depleted thin-body FnFET process", 2003. https://scholarworks.rit.edu/theses/7314/
[27] Wong, H.-S., K. K. Chan, and Y. Taur, "Self-aligned (top and bottom) double-gate MOSFET with a 25 nm thick silicon channel", *IEEE International Electron Devices Meeting*, pp. 427–430, 1997.
[28] Monfray, S., et al., "50 nm-gate all around (GAA)-silicon on nothing (SON)-devices: A simple way to co-integration of GAA transistors within bulk MOSFET process", *IEEE Symposium on VLSI Technology*, pp. 108–109, 2002.
[29] Oh, S.-H., et al., "50 nm vertical replacement-gate (VRG) pMOSFETs", *IEEE International Electron Devices Meeting*, pp. 65–68, 2000.
[30] Boucart, K., and A. M. Ionescu, "Length scaling of the double gate tunnel FET with a high-k gate dielectric", *Solid-State Electronics*, vol. 51, no. 11, pp. 1500–1507, 2007.
[31] Colinge, J.-P., C.-W. Lee, I. Ferain, N. D. Akhavan, R. Yan, P. Razavi, R. Yu, A. N. Nazarov, and R. T. Doria, "Reduced electric field in junctionless transistors", *Applied Physics Letters*, vol. 96, no. 7, p. 073510, 2010.
[32] Chou, S., "Intel news release", *Technical Reports.* https://www.intel.com/pressroom/archive/releases/2003/20030612tech.htm
[33] Hu, C., "TSMC's Hu sees earlier entry for FinFET", *Technical Reports.* https://www.eetimes.com/tsmcs-hu-sees-earlier-entry-for-finfet/#
[34] Keyes, R. W., "High-mobility FET in strained silicon", *IEEE Transactions on Electron Devices*, vol. 33, no. 6, pp. 863–863, 1986.
[35] Chaudhry, A., and M. J. Kumar, "Controlling short-channel effects in deep-submicron SOI MOSFETs for improved reliability: A review", *IEEE Transactions on Device and Materials Reliability*, vol. 4, no. 1, pp. 99–109, 2004.
[36] Celler, G., and S. Cristoloveanu, "Frontiers of silicon-on-insulator", *Journal of Applied Physics*, vol. 93, no. 9, pp. 4955–4978, 2003.
[37] Cristoloveanu, S., "From SOI basics to nano-size", *Nanotechnology for Electronic Materials and Devices*, pp. 67, 2010.
[38] Hu, C., "SOI and nanoscale MOSFETs", *Device Research Conference, IEEE*, pp. 3–4, 2001.
[39] Hsiao, T. C., and J. C. Woo, "Subthreshold characteristics of fully depleted submicrometer SOI MOSFET's", *IEEE Transactions on Electron Devices*, vol. 42, no. 6, pp. 1120–1125, 1995.
[40] Colinge, J. P., "Multiple-gate SOI MOSFETs", *Solid-State Electronics*, vol. 48, no. 6, pp. 897–905, 2004.
[41] Colinge, J.-P., et al., *FinFETs and other Multi-gate Transistors,* Springer, 2008. https://www.springer.com/gp/book/9780387717517

[42] Denton, J. P., and G. W. Neudeck, "Fully depleted dual-gated thin-film SOI P-MOSFETs fabricated in SOI islands with an isolated buried polysilicon backgate", *IEEE Electron Device Letters*, vol. 17, no. 11, pp. 509–511, 1996.

[43] Suzuki, K., and T. Sugii, "Analytical models for n+-p+ double-gate SOI MOSFET's", *IEEE Transactions on Electron Devices*, vol. 42, no. 11, pp. 1940–1948, 1995.

[44] Suzuki, K., T. Tanaka, Y. Tosaka, H. Horie, and T. Sugii, "High-speed and low-power n+-p+ double-gate SOI CMOS", *IEICE Transactions on Electronics*, vol. 78, no. 4, pp. 360–367, 1995.

[45] Wong, H.-S., D. Frank, and P. Solomon, "Device design considerations for doublegate, ground-plane, and single-gated ultra-thin SOI MOSFET's at the 25 nm channel length generation", *IEEE International Electron Devices Meeting*, Dec. 1998, pp. 407–410.

[46] Balestra, F., S. Cristoloveanu, M. Benachir, J. Brini, and T. Elewa, "Double-gate silicon on insulator transistor with volume inversion: A new device with greatly enhanced performance", *IEEE Electron Device Letters*, vol. 8, no. 9, pp. 410–412, 1987.

[47] Khandelwal, S., et al., "BSIM-IMG: A compact model for ultrathin-body SOI MOSFETs with back-gate control", *IEEE Transactions on Electron Devices*, vol. 59, no. 8, pp. 2019–2026, 2012.

[48] Wind, S., D. Frank, and H.-S. Wong, "Scaling silicon MOS devices to their limits", *Microelectronic Engineering*, vol. 32, no. 1, pp. 271–282, 1996.

[49] Sun, X., V. Moroz, N. Damrongplasit, C. Shin, and T.-J. K. Liu, "Variation study of the planar ground-plane bulk MOSFET, SOI FinFET, and trigate bulk MOSFET designs", *IEEE Transactions on Electron Devices*, vol. 58, no. 10, pp. 3294–3299, Oct. 2011.

[50] Park, S. H., Y. Liu, N. Kharche, M. S. Jelodar, G. Klimeck, M. S. Lundstrom, and M. Luisier, "Performance comparisons of III_V and strained-Si in planar FETs and nonplanar FinFETs at ultrashort gate length (12 nm)", *IEEE Transactions on Electron Devices*, vol. 59, no. 8, pp. 2107–2114, 2012.

[51] Manoj, C., and V. R. Rao, "Impact of high-k gate dielectrics on the device and circuit performance of nanoscale nFETs", *IEEE Electron Device Letters*, vol. 28, no. 4, pp. 295–297, 2007.

[52] Agrawal, S., and J. G. Fossum, "On the suitability of a high-gate dielectric in nanoscale nFET CMOS technology", *IEEE Transactions on Electron Devices*, vol. 55, no. 7, pp. 1714–1719, 2008.

[53] Pal, P. K., B. K. Kaushik, and S. Dasgupta, "Investigation of symmetric dual-(k) spacer trigate FinFETs from delay perspective", *IEEE Transactions on Electron Devices*, vol. 61, no. 11, pp. 3579–3585, 2014.

[54] Pal, P. K., B. K. Kaushik, and S. Dasgupta, "Asymmetric dual-spacer trigate nFET device circuit codesign and its variability analysis", *IEEE Transactions on Electron Devices*, vol. 62, no. 4, pp. 1105–1112, 2015.

[55] Mathew, L., et al., "Inverted-T channel FET (ITFET)-fabrication and characteristics of vertical horizontal, thin body, multi-gate, multi-orientation devices, ITFET SRAM bit-cell operation: A novel technology for 45nm and beyond CMOS", *IEEE International Electron Devices Meeting*, pp. 713–716, 2005.

[56] Zhang, W., J. G. Fossum, and L. Mathew, "The ITFET: A novel FinFET-based hybrid device", *IEEE Transactions on Electron Devices*, vol. 53, no. 9, pp. 2335–2343, 2006.

[57] Zhang, W., J. G. Fossum, and L. Mathew, "A hybrid NFET/SOI MOSFET", *IEEE International SOI Conference*, pp. 151–153, 2005.

[58] Fahad, H. M., A. M. Hussain, G. T. Sevilla, and M. Hussain, "Wavy channel transistor for area efficient high performance operation", *Applied Physics Letters*, vol. 102, no. 13, p. 134109, 2013.

[59] Fahad, H. M., C. Hu, and M. M. Hussain, "Simulation study of a 3-D device integrating FinFET and UTBFET", *IEEE Transactions on Electron Devices*, vol. 62, no. 1, pp. 83–87, 2015.

[60] Sze, S. M., *Physics of Semiconductor Devices*, third edition, John Wiley & Sons, 2009. https://onlinelibrary.wiley.com/doi/book/10.1002/0470068329
[61] Vigna, B., "More than Moore: Micro-machined products enable new applications and open new markets", *IEEE International Electron Devices Meeting*, pp. 8, 2005.
[62] Kahng, A. B., "Scaling: More than Moore's law", *IEEE Design & Test of Computers*, vol. 27, no. 3, pp. 86, 87, 2010.
[63] Flandre, D., L. Ferreira, P. Jespers, and J.-P. Colinge, "Modelling and application of fully depleted SOI MOSFETs for low voltage, low power analogue CMOS circuits", *Solid-State Electronics*, vol. 39, no. 4, pp. 455–460, 1996.
[64] Razavi, B., "CMOS technology characterization for analog and RF design", *IEEE Journal of Solid-State Circuits*, vol. 34, no. 3, pp. 268–276, 1999.
[65] Kilchytska, V., et al., "Influence of device engineering on the analog and RF performances of SOI MOSFETs", *IEEE Transactions on Electron Devices*, vol. 50, no. 3, pp. 577–588, 2003.
[66] Lazaro, A., and B. Iniguez, "RF and noise performance of double gate and single gate SOI", *Solid-State Electronics*, vol. 50, no. 5, pp. 826–842, 2006.
[67] Raskin, J.-P., T. M. Chung, V. Kilchytska, D. Lederer, and D. Flandre, "Analog/RF performance of multiple gate SOI devices: Wideband simulations and characterization", *IEEE Transactions on Electron Devices*, vol. 53, no. 5, pp. 1088–1095, 2006.
[68] Mohankumar, N., B. Syamal, and C. Sarkar, "Influence of channel and gate engineering on the analog and RF performance of DG MOSFETs", *IEEE Transactions on Electron Devices*, vol. 57, no. 4, pp. 820–826, Apr. 2010.
[69] Sarkar, A., A. K. Das, S. De, and C. K. Sarkar, "Effect of gate engineering in double-gate MOSFETs for analog/RF applications", *Microelectronics Journal*, vol. 43, no. 11, pp. 873–882, 2012.
[70] Sharma, R. K., and M. Bucher, "Device design engineering for optimum analog/RF performance of nanoscale DG MOSFETs", *IEEE Transactions on Nanotechnology*, vol. 11, no. 5, pp. 992–998, 2012.
[71] Matsuzawa, A., "High quality analog CMOS and mixed signal LSI design", *IEEE International Symposium in Quality Electronic Design*, pp. 97–104, 2001.
[72] Silveira, F., D. Flandre, and P. Jespers, "A g_m/ID based methodology for the design of CMOS analog circuits and its application to the synthesis of a silicon-on-insulator micropower OTA", *IEEE Journal of Solid-State Circuits*, vol. 31, no. 9, pp. 1314–1319, 1996.
[73] Yu, B., C. H. Wann, E. D. Nowak, K. Noda, and C. Hu, "Short-channel effect improved by lateral channel-engineering in deep-submicronmeter MOSFET's", *IEEE Transactions on Electron Devices*, vol. 44, no. 4, pp. 627–634, 1997.
[74] Chen, C., et al., "High-performance fully-depleted SOI RF CMOS", *IEEE Electron Device Letters*, vol. 23, no. 1, pp. 52–54, 2002.
[75] Boots, H. M., G. Doornbos, and A. Heringa, "Scaling of characteristic frequencies in RF CMOS", *IEEE Transactions on Electron Devices*, vol. 51, no. 12, pp. 2102–2108, 2004.
[76] Dambrine, G., C. Raynaud, D. Lederer, M. Dehan, O. Rozeaux, M. Vanmackelberg, F. Danneville, S. Lepilliet, and J.-P. Raskin, "What are the limiting parameters of deep-submicron MOSFETs for high frequency applications?" *IEEE Electron Device Letters*, vol. 24, no. 3, pp. 189–191, 2003.
[77] Priji¢, Z., Z. Pavlovi¢, S. Risti¢, and N. Stojadinovi¢, "Zero-temperature-coefficient (ZTC) biasing of power VDMOS transistors", *Electronics Letters*, vol. 29, no. 5, pp. 435–437, 1993.
[78] Shoucair, F., "Analytical and experimental methods for zero-temperature-coeffcient biasing of MOS transistors", *Electronics Letters*, vol. 25, no. 17, pp. 1196–1198, 1989.
[79] Artola, L., et al., "SEU prediction from set modeling using multi-node collection in bulk transistors and SRAMs down to the 65 nm technology node", *IEEE Transactions on Nuclear Science*, vol. 58, no. 3, p. 1338, 2011.

[80] Mitra, S., P. Sanda, and N. Seifert, "Soft errors: Technology trends, system effects, and protection techniques", *13th IEEE International On-Line Testing Symposium (IOLTS 2007)*, p. 4, 2007.
[81] Vial, C., J. Palau, J. Gasiot, M. Calvet, and S. Fourtine, "A new approach for the prediction of the neutron-induced SEU rate", *IEEE Transactions on Nuclear Science*, vol. 45, no. 6, pp. 2915–2920, 1998.
[82] www.tsmc.com/english/dedicatedFoundry/technology/logic/l_7nm
[83] www.ieee.org/ns/periodicals/EDS/EDS-JULY-2020-HTML/index.html
[84] https://spectrum.ieee.org/semiconductors/devices/the-nanosheet-transistor-is-the-next-and-maybe-last-step-in-moores-law

2 High-Performance Tunnel Field-Effect Transistors (TFETs) for Future Low Power Applications

Ribu Mathew, Ankur Beohar, and Abhishek Kumar Upadhyay

CONTENTS

2.1 INTRODUCTION

From inception, the advent of electronics industry has been propelled by advancements made in the field of integrated circuit (IC) technology. In recent years, continuous miniaturization along with batch fabrication of ICs has resulted in high performance to cost index of modern electronics system. This tremendous growth in the semiconductor IC industry has been due to the expansion in the material set, novel device design, and innovations in circuit and architectural levels. The silicon (Si)-based metal oxide semiconductor field-effect transistor (MOSFET) has been the cornerstone device for the growth of the IC industry, especially for commercial electronics applications.

Over the years, continuous scaling of Si MOSFETs has resulted in an improvement in performance to cost index. However, in the sub-nano meter technology regime, the performance of MOSFET devices has been limited due to short-channel effects (SCEs) like subthreshold conduction, drain-induced barrier lowering (DIBL), and high leakage current, to mention a few. Even though innovative MOSFET designs based on gate and channel engineering like double gate (DG) MOSFET, dual material

DOI: 10.1201/9781003200987-2

gate (DMG) MOSFET, dual material surrounding gate (DMSG), etc. structures have been reported, a major performance limitation of such devices has been the limit of scale supply voltage due to subthreshold swing of value 60mV/decade at room temperature. A solution to continue scaling and improve the performance of ICs is an alternative device, that is, a tunnel field-effect transistor (TFET). Compared to a conventional MOSFET device, a TFET has advantages in terms of high ON current (I_{ON}) to OFF current (I_{OFF}) ratio, faster switching times, improved subthreshold swing (SS), thereby low leakage current and low power consumption. The aforementioned characteristics of TFET devices provide them an edge over conventional MOSFET for high-speed and low power applications [1–5]. In addition, in recent years, in the field of healthcare, TFET-based biosensors have been extensively explored as an alternative to micro/nano electro-mechanical systems (MEMS/NEMS)-based chemical/biological sensors [6–15]. Compared to MEMS/NEMS sensors, TFET-based sensors hold advantages such as better integration with CMOS process flow, simplicity of fabrication, and relatively mechanical stability, to mention a few.

This treatise encompasses examples of TFET devices with various configurations of TFET devices for numerous applications. In this chapter, we elucidate silicon (Si), germanium (Ge) and silicon-germanium (SiGe) materials-based TFET devices. The focus of this chapter is to explain the basic device operation of TFET with mathematical models. Further, a comparison of various TFET devices considering their geometry and material selection is performed. In the later sections, various device configurations and operations of Si TFET, Ge TFET, and SiGe TFET are analyzed, followed by the design challenges and future.

2.2 TFET DEVICE MODELS AND OPERATIONS

A typical TFET device is a p-i-n diode that behaves like a transistor under reverse bias condition. A typical hetero gate dielectric cylindrical GAA SiGe-based TFET with a high-density delta (HDD) layer is inserted across the source-channel junction. The 3-D structure of the proposed device along with the cross-sectional view is shown in Figure 2.1. In a generic TFET structure, an intrinsic (i) region is sandwiched between two oppositely doped regions (n and p). Unlike conventional MOSFET devices in which the electrical conduction is based on drift-diffusion of carriers, the underlying physics of conduction in TFET devices is band-to-band tunneling (BTBT) of carriers across device junction. When operated in reverse bias condition, electrical conduction takes place due to BTBT phenomenon controlled by the external applied gate voltage. The tunneling of majority charge carriers through source channel interface constitutes the device current.

The low tunneling current is a major concern in silicon (Si)-based TFET due to the relatively high bandgap and low tunneling effective mass (m^*). It is well known that the BTBT current (I_{BTBT}) has lower dependency on the applied external electric field, which may cause lower transconductance. To improve the ON current (I_{ON}) of a TFET device, the following approaches can be used:

- Use of lower bandgap material in the source/channel region
- Lower equivalent oxide thickness (EOT) of dielectric material
- More abrupt doping profile in the source region

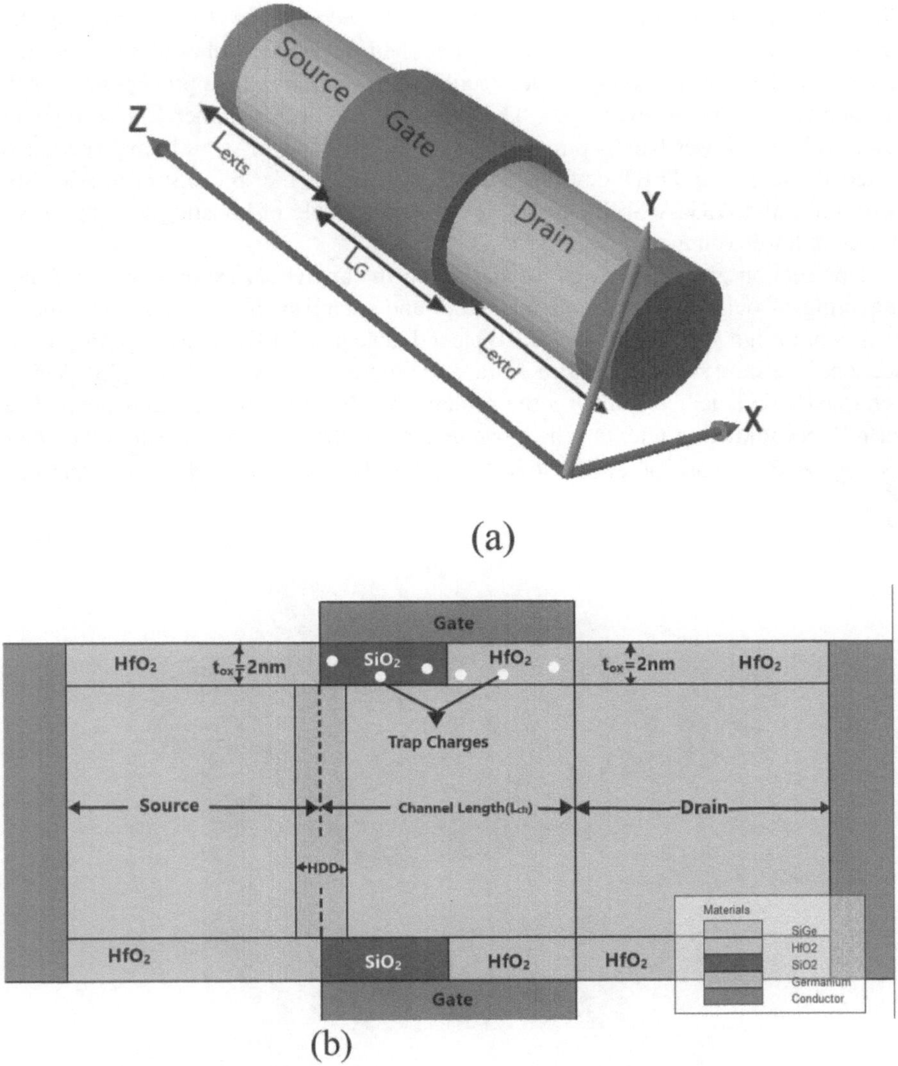

FIGURE 2.1 Cyl.GAA Hetero gate TFET Device: (a) 3-D Device View and (b) 2-D Cross-Sectional View

The lower bandgap material offers the lower tunnel barriers and increases the BTBT generation rate, which is an exponential function of the bandgap. For this purpose, the use of SiGe or Ge in the entire active region of the device area will increase the parameter I_{ON}. Similarly, lower EOT offers high I_{ON} by establishing the higher coupling between the gate voltage (V_G) and the channel potential (V_{CH}). However, recently the thickness of gate dielectric materials has been chosen to be less, so lower EOT may result in indirect tunneling leakage and thereby leakage current. To overcome this issue, high-k dielectric materials are used as gate dielectric.

Similarly, the abrupt source doping also helps to increase the I_{ON} by reducing the tunneling barrier width [16]. The aforementioned three approaches along with gate oxide and SOI layer thickness reduction and improved fabrication process have been considered to improve the I_{ON} of TFET devices. To obtain higher ON currents at lower voltages, lower bandgap materials like Ge [17] and InAs [18] have been also tested for realizing TFET devices. The heterojunction like SiGe/Si [19], AlGaSb/InAs [20], and AlGaAsSb/InGaAs [21] are also capable of boosting I_{ON} of TFET device at lower voltages.

This section gives a detailed discussion on the analytical expressions for Zener tunneling in bulk and 1-D semiconductors and establish the principles for tunnel transistor design [22]. Schematic of a typical double gate TFET along with the cross-sectional view are shown in Figure 2.2 (a) and Figure 2.2 (b), respectively [23]. A conventional TFET device works on the principle of Zener tunneling mechanism. The term Zener tunneling refers to the movement of electrons from one band to another energy band. In case of TFET, $P^{+}N^{+}$ junction is heavily doped. Thus, electrons

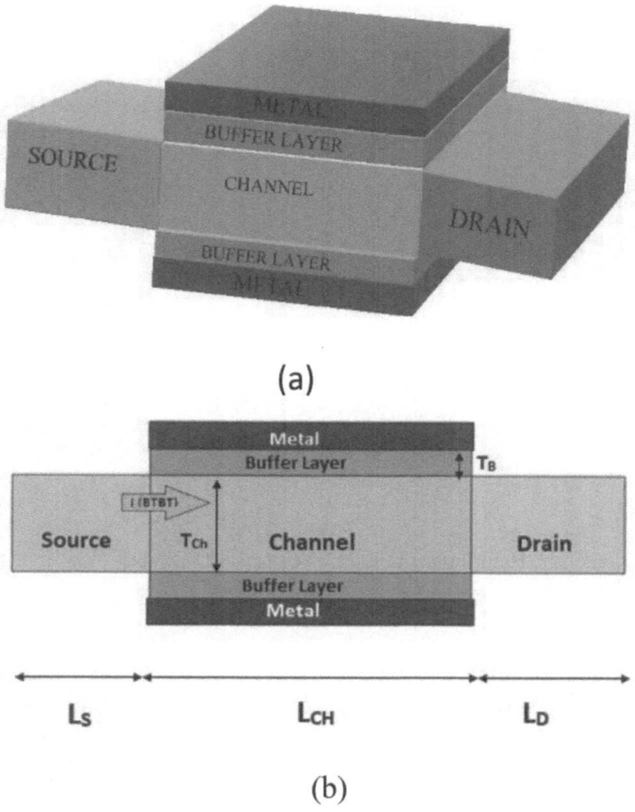

FIGURE 2.2 (a) The 3-D View and (b) Cross-Sectional View of Double Gate (DG) TFET (Reproduced from Ref. [23].)

from the valance band of P^+ semiconductor can tunnel to the conduction band of N^+ semiconductor under reverse biased condition. The tunneling probability of electrons is calculated by the Wentzel-Kramers-Brillouin (WKB) approximation. Here the electrons from the valance band of P^+ tunnel through the forbidden energy gap through the conduction band of N^+ region. This tunneling mechanism where the electrons pass through the triangular-shaped potential barrier with height is equal to or greater than the bandgap and the slope of potential barrier is equal to the electron charge (q) times applied electric field $(\mathfrak{I})$.

In bulk semiconductor, the longitudinal (E_x) and transverse $(E_\perp)$ energy is additively faced by charge carriers. Longitudinal energy $\left(E_x = \hbar^2 k_x^2 / 2m_R^*\right)$ corresponds to the tunneling direction, while transverse energy $\left(E_\perp = \hbar^2\left(k_y^2 + k_z^2\right) / 2m_R^*\right)$ is conserved during the entire tunneling process. Here $\hbar$ is the reduced Planck constant, and the reduced effective mass $m_R^* = \left(1/m_e^* + 1/m_h^*\right)^{-1}$, m_e^* and m_h^* is the effective mass of electron and hole, respectively.

The expression for the wave vector in the direction of tunneling is [23]

$$K_x(x) = \sqrt{\frac{2m_R^*}{\hbar^2}\left(E_x - U(x)\right)} \tag{2.1}$$

where $U(x)$ is the potential energy profile. In the tunnel barrier, the $U(x) = E + q\mathfrak{I}$, in the region $0 < x < d$, where $d = E_G / q\mathfrak{I}$. The mathematical model of the wave vector is as follows:

$$K_x(x) = \sqrt{\frac{2m_R^*}{\hbar^2}\left(-q\mathfrak{I}x - E_\perp\right)} \tag{2.2}$$

The *WKB* approximation is used to calculate the tunneling probability of charge carrier [24].

$$T_{WKB} \approx exp\left(-2\int_0^d \left|K_x(x)\right| dx\right) \tag{2.3}$$

Now, putting the $K_x(x)$ from (2.2) in (2.3), we have

$$T_{WKB} \approx exp\left(-\frac{4\sqrt{2m_R^*}\left(E_G + E_\perp\right)^{3/2}}{3q\hbar\mathfrak{I}}\right)$$

$$\approx exp\left(-\frac{4\sqrt{2m_R^*}\left(E_G\right)^{3/2}}{3q\hbar\mathfrak{I}}\right) exp\left(-\frac{E_\perp}{\mathcal{R}}\right) \tag{2.4}$$

where $\mathcal{R} = \left(q\hbar\mathfrak{I}\right) / 2\sqrt{2m_R^* E_G}$ is used to determine the impact of transverse-energy-state carriers on the tunneling magnitude.

The Zener tunneling current density (J_{3D}) in bulk semiconductor is calculated by the charge flux times the tunneling probability from Fermi energy at the P^{+} side to the N^{+} side [25]

$$J_{3D} = \int\int q v_G \rho_{x\perp} dk_x 2\pi k_{\perp} dk_{\perp} \times (f_V - f_C) T_{WKB} \tag{2.5}$$

where v_G is the group velocity, $\rho_{x\perp}$ is the density of states in the tunneling direction, and in the transverse direction, f_V and f_C are the Fermi-Dirac distribution at the valence band of the P^{+} side and conduction band of the N^{+} side, respectively.

For mathematical simplicity, here it has assumed that $f_V - f_C \approx 1$. Then (2.5) will become

$$J_{3D} \approx \left(\frac{q m_R^{*}}{2\pi^2\hbar^2}\right) \int\int T_{WKB}^{1D} exp\left(\frac{-E_{\perp}}{\mathcal{R}}\right) dE_{\perp} dE_x \tag{2.6}$$

Now integrating E_x form 0 to qV_R and $E_{\perp}$ from 0 to $qV_R - E_x$, we have the expression for derived current density, which is applicable for both direct and indirect semiconductors

$$J_{3D} = \frac{\sqrt{2m_R^{*}} q^3 \Im V_R}{8\pi^2\hbar^2 E_G^{1/2}} exp\left(-\frac{4\sqrt{2m_R^{*}} E_G^{3/2}}{3q\Im\hbar}\right) \tag{2.7}$$

where symbols have their usual meanings.

In case of an ultra-thin channel material, 1-D Zener tunneling takes place. In this case, the transverse energy is quantized and included in the increased bandgap, that is, substituting $E_{\perp} = 0$ in (2.4), we have the new equation for the tunneling probability, which is given by [22]

$$T_{WKB}^{1D} \approx exp\left(-\frac{4\sqrt{2m_R^{*}}(E_G)^{3/2}}{3q\hbar\Im}\right) \tag{2.8}$$

The 1-D Zener tunneling current is also calculated by following a similar approach as in bulk Zener tunneling.

$$I_{1D} = \frac{q^2}{\pi\hbar} T_{WKB}^{1D} V_T ln\left(\frac{1}{2}\left(1 + cosh\frac{V_R}{V_T}\right)\right)$$

If $V_R \gg V_T$, the tunneling current becomes $I_{1D} = q^2 / \pi\hbar T_{WKB}^{1D} (V_R - V_T) ln(2.4)$. Furthermore, if $T = 0$, the tunneling is directly proportional to the applied reverse bias, that is, $I_{1D} = \frac{q^2}{\pi\hbar} T_{WKB}^{1D} (V_R) ln(2.4)$.

2.3 MULTI-GATE TFETs

As discussed in earlier sections, the design of analog/RF devices for ultra low power circuit applications has become increasingly challenging with the rigorous down-scaling in standard planar digital CMOS technologies in the deep sub-micrometer [26, 27]. However, the novel device architecture like multi-gate device (MuGFET) and high-k dielectric material will be used to improve the electrical properties of analog/RF device, which will also influence the circuit performance [28]. The MuGFETs have a strong potential to extend the *CMOS* scaling into the sub-30 nm regime [29]. They offer superior electrostatic gate control over the channel charge and offer low screening length due to multiple gates with suppressed *SCEs* and leakage current. The optimized structure also helps to mitigate several other issues of nanoscale MOSFETs, for example, mobility degradation, random dopant fluctuation, and compatibility with mid-gap material gate [30]. Further, strained silicon, metal gate, and high-k dielectric gate material can also enhance the driving current of MOS device. The screening length can also be reduced by decreasing the thickness of gate oxide or by using the low bandgap material. The circuit performance can also be improved by the use of novel gate stack material, reduced parasitic capacitance, and hole mobility improvement. Therefore, the MuGFETs are strong candidates for replacing conventional single gate MOSFET in the future.

2.4 CLASSIFICATION OF TFETs BASED ON MATERIAL

This treatise encompasses examples of innovative designs of TFET with careful material selection and geometrical design to maximize performance to cost index ratio. In this section, we elucidate TEFTs based on silicon (Si), germanium (Ge), and silicon-germanium (SiGe) materials with focus on their device structure, physics of operation, and device characteristics.

Various TFET devices reported in the literature along with their features are summarized in Table 2.1.

SI-TEFT: Source inside TFET; DI-TEFT: Drain inside TFET; N_S: Source doping concentration; N_D: Drain doping concentration; N_{ch}: Channel doping concentration; t_{si}: thickness of the channel.

2.5 Si TFETs

The conventional design and operation of a TFET device has been explained in section 2. In this section, we focus on various Si-based TFET devices and their features.

This treatise encompasses examples of Si-TFET designs that depict better performance to cost index than conventional TFET design. This improvement in performance metrics of Si-TFET has been accomplished by judicious material selection and geometrical optimization coupled with innovative designs. For instance, Kumari et al. [31] reported a ring tunnel field-effect transistor (R-TFET) with two configurations: (i) source inside TFET (SI-TFET) and (ii) drain inside TFET (DI-TFET). The virtual device structure of R-TFET is shown in Figure 2.3.

TABLE 2.1

Summary of Si, Ge, and SiGe Material-Based TFET Device Features from the Literature

Sr. No.	Ref no.	Device	Technology	Material	Geometry	Features
1	[31] Kumari et al.	Ring field-effect transistor (R-TFET)	22 nm, 32 nm, 45 nm, 65 nm	Si $N_s = 10^{20}$ cm^{-3} (Boron), $N_{ch} = 10^{17}$ cm^{-3} (Arsenic), and $N_d = 5\times10^{18}$ cm^{-3} (Arsenic),	Ring field-effect transistor (R-TFET) (i) SI-TFET, and (ii) DI-TFET tsi= 10 nm	- better immune to short-channel effects - high ON current compared to DG-TFET and SI-TFET
2	[32] Ma et al.	Low temperature polycrystalline silicon tunnel field-effect transistor (LTPS-TFET)	1μm, 2μm, 5μm, 10μm, 20μm	Polycrystalline silicon Source: boron at a dose of 5×10^{15} cm^{-2} Drain: phosphorus at a dose of 5×10^{15} cm^{-2}	Gate oxide: 50 nm-thick SiO_2	- Suitable to realize modules of 3-D ICs
3	[33] Liu et al.	High Schottky barrier bidirectional tunnel field-effect transistor (HSB-BTFET)	4 nm	Si doping concentration of p++ region of TFET = 10^{20} cm^{-3}	High Schottky barrier-TFET	- Source and drain region are symmetric and interchangeable
4	[34] Liu et al.	Si nanowire (NW) Gate-all-around (GAA) TFET	Gate length = 200 nm, radius of SiN = 35 nm	Si, Gate dielectric: SiO_2 Source: p-type 1×10^{19} cm^{-3} Drain: n-type 1×10^{19} cm^{-3}, Channel doping: 1×10^{17} cm^{-3}	Gate-all-around (GAA) TFET	- Gate-source (GS) overlap structure has better performance than GS underlap. - Source doping gradient (SDG) has negligible impact on performance.
5	[35] Ghosh et al.	All Ge-pTFET GaAsP-Ge-TFET SiGe-Ge-TFET InAlAs-Ge-TFET	50 nm	Source, body, drain - all Ge Source -GaAsP, body, drain-Ge Source-SiGe, body, drain-Ge Source-InAlAs, body, drain-Ge	Double-gate conventional	Type-II staggered heterostructure

Sr. No.	Ref no.	Device	Technology	Material	Geometry	Features
6	[36] Datta et al.	Ge-source TFET	5μm	Source-Ge, body, drain-Si	Single-gate conventional	linearity, harmonic distortion, analog performance dependency on temperature
7	[37] Vanlalawpuia et al.	Ge-Source Vertical TFET	30 nm	Source-Ge, body, drain-Si	Single-gate vertical	Delta-doped layer
8	[38] Paras et al.	Ge-source dual metal gate vertical TFET (VGeDMG)	20 nm	Source-Ge, body, drain-Si	Dual-gate vertical	Applicable to low-power device with enhancement in DC and RF characteristics
		Si-source dual metal gate vertical TFET (VGeDMG)	20 nm	Source, body, drain-all Si		
		Si-source dual metal gate TFET	50 nm		Dual-gate conventional	
9	[39] Li et al.	Ge-pocket TFET	20 nm	Source-SiGe, body, drain-Si	Dual-gate conventional	Insertion of Ge-pocket in between source and channel junction
10	[40] Kato et al.	p-channel TFET based on bilayer structures	100 nm	• SiGe • ZnSnO • SiGe-on-insulator • p-Ge • n-ZnO	• Bilayer TFET hetero-structure • 2-D	• It is not simple to deposit a high-quality p-IV layer on an n-OS material.
11	[41] Kumar et al.	Charge-plasma-based engineered dual metal-heterogeneous gate Si-Si0.55Ge0.45	25 nm	• $ZrSiO_4$ • SiO_2 • SiGe–Si, GaAs, InAs, and AlGaAs • SiGe	• Dopingless gate-all-around (GAA) nanowire tunnel field-effect transistor (NWTFET) • 3-D	• High-k dielectrics in thin layers matching the Si lattice are difficult to produce.
12	[42] J. Lee et al.	P-type SiGe channel tunnel field-effect transistor	20 nm	• SiGe nanowire • Al_2O • HfO_2	• 2-D	• The TFET with the concentration-graded SiGe channel can improve drive current due to a smaller bandgap.

TABLE 2.1 *(Continued)*

Summary of Si, Ge, and SiGe Material-Based TFET Device Features from the Literature

Sr. No.	Ref no.	Device	Technology	Material	Geometry	Features
13	[43] Mayer et al.	Si1-xGexOI and GeOI substrates on CMOS compatible tunnel FET	100 nm	• HfO_2 • GeOI • TiN	• 2-D	• Leading to parasitic conduction in the OFF state
14	[44] Bhuwalka et al.	P-channel tunnel field-effect transistors	50 nm	• SiGe on insulator • Polysilicon • Strained SiGe	• 2-D	• Further propose a lateral tunnel FET on SiGe-on-insulator with high on currents.
15	[45] Bae et al.	Ge/Si hetero-junction tunneling field-effect transistors	30 nm	• Ge • Strained Si • Source contact Ni • Al contact pads • Al_2O_3	• 2-D	• Ultra low power applications.
16	[46] Walke et al.	$Si/Si_{0.55}Ge_{0.45}$ Heterojunction tunnel FET	70 nm	• N-type Si/SiGe heterojunction • + SiGe source capped • SiO_2 / HfO_2(1.8 nm)/TiN (2 nm)/P-doped	• 2-D	• The device shows an increase in tunneling current with gate length.
17	[47] Shafi et al.	SiGe tunnel FET	54 nm	• SiGe • SiO_2 • Intrinsic silicon	• 2-D	• Lower energy bandgap material in source region of TFET. • Trans-conductance to drain current ratio is 28% higher in CP-TFET than conventional TFET.
18	[48] Patel et al.	N-channel tunnel FET with strained SiGe layer	30 nm	• Si • SiGe • Strained SiGe • SiO_2 • SiN_4	• 2-D	• The proposed device is nearly free of SCE and DIBL and can be scaled up to channel lengths of 30 nm.

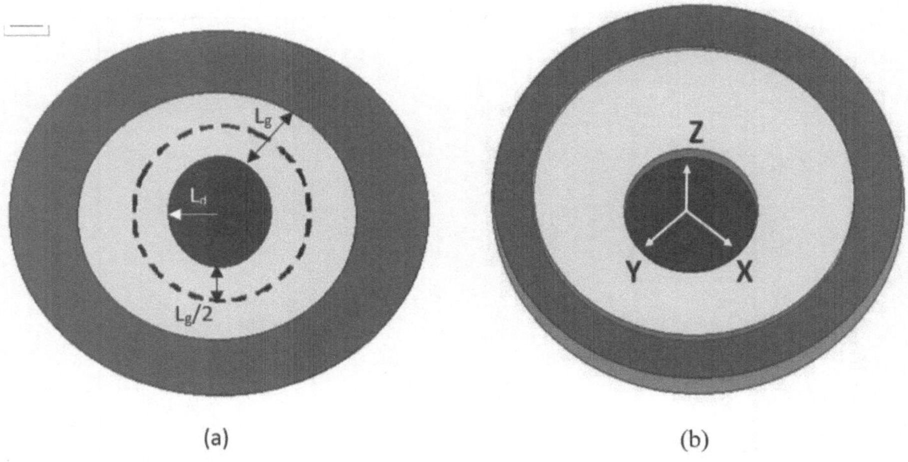

FIGURE 2.3 (A) A 3-D View and (b) a 2-D Cross-Sectional View (across Z-axis) of an R-TFET

(Redrawn from Ref. [31].)

I_{DS}-V_{GS} characteristics of the DG-TFET, DI-TFET, and SI-TFET devices at drain source (V_{DS}) = 0.5 V is depicted in Figure 2.4. It is evident that DI-TFET has better linear characteristics at higher V_{GS}.

Liu et al. [33] reported a high Schottky barrier bidirectional tunnel field-effect transistor (HSB-BTFET) device with symmetric and interchangeable source and drain region. The top view and cross-sectional view (across AA) of the HSB-BTFET is shown in Figure 2.5. In the device, Schottky contacts are formed at source/drain interface. The main control gate is a pair of brackets formed on both lateral sides of source and drain controlling three sides of Silicon. On the other hand, the assistant gate controls the central part of Silicon body.

The transfer characteristics of the HSB-BTFET along with Schottky barrier (SB)-MOSFET and TFET are shown in Figure 2.6. It is found that the reverse bias transfer characteristics of SB-MOSFET are poor as there is enhanced leakage current for larger V_{DS}. The premise is low for an SB-MOSFET. Further, it is seen that the forward current of TFET is relatively small. It is observed that the HSB-BTFET depicts higher ON current and sharper SS compared to TFET and SB-MOSFET.

Liu et al. [34] reported a silicon nano wire (NW) gate-all-around (GAA) TFET as shown in Figure 2.7. The impact of gate-source (GS) overlap and underlap along with source doping gradient was investigated using numerical simulations. The material and geometrical features have been summarized in Table 2.1. It was found that the BTBT occurs at the edge of gate and source end instead of source and channel interface. Further, it was observed that GS overlap TFET depicts better performance than other combinations of overlap structures because BTBT occurs at the source-end gate edge. The premise is also due to the fact that high space charge density and sharp potential change occur in GS overlap TFET structures. Further, through simulations it is observed that for a GS underlap structure, the SDG determines the doping

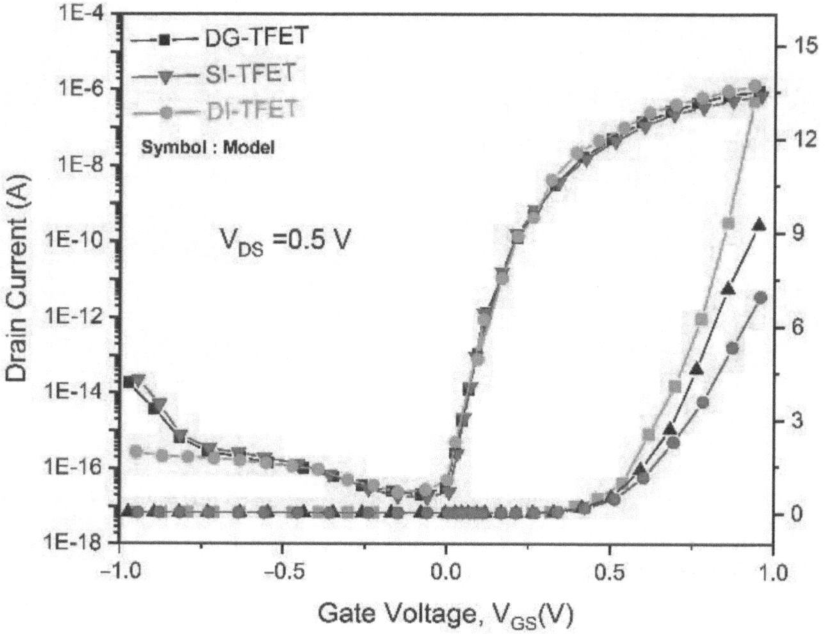

FIGURE 2.4 Characteristics of Three Devices DG-TFET, DI-TFET, and SI-TFET: I_{DS}-V_{GS} at V_{DS} = 0.5 V

(Reproduced from Ref. [31].)

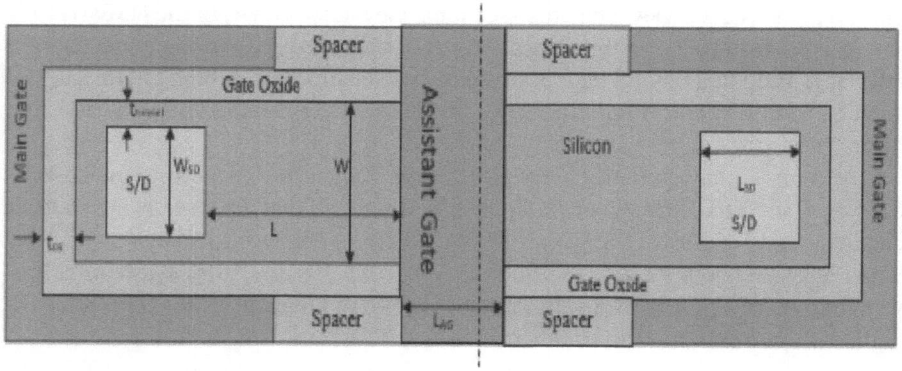

FIGURE 2.5 HSB-BTFET Device: Top View

(Redrawn from Ref. [33].)

concentration of BTBT region. For instance, for a GS underlap TFET, larger SDG results in higher doping concentration of BTBT region, thereby improving the device performance. The premise is evident from the transfer characteristics of the device for various geometrical configurations as shown in Figure 2.7 (b).

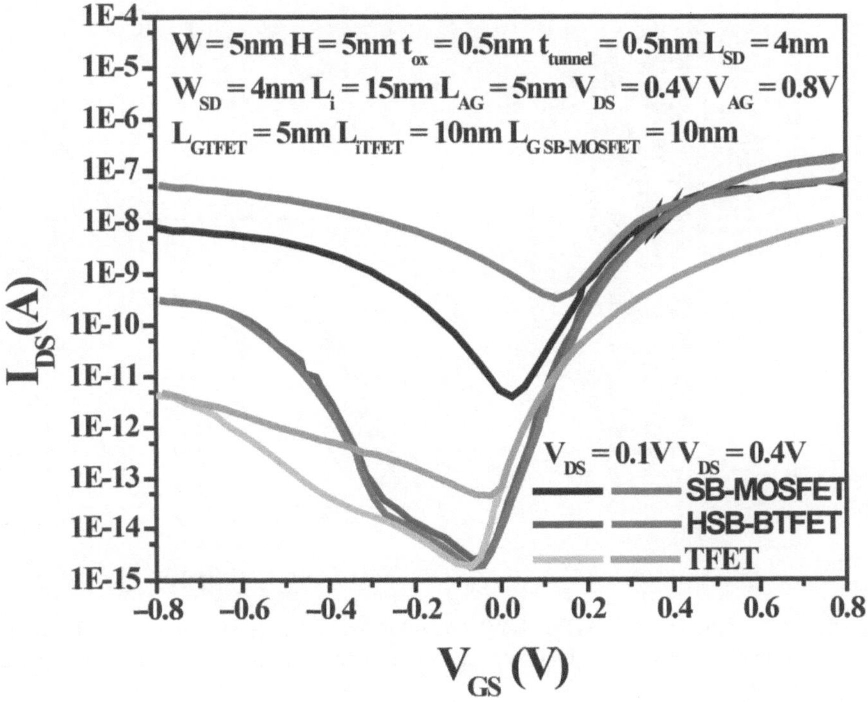

FIGURE 2.6 Transfer Characteristics of the HSB-BTFET along with SB-MOSFET and TFET

(Reproduced from Ref. [33].)

Various Si-based TFET devices reported in the literature along with their features are summarized in Table 2.2.

Abbreviations: SI-TEFT: source inside TFET; DI-TEFT: drain inside TFET; DIBT: drain-induced barrier thinning; TFT: Thin-film transistor; SS: subthreshold swing; t_{tunnel}: intrinsic tunneling region thickness between gate oxide and source/drain contact; L_{SD}: source/drain contacts length; W_{SD}: source/drain contacts width; Li: intrinsic silicon region length between the source/drain contact and the assistant gate contact; H: silicon body height; L_{AG}: assistant gate length; W: silicon body width.

2.6 Ge TFETs

In this section, we will elucidate the details related to Ge TFET device. The higher effective mass and larger bandgap of silicon (Si) material cause the lower ON current (I_{on}) and switching speed of TFET device. To mitigate this concern, Germanium (Ge) is used as substrate due to the smaller bandgap. In addition, the higher value of ON current (I_{on}) can also be achieved.

In TFETs, the higher doping and lower bandgap at the source-body tunnel junction will affect the ON current (I_{on}) of the device. This is because of the lower bandgap

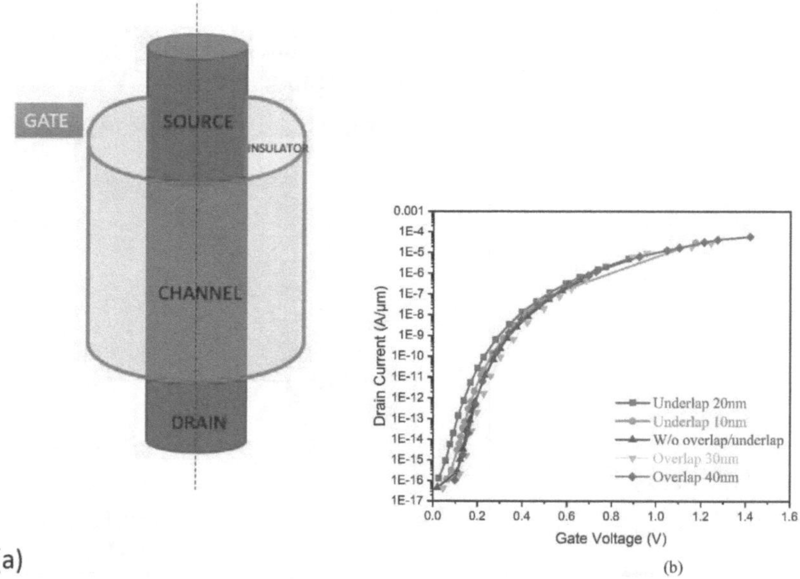

FIGURE 2.7 Silicon Nano Wire (NW) Gate-All-Around (GAA) TFET: (a) 3-D Model and (b) Drain Current (I_{DS}) vs Gate Voltage (V_G) of the Device for Various Geometrical Configurations

(Reproduced from Ref. [34].)

results in the lower tunneling barrier at source-channel junction, causing an increase in the charge carrier tunneling rate, which boosts the ON current (I_{on}).

By utilizing the material properties of germanium, there are researchers that offer better performance of TFET in terms of I_{ON}.

Ghosh et al. [35] did a comparative analysis on the analog performance of the Ge-body-based pTFETs with different compound semiconductors (InAlAs, SiGe, and GaAsP) as source materials. Their study confirms that the performance of this kind of pTFETs with different compound semiconductor source materials offers comparatively superior analog/RF performance. Moreover Datta et al. claimed that the Ge-source TFET offers lower harmonic distortion (HD). The HD arises due to the nonlinear characteristics of the device. It is an important concern in analog-/mixed-signal applications [36].

Vanlalawpuia et al. [37] claimed that the δ-doped germanium-source vertical TFET (Ge-source vTFET) is a potential candidate for the low-power application. For this device structure, higher I_{ON} is obtained due to higher charge carrier tunneling in a direction parallel to the gate electric field. The δ-doped source further reduces the I_{OFF}.

TFET with substrate material as silicon has been explored to overcome the SCEs of the conventional MOSFETs. However, TFET device also suffers from limitations such as ambipolar behavior, low ON current, and limited OFF current. A possible alternative at the material level is germanium (Ge) that due to its low bandgap can reduce tunneling width and increase ON current at the source-to-channel junction of a

TABLE 2.2
Summary of Si TFET Device Parameters

Sr. no.	Ref no.	Simulation model/fabrication	Device parameters Gate length (Lg), I_{on}, I_{off}, g_m, f_T, f_{max}
1	[31] Kumari et al.	Simulation Dynamic nonlocal path band-to-band tunneling model, The Kane's BTBT model with parameters ($A = 4\times10^{14}$ $cm^{-3}s^{-1}$ and $B = 1.9\times10^{7}$ V/cm), Masetti and Highk-Lombardi models, Bandgap narrowing model, Shockley-Read-Hall (SRH) with doping-dependent lifetime model, and Auger recombination model	Lg = 22 nm SI-TFET: I_{ON} = 10.5μA, $I_{ON}/I_{OFF} = 2.1\times10^9$, DIBT = 152mV/V, g_m = 48.2μS, SS = 52.2mV/decade DI-TFET: I_{ON} = 18.7 μA, $I_{ON}/I_{OFF} = 1.1\times10^9$, DIBT = 89mV/V, g_m = 94.9μS, SS = 48.6mV/decade
2	[32] Ma et al.	Fabrication LTPS-TEFT LTPS-TFT	LTPS-TEFT: W/L = 100μm/10μm, V_{TH} = 3.456V, SS = 0.688 V/decade, $I_{ON}/I_{min} = 2.89\times10^6$ LTPS-TFT: W/L = 100μm/10μm, V_{TH} = 2.581V, SS = 0.359 V/decade, $I_{ON}/I_{min} = 6.04\times10^6$
3	[33] Liu et al.	Simulation SRH recombination model, Auger recombination model, Fermi-Dirac statistic model, mobility models, bandgap narrowing model, a standard band to band tunneling model, Fowler-Nordheim tunneling model	T_{tunnel} = 0.5 nm, t_{ox} = 0.5 nm, W_{SD} = 4 nm, W = 5 nm, L_{SD} = 4 nm, L_{AG} = 5 nm, L_i = 5 nm, H = 5 nm $I_{foward}/I_{reverse} = 10^3$, $I_{ON}/I_{OFF} > 10^7$, avg. SS = 42mV/decade
4	[34] Liu et al.	Simulation Non-local BTBT model, bandgap narrowing model, mobility model	V_{DS} = 1.2V, V_{GS} = 1.2V, gate-source (GS) GS underlap = 10 nm, I_{DS} (μA/μm) = 28 GS underlap = 20 nm, I_{DS} (μA/μm) = 10 w/o overlap or underlap, I_{DS} (μA/μm) = 22 GS overlap = 30 nm, I_{DS} (μA/μm) = 22 GS overlap = 50 nm, I_{DS} (μA/μm) = 23

TFET device. Literature encompasses examples of Ge substrate-based TFET devices. For instance, Beohar et al. [49] proposed drain-underlap (DU) cylindrical (Cyl) gate-all-around (GAA) TFETs based on a Ge source. In this work, the fringing field effects were investigated. Further, hetero-spacer dielectric was incorporated over the source region of the device. Various characteristics such as direct current and RF characteristics were investigated that are responsible for I_{ON}, I_{OFF}, subthreshold swing (SS), C_{gs}, C_{gd}, and cutoff frequency. It was observed that the parameter C_{gd} is minimized due to the combined effects of steep band-to-band tunneling (BTBT) and simultaneous reduction in I_{OFF}. The device characteristic (I_{DS} Vs V_{GS}) is shown in Figure 2.8. It was reported that high depletion width of drain-to-channel junction results in low OFF current for a Ge-GAA-HTS due to the presence of high-k dielectric spacer.

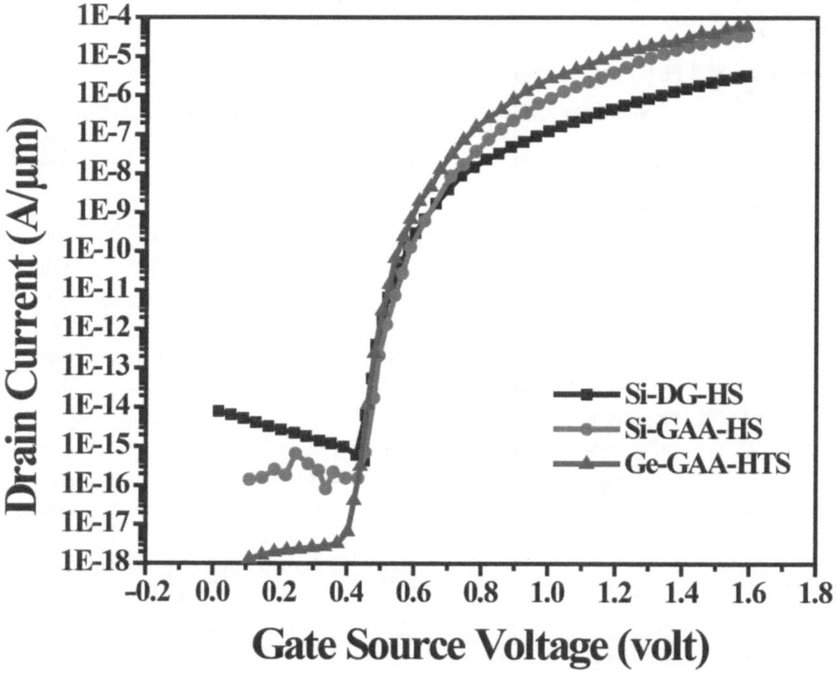

FIGURE 2.8 Transfer Characteristics of Different Topologies of TFET Devices (Reproduced from Ref. [49].)

Kao et al. [50] investigated the theoretical analysis for three different materials: silicon, germanium, and Si_{1-x}-Ge_x. This work focused on the theoretical investigation with Kane's band-to-band tunneling (BTBT) model using parameters A and B at various tunneling directions (100, 110, 111). The goal of this work was to find the importance of Ge mole fraction on BTBT parameters. Furthermore, from the calculated results of BTBT generation rate for a uniform electric field, the dominance of direct tunneling over indirect tunneling for Ge was observed.

Krishnamohan et al. [51] investigated a double-gate, strained-Ge, heterostructure TFET. High ON current and lower subthreshold slope was achieved due to the presence of low bandgap material (strained-Ge) as well as dual gate structure. Furthermore, the impact of quantum transport model and non-local BTBT were investigated to solve the ambipolar behavior issues of the device. Three different topologies (underlapped drain, lower drain doping, and lateral heterostructure) were investigated. The On current of the device was found to be 300μA/μm primarily because of the low bandgap material s-Ge as well as the electrostatics of the dual gate structure (Figure 2.9).

Nayfeh et al. [52] proposed a heterojunction TFET with strained silicon (for channel and drain region) and strained germanium (for source region) as shown in Figure 2.10. Initial investigation was performed for strained SiGe diodes with tunneling model and then extended for the heterojunction TFET. It was found that the

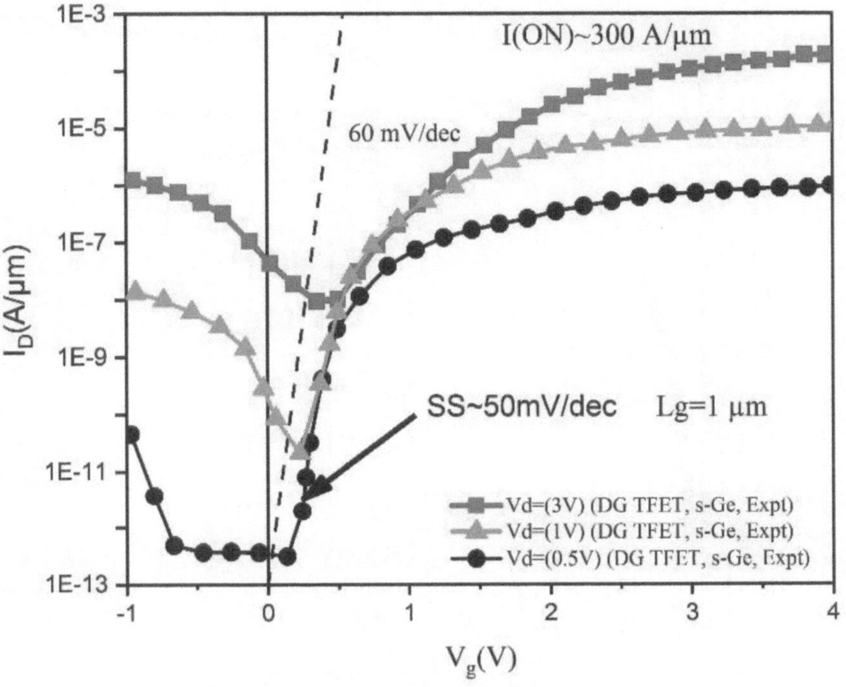

FIGURE 2.9 Transfer Characteristics at Different Drain Voltages

(Reproduced from Ref. [51].)

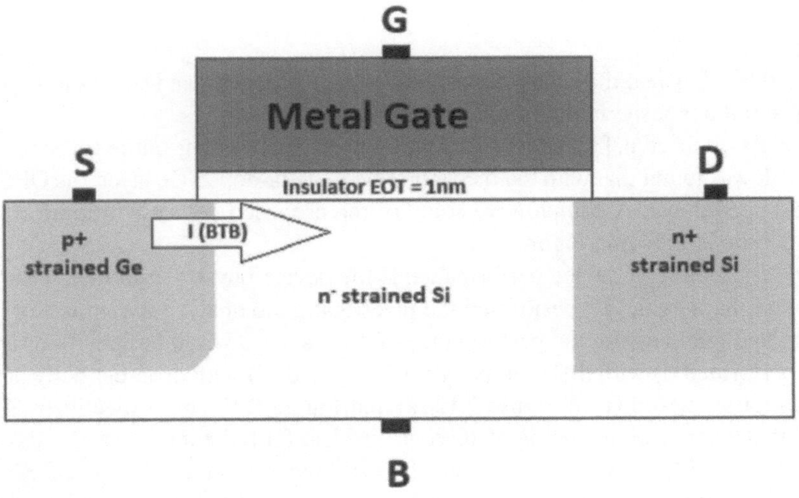

FIGURE 2.10 Schematic of Heterojunction TFET (HTFET)

(Redrawn from Ref. [52].)

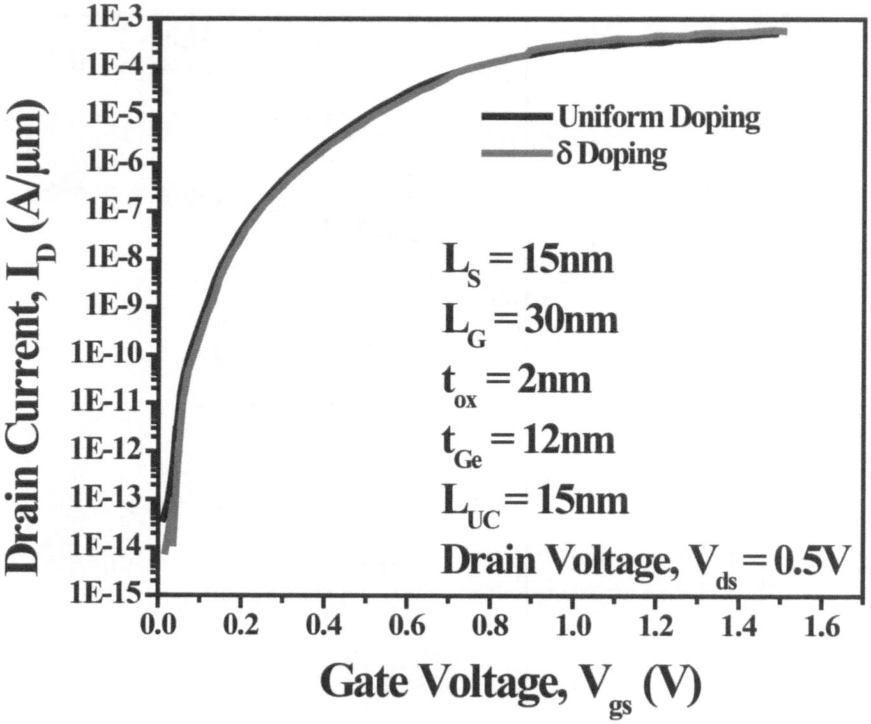

FIGURE 2.11 Comparative Study of the Vertical TFET Device with Uniform and Delta Doping Profiles

(Reproduced from Ref. [37].)

reported HTFET had the capability for low-voltage operation and steep subthreshold swing simultaneously maintaining the high driving current.

Vanlalawpuia et al. [37] proposed a delta-doped germanium source vertical TFET device. It was found that with the incorporation of delta-doped Ge layer, the OFF current dropped down. A comparative study of the proposed device with uniform and delta doping is shown in Figure 2.11.

Another aspect of device performance is the device physical parameters such as thickness. Kim et al. [53] performed the investigation to analyze the impact of body doping and thickness on the performance of a Ge-source-based device. Two studies were performed (i) with different body thickness and (ii) with different body doping concentration as shown in Figure 2.12 (a) and Figure 2.12 (b), respectively. It was found that average body doping of 10^{18} cm^{-3} is ideal for reducing I_{OFF} and rising I_{ON}.

Lee et al. [54] reported the impact of work function variation on a Ge-Source-based TFET. It was found that work-function varied threshold voltage and the device characteristics were a function of the grain size, gate width, and equivalent oxide thickness.

Various Ge-based TFET devices reported in the literature along with their features are summarized in Table 2.3.

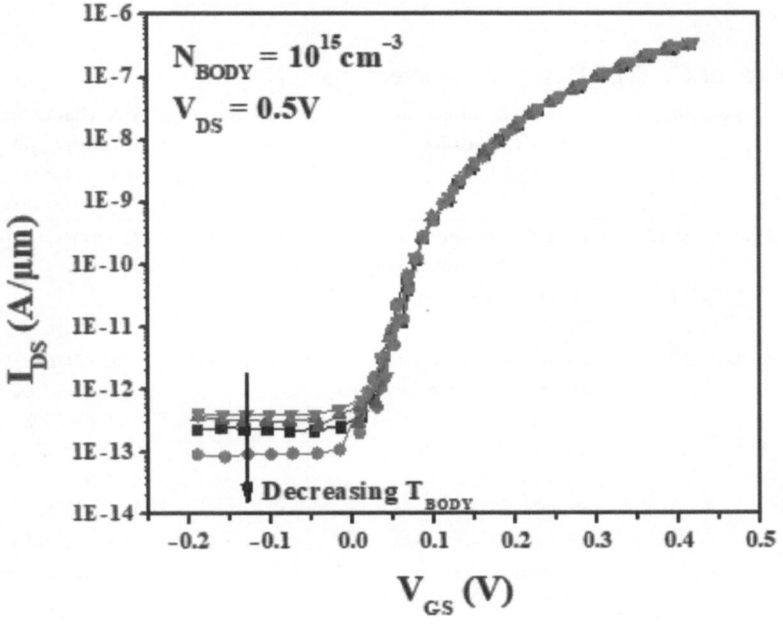

(a)

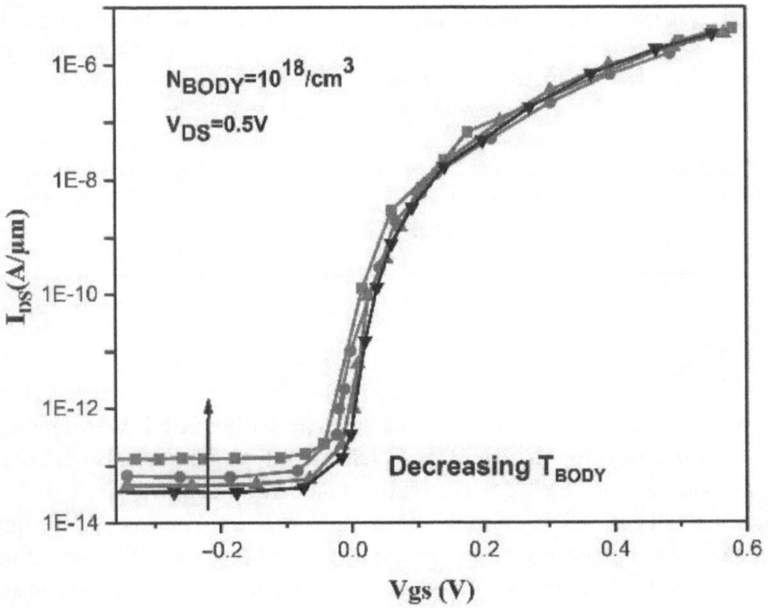

(b)

FIGURE 2.12 Transfer Characteristics for Different Body Thickness (20, 30, 50, and 100 nm): (a) Body Doping = 10^{15} cm^{-3} and (b) Body Doping = 10^{18} cm^{-3}

(Reproduced from Ref. [53].)

TABLE 2.3
Summary of Ge TFET Device Parameters

Sr. no.	Reference	Simulation model/ Fabrication	Mode	Device characteristics I_{ON}(A/µm), I_{OFF}(A/µm), g_m(mS), f_t(GHz), f_{max}(GHz), SS (mV/dec.)
1	Ghosh et al. [35]	Drift diffusion carrier transport, Masetti mobility model, Kane's non-local BTBT	2-Dimensional	7×10^{-5}, 6×10^{-12}, 0.09, 15, 180, 20.85
2	Datta et al. [36]	Non-local BTBT for source, bandgap narrowing for drain. Lombardi CVT for temperature dependency, SRH and Auger for recombination	2-Dimensional	10^{-5} for all temperatures, 10^{-9} at T = 450K 10^{-12} at T = 300K, 0.035, N/A, N/A, 100 at T = 450K 80 at T = 300K
3	Vanlalawpuia et al. [37]	Bandgap narrowing, non-local BTBT for comparison with Si-based TFET. Masetti mobility model for carrier mobility, SRH for recombination	2-Dimensional	2.76×10^{-5}, 1.2×10^{-14}, N/A, N/A, N/A, 21.2
4	Paras et al. [38]	Drift diffusion transport, SRH recombination, trap assisted tunneling, Fermi-Dirac statistics, non-local BTBT, bandgap narrowing	2-Dimensional	4.8×10^{-3}, 4.80×10^{-15}, 2.4, 122, N/A, 26
5	Li et al. [39]	Doping-dependent mobility, interface mobility degradation, high-field saturation	2-Dimensional	4.8×10^{-7}, 10^{-11}, N/A, N/A, 30

2.7 SiGe TFETs

In this section, we elucidate the features of SiGe-based TFET with emphasis on device structure, working principle, and characteristics. This treatise encompasses numerous examples of TFET devices realized on SiGe substrate.

K. Bhuwalka et al. [59] proposed silicon tunnel FETs by means of 2-D tunnels, a modern bandgap engineering-based device design to boost the efficiency of the tunnel FET with better I_{on} and V_T. The structure reported has been realized on strained SiGe layer formed over silicon-on-insulator (SOI) substrate with SiGe electrodes forming source and drain regions. A typical schematic representation of a TFET with SiGe on insulator is shown in Figure 2.13. The structure boosts the mobility of hole, thereby increasing the drain current. The device is realized with a boron doped layer of 25 nm (6–8) × $10^{19}cm^{-3}$ to form the p^+ source field with 70 and 25 nm i-silicon

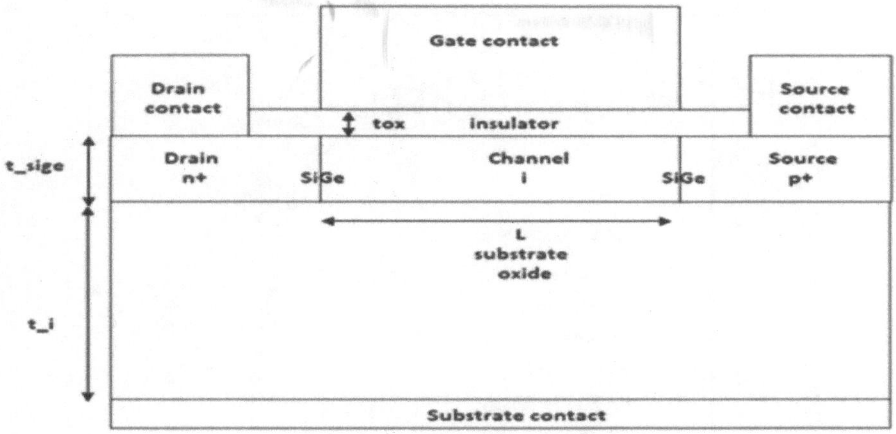

FIGURE 2.13 Schematic Representation of a Tunnel FET with SiGe on an Insulator [59]

layers forming the channel for the two devices. A silicon sheet of 250 nm thickness with 8×10^{19} cm^{-3} doped arsenic is grown on top to form the drain electrode. Further, a 4.5 nm layer of silicon oxide is thermally grown with wet oxidation at 800°C for 5 min after mesa etching. During epitaxial growth, a maximum temperature of 775°C is maintained and the two devices are processed under identical conditions.

K.-W. Jo et al. [60] reported the effect of SiGe layer thickness on strained Ge-on-insulator pMOSFET realized with Ge condensation method. The epitaxially grown Si (10 nm)/$Si_{0.75}Ge_{0.25}$(40 or 60 nm) stack on a (100)-oriented silicon-on-insulator (SOI) wafer (10 nm-thick) with a buried oxide (BOX) 25 nm-thick as a starting substrate was used to realize the device. It encompasses the intermixing annealing. It may be noted that to minimize the pressure relaxation in the long Ge fraction, high cooling time was performed.

The compressive strain is measured as a function of Ge fraction in SGOI layers. It contrasts the effects of 40-nm-thick SiGe and 60-nm-thick SiGe with 4h slow- and fast-cooling process. Thickness of final GOI films realized from 60-nm-thick SiGe and 40-nm-thick SiGe initial substrates was found to be 15 nm and 10 nm, respectively. It was reported that the thinner (40 nm) SiGe substrate has higher compressive strain than the thick (60 nm) one in the Ge fraction.

The transfer characteristics of GOI p-MOSFETs for 60- and 40-nm-thick SiGe samples with back gate operation is shown in Figure 2.14 (a). It was found that in comparison with the other structures, the 10-nm-thick GOI p-MOSFET obtained from the 40-nm-thick SiGe sample with slow cooling depicted higher Id, lower I_{OFF}, and better device characteristics. For this device, a high ON/OFF current proportion of 7.2×10^5 was reported. It was found that in the thinner SiGe sample with slow cooling, the improvement in the transfer characteristics is primarily due to the lower concentration of defects and dislocations by strain relaxation mitigation. The output characteristics of the 10-nm-thick GOI MOSFET with a compressive strain of 1.75% are shown in Figure 2.14 (b). Standard testing operation of the Ge pMOSFET was verified with a high-quality 10 nm thickness of the GOI tube.

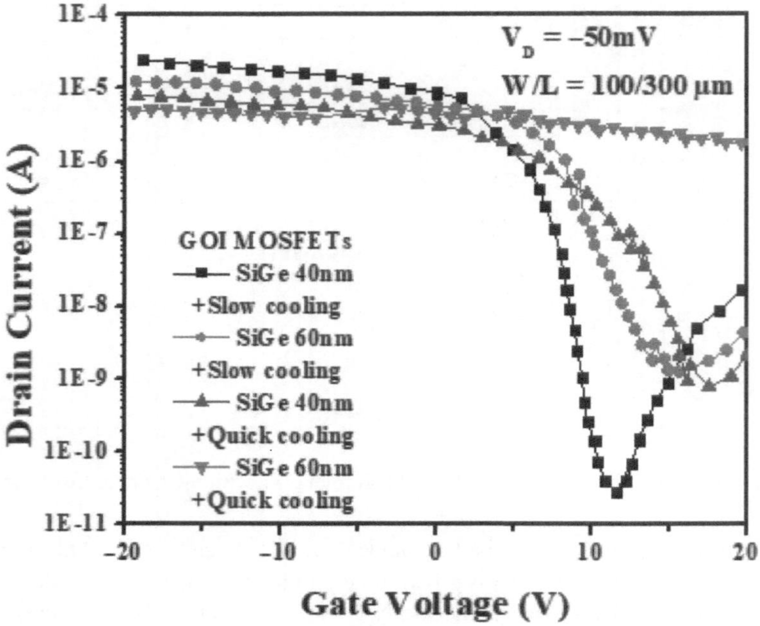

(a)

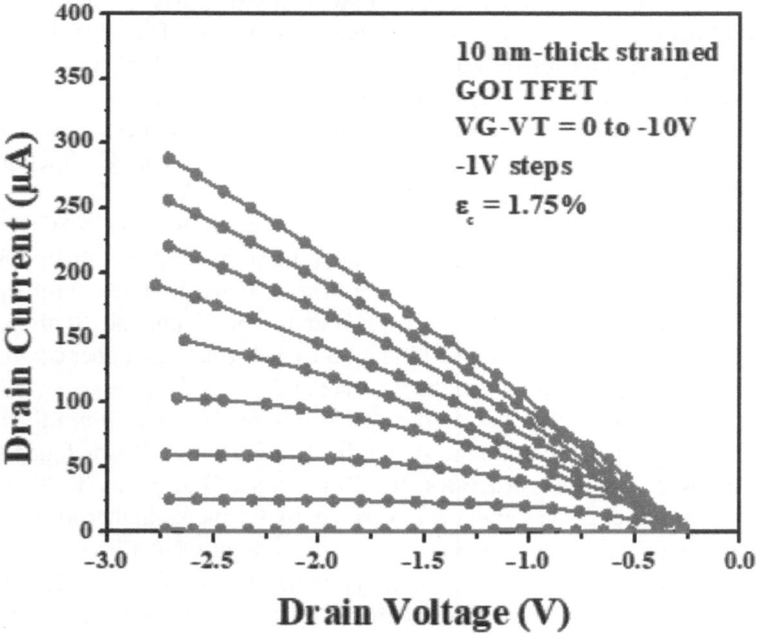

(b)

FIGURE 2.14 (a) UTB GOI pMOSFETs Characteristics Fabricated from 40 nm or 60 nm Thick SiGe Substrates and (b) Characteristics of 10-nm Thick sGOI Substrate pMOSFET [60]

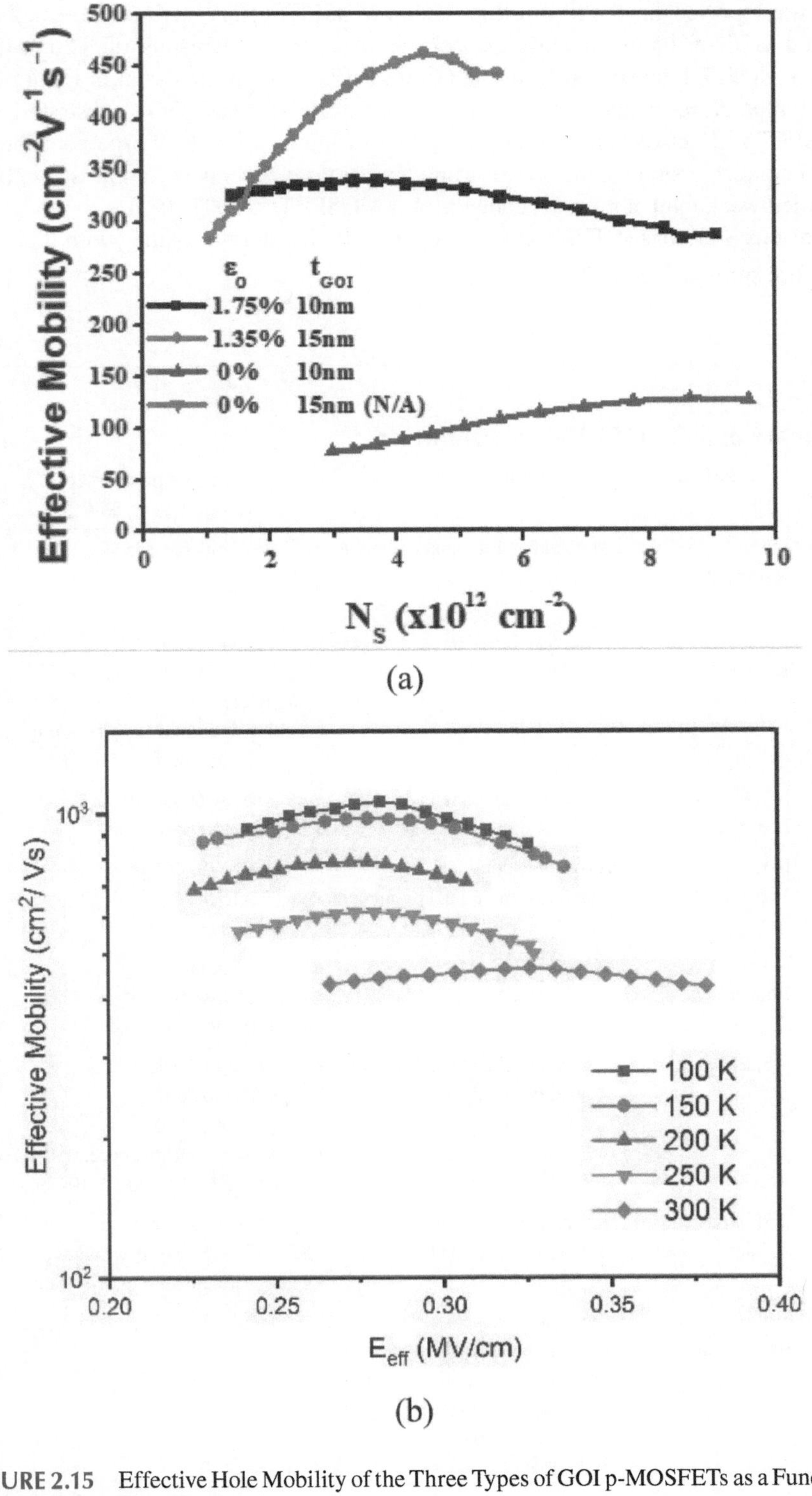

FIGURE 2.15 Effective Hole Mobility of the Three Types of GOI p-MOSFETs as a Function of (a) Ns and (b) E_{eff} [60]

It was reported that the 10-nm-thick GOI p-MOSFET offers 1.8 times more mobility around GOI of 10 nm against the conventional structure results of 9.2-nm-thick GOI p-MOSFET fabricated from the 60-nm-thick-SiGe layer as seen in Figure 2.15 (a). Temperature dependence of the 10-nm-thick strained GOI substrate-based MOSFET's effective hole mobility is shown in Figure 2.15 (b). Large contribution of phonon dispersion significant to strong temperature dependence contrasts with the recorded weak temperature dependence of p-MOSFETs of UTB GOI.

Various SiGe-based TFET devices reported in the literature along with their features are summarized in Table 2.4.

TABLE 2.4
Summary of SiGe TFET Device Parameters

S. no.	Ref no.	Simulation model	Device parameters L_{Ch} I_{ON} I_{OFF} g_m SS $C_{gs}C_{gd}$
1	[40] Kato et al.	Vertical band-to-band tunneling model	• SS = ~91 mV/decade • $\frac{W}{L}=\frac{10}{15}\mu m$
2	[41] Kumar et al.	Shockley-Read-Hall and Auger models	• I_{ON} = 5.54 μA/μm • I_{OFF} = 0.1 μA/μm • negligible ambipolar current (10 – 19 A/μm) • $L_{Ch/SiGe}$ = 25 nm
3	[42] Lee et al.	Band-to-band tunneling model	• SS = below 60 mV/dec • $I_{ON}/I_{OFF} = 10^6$
4	[43] Mayer et al.	A dynamic non-local band-to-band tunneling model	• I_{OFF} = ~10–100 fA/μm • V_{GD} = +1.15 V • L_G = 100 nm
5	[44] Bhuwalka et al.	Kane's model	• I_{ON} = 1 mA/mm • I_{OFF} = 250fA/mm • V_{DS} = 0.7 V
6	[45] Bae et al.	A gate modulated band-to-band tunneling (BTBT)	• SS = 60.6 mV/dec • I_{ON} = 82.3 nA/μm • $I_{ON}/I_{OFF} = 6.8 * 10^6$ • Ge-source impurity concentrations = $1 * 10^{18}$, $1 * 10^{19}$, and $1 * 10^{20}$ cm^{-3}
7	[55] A. Vandooren et al.	Dynamic nonlocal trap-assisted Schenk BTBT model	• I_{ON} = 0.1–1 μA/μm • SS = 100–150 mV/dec range • gate-to-drain underlap of ~40 nm
8	[56] Hanna et al.	• A dynamic non-local band-to-band (BTB) tunneling model is utilized in conjunction with Shockley-Reed-Hall (SRH) recombination model.	• I_{ON} = 244 μA/μm I_{ON} = 83 μA/μm

TABLE 2.4 *(Continued)*
Summary of SiGe TFET Device Parameters

S. no.	Ref no.	Simulation model	Device parameters L_{Ch} I_{ON} I_{OFF} g_m SS $C_{gs}C_{gd}$
9	[57] Ferhati at al.	• Non-local band-to-band quantum tunneling model • Models for carrier's recombination effect Shockley-Read-Hall (SRH), Auger and surface recombination) are adopted.	• $I_{OFF} = 1.6 \times 10^{-13}$ A • Signal-to-noise ratio = 3.5×10^7 • $V_{ds}(V) = 1$ • $V_{gs}(V) = 0.4$ • g_m = 48 mV/dec
10	[58] Wang et al.	Dynamic non-local band-to-band tunneling model is used. ($A = 1.46 \times 10^{17}$ & $B = 3.59 \times 10^6$)	• SS = 58 mV/dec • V_{gs} = 0 to 0.46 V • charge density = $\sim 10^{11}$ q/cm^2 • $I_{ON}/I_{OFF} = \sim 10^8$ • off-leakage current = 2.05×10^{-14} A/μm

2.8 CHALLENGES AND FUTURE TRENDS

Hot carrier stress (HCS) and positive/negative gate bias stress (P/NGBS) instability are critical issues in TFET devices. Various device parameters such as rho, intrinsic resistance, and transport delay degrades due to NQS effect in Ge-pTFETs compared to other pTFETs. The transconductance is observed to be less for TFET in comparison to the same technology node MOSFET. Similarly, the transmit time for the TFET based on Ge is less compared to Si and SiGe. This results in high switching speed for TFET device based on Ge. With low delta-doping, there is a reduction in inner electric field across source/channel junction that results in rising SS_{avg}. The delta-doped Ge-source TFET is a potential candidate for ultra low power and high-performance technology. However, it produces depletion zones across gate source fringing field.

TFET based on SiGe is found to be better material for futuristic analog/RF applications perspective. Si-based conventional devices such as MOSFTEs, TFETs, and SoC can be substituted with SiGe-based devices. The devices will also have better reliability due to high thermal stability of SiGe. Moreover, in terms of bandgap properties, SiGe has less bandgap compared to Si. This improves the ON current (I_{ON}) and results in the enhancement of switching speed with low subthreshold swing (SS) useful for optoelectronic-based applications such as personal computers, high-speed telecommunications, long-distance fiber communication, local area network (LAN), and chip-to-chip communication, to cite a few. Although silicon tends to be the most dominating material for electronic devices, silicon's indirect bandgap property limits its application for electronic-based optical devices. Many of high-performance devices are aided with III-V compound semiconductor-based transistors due to direct bandgap feature and high quantum effect. However, silicon-based optoelectronic devices tend to be a much better option, due to some limitations such as poor thermal property, high cost, and incompatibility with silicon. With the integration of germanium and combining it as SiGe, the bandgap lowers down and with low cost, better thermal properties as well as mechanical properties predict SiGe-based devices to be better option for electronic-based optical applications. Although the ambipolarity of TFET based on low bandgap such as SiGe and Ge is a major challenge that causes increment in switching leakage in a TFET-based circuit like inverter, by careful design limitations induced due to such factors can be improved.

Silicon-Germanium emerges to be a substitute for conventional silicon-based TFETs. Silicon-Germanium has better optical properties, which are useful for silicon-based photonic applications. SiGe can also be combined with Si for realizing heterojunctions. SiGe was initially used to realize bipolar devices, but in recent times, metal oxide semiconductor (MOS) devices have been also realized with SiGe. Fabrication of SiGe-based devices involves technologies such as low-pressure chemical vapor deposition (LPCVD), molecular beam epitaxy (MBE) as well as ultra-high vacuum-chemical vapor deposition (UHV-CVD). The most prominent role is played by the temperature (500–7,000°C) at which epitaxial layer growth (SiGe) takes place. Furthermore, these techniques result in growth of the layers without distorting the doping concentrations for other structures present within the silicon wafer. Moreover, with the use of SiGe material, tunneling barrier width for the device junctions can be reduced. This results in enhanced rate of flow of majority charge carriers and further improves the switching speed of the device. SiGe TFET implemented on the fully depleted SoI (silicon germanium on insulator) chip with CMOS transistors for a wide range of possible applications in low power analog/RF and SoC applications.

REFERENCES

[1] Rajan, C., Patel, J., Sharma, D., Behera, A. K., Lodhi, A., Lemtur, A., & Samajdar, D. P. (2020). Implementation of Σ Δ ADC using electrically doped III-V ternary alloy semiconductor nano-wire TFET. *Micro & Nano Letters, 15*(4), 266–271.

[2] Singh, A. K., Tripathy, M. R., Singh, P. K., Baral, K., Chander, S., & Jit, S. (2020). Deep insight into DC/RF and linearity parameters of a novel back gated ferroelectric TFET on SELBOX substrate for ultra low power applications. *Silicon*, 1–11.

[3] Guha, S., Pachal, P., Ghosh, S., & Sarkar, S. K. (2020). Analytical model of a novel double gate metal-infused stacked gate-oxide tunnel field-effect transistor (TFET) for low power and high-speed performance. *Superlattices and Microstructures, 146*, 106657.

[4] Vidhyadharan, S., & Dan, S. S. (2021). An efficient ultra-low power and superior performance design of ternary half adder using CNFET and gate-overlap TFET devices. *IEEE Transactions on Nanotechnology, 20,* 365–376.

[5] Joshi, S., Dubey, P. K., & Kaushik, B. K. (2020). Photosensor based on split gate TMD TFET using photogating effect for visible light detection. *IEEE Sensors Journal, 20*(12), 6346–6353.

[6] Mathew, R., & Sankar, A. R. (2018). A review on surface stress-based miniaturized piezoresistive SU-8 polymeric cantilever sensors. *Nano-micro Letters, 10*(2), 1–41.

[7] Mathew, R., & Sankar, A. R. (2015). Design of a triangular platform piezoresistive affinity microcantilever sensor for biochemical sensing applications. *Journal of Physics D: Applied Physics, 48*(20), 205402.

[8] Mathew, R., & Sankar, A. R. (2018). Impact of isolation and immobilization layers on the electro-mechanical response of piezoresistive nano cantilever sensors. *Journal of Nanoscience and Nanotechnology, 18*(3), 1636–1647.

[9] Mathew, R., & Sankar, A. R. (2018). Optimization of a nano-cantilever biosensor for reduced self-heating effects and improved performance metrics. *Journal of Micromechanics and Microengineering*, *28*(8), 085012.

[10] Mathew, R., & Ravi Sankar, A. (2017). In silico modeling and investigation of self-heating effects in composite nano cantilever biosensors with integrated piezoresistors. *AIP Advances*, *7*(3), 035108.

[11] Mathew, R., & Sankar, A. R. (2018). Piezoresistive composite silicon dioxide nanocantilever surface stress sensor: Design and optimization. *Journal of Nanoscience and Nanotechnology*, *18*(5), 3387–3397.

[12] Mathew, R., & Sankar, A. R. (2019). Influence of surface layer properties on the thermo-electro-mechanical characteristics of a MEMS/NEMS piezoresistive cantilever surface stress sensor. *Materials Research Express*, *6*(8), 086304.

[13] Mathew, R., & Sankar, A. R. (2020). Temperature drift-aware material selection of composite piezoresistive micro-cantilevers using Ashby's methodology. *Microsystem Technologies*, 1–14.

[14] Mathew, R., & Sankar, A. R. (2016). Numerical study on the influence of buried oxide layer of SOI wafers on the terminal characteristics of a micro/nano cantilever biosensor with an integrated piezoresistor. *Biomedical Physics & Engineering Express*, *2*(5), 055012.

[15] Mathew, R., & Sankar, A. R. (2018, Oct.). Temperature induced inaccuracy in composite piezoresistive micro/nano cantilever chemical/biological sensors. In *2018 IEEE SENSORS* (pp. 1–4). IEEE.

[16] Choi, W. Y., Park, B. G., Lee, J. D., & Liu, T. J. K. (2007). Tunneling field-effect transistors (TFETs) with subthreshold swing (SS) less than 60 mV/dec. *IEEE Electron Device Letters*, *28*(8), 743–745.

[17] Bowonder, A., Patel, P., Jeon, K., Oh, J., Majhi, P., Tseng, H. H., & Hu, C. (2008, May). Low-voltage green transistor using ultra shallow junction and hetero-tunneling. In *Extended Abstracts-2008 8th International Workshop on Junction Technology (IWJT'08)* (pp. 93–96). IEEE.

[18] Luisier, M., & Klimeck, G. (2009). Atomistic full-band design study of InAs band-to-band tunneling field-effect transistors. *IEEE Electron Device Letters*, *30*(6), 602–604.

[19] Nayfeh, O. M., Chleirigh, C. N., Hennessy, J., Gomez, L., Hoyt, J. L., & Antoniadis, D. A. (2008). Design of tunneling field-effect transistors using strained-silicon/strained-germanium type-II staggered heterojunctions. *IEEE Electron Device Letters*, *29*(9), 1074–1077.

[20] Koswatta, S. O., Koester, S. J., & Haensch, W. (2009, Dec.). 1D broken-gap tunnel transistor with MOSFET-like on-currents and sub-60mV/dec subthreshold swing. In *2009 IEEE International Electron Devices Meeting (IEDM)* (pp. 1–4). IEEE.

[21] Wang, L., Yu, E., Taur, Y., & Asbeck, P. (2010). Design of tunneling field-effect transistors based on staggered heterojunctions for ultralow-power applications. *IEEE Electron Device Letters*, *31*(5), 431–433.

[22] Seabaugh, A. C., & Zhang, Q. (2010). Low-voltage tunnel transistors for beyond CMOS logic. *Proceedings of the IEEE*, *98*(12), 2095–2110.

[23] Hurkx, G. A. M. (1989). On the modelling of tunnelling currents in reverse-biased p-n junctions. *Solid-State Electronics*, *32*(8), 665–668.

[24] Landau, L. D., & Lifshitz, E. M. (1958). *Quantum Mechanics*. Reading: Addison Wesley Press, p. 174.

[25] Moll, J. L. (1964). *Physical and Quantum Electronics*. New York: McGraw-Hill, pp. 249–253.

[26] Narasimhulu, K., Desai, M. P., Narendra, S. G., & Rao, V. R. (2004). The effect of LAC doping on deep submicrometer transistor capacitances and its influence on device RF performance. *IEEE Transactions on Electron Devices*, *51*(9), 1416–1423.

[27] Bangsaruntip, S., Cohen, G. M., Majumdar, A., & Sleight, J. W. (2010). Universality of short-channel effects in undoped-body silicon nanowire MOSFETs. *IEEE Electron Device Letters, 31*(9), 903–905.
[28] Colinge, J.-P., & others. (2008). *FinFETs and other Multi-gate Transistors*, vol. 73. New York: Springer.
[29] Francis, P., Terao, A., Flandre, D., & Van de Wiele, F. (1994). Modeling of ultrathin double-gate nMOS/SOI transistors. *IEEE Transactions on Electron Devices, 41*(5), 715–720.
[30] Young, K. K. (1989). Short-channel effect in fully depleted SOI MOSFETs. *IEEE Transactions on Electron Devices, 36*(2), 399–402.
[31] Kumari, T., Singh, J., & Tiwari, P. K. (2020). Investigation of ring-TFET for better electrostatics control and suppressed ambipolarity. *IEEE Transactions on Nanotechnology, 19*, 829–836.
[32] Ma, W. C. Y., Hsu, H. S., & Wang, H. C. (2020). Various reliability investigations of low temperature polycrystalline silicon tunnel field-effect thin-film transistor. *IEEE Transactions on Device and Materials Reliability, 20*(4), 775–780.
[33] Liu, X., Ma, K., Wang, Y., Wu, M., Lee, J. H., & Jin, X. (2020). A novel high Schottky barrier based bilateral gate and assistant gate controlled bidirectional tunnel field effect transistor. *IEEE Journal of the Electron Devices Society, 8*, 976–980.
[34] Liu, K. M., & Cheng, C. P. (2020). Investigation on the effects of gate-source overlap/underlap and source doping gradient of N-type si cylindrical gate-all-around tunnel field-effect transistors. *IEEE Transactions on Nanotechnology, 19*, 382–389.
[35] Ghosh, S., Koley, K., Saha, S. K., & Sarkar, C. K. (2020). Heterostructure Ge-Body pTFETs for analog/RF applications. *IEEE Journal of the Electron Devices Society, 8*, 1202–1209.
[36] Datta, E., Chattopadhyay, A., Mallik, A., & Omura, Y. (2020). Temperature dependence of analog performance, linearity, and harmonic distortion for a Ge-source tunnel FET. *IEEE Transactions on Electron Devices, 67*(3), 810–815.
[37] Vanlalawpuia, K., & Bhowmick, B. (2019). Investigation of a Ge-source vertical TFET with delta-doped layer. *IEEE Transactions on Electron Devices, 66*(10), 4439–4445.
[38] Paras, N., & Chauhan, S. S. (2019). A novel vertical tunneling based Ge-source TFET with enhanced DC and RF characteristics for prospect low power applications. *Microelectronic Engineering, 217*, 111103.
[39] Li, W., & Woo, J. C. (2018). Optimization and scaling of Ge-pocket TFET. *IEEE Transactions on Electron Devices, 65*(12), 5289–5294.
[40] Kato, K., Jo, K. W., Matsui, H., Tabata, H., Mori, T., Morita, Y., & Takagi, S. (2020). P-channel TFET operation of bilayer structures with type-II heterotunneling junction of oxide-and group-IV semiconductors. *IEEE Transactions on Electron Devices, 67*(4), 1880–1886.
[41] Kumar, N., & Raman, A. (2019). Design and investigation of charge-plasma-based work function engineered dual-metal-heterogeneous gate Si-Si 0.55 Ge 0.45 GAA-cylindrical NWTFET for ambipolar analysis. *IEEE Transactions on Electron Devices, 66*(3), 1468–1474.
[42] Lee, J., Lee, R., Kim, S., Lee, K., Kim, H. M., Kim, S., . . . & Park, B. G. (2020). Surface Ge-rich p-type SiGe channel tunnel field-effect transistor fabricated by local condensation technique. *Solid-State Electronics, 164*, 107701.
[43] Mayer, F., Le Royer, C., Damlencourt, J. F., Romanjek, K., Andrieu, F., Tabone, C., . . . & Deleonibus, S. (2008, Dec.). Impact of SOI, Si 1-x Ge x OI and GeOI substrates on CMOS compatible tunnel FET performance. In *2008 IEEE International Electron Devices Meeting* (pp. 1–5). IEEE.
[44] Bhuwalka, K. K., Born, M., Schindler, M., Schmidt, M., Sulima, T., & Eisele, I. (2006). P-channel tunnel field-effect transistors down to sub-50 nm channel lengths. *Japanese Journal of Applied Physics, 45*(4S), 3106.

[45] Bae, T. E., Kato, K., Suzuki, R., Nakane, R., Takenaka, M., & Takagi, S. (2018). Influence of impurity concentration in Ge sources on electrical properties of Ge/Si hetero-junction tunneling field-effect transistors. *Applied Physics Letters, 113*(6), 062103.

[46] Walke, A. M., Vandooren, A., Rooyackers, R., Leonelli, D., Hikavyy, A., Loo, R., . . . & Thean, A. V. Y. (2014). Fabrication and analysis of a $Si/Si_{0.55}Ge_{0.45}$ heterojunction line tunnel FET. *IEEE Transactions on Electron Devices, 61*(3), 707–715.

[47] Shafi, N., Sahu, C., & Periasamy, C. (2018). Virtually doped SiGe tunnel FET for enhanced sensitivity in biosensing applications. *Superlattices and Microstructures, 120*, 75–89.

[48] Patel, N., Ramesha, A., & Mahapatra, S. (2008). Drive current boosting of n-type tunnel FET with strained SiGe layer at source. *Microelectronics Journal, 39*(12), 1671–1677.

[49] Beohar, A., Yadav, N., Shah, A. P., & Vishvakarma, S. K. (2018). Analog/RF characteristics of a 3D-Cyl underlap GAA-TFET based on a Ge source using fringing-field engineering for low-power applications. *Journal of Computational Electronics, 17*(4), 1650–1657.

[50] Kao, K. H., Verhulst, A. S., Vandenberghe, W. G., Soree, B., Groeseneken, G., & De Meyer, K. (2011). Direct and indirect band-to-band tunneling in germanium-based TFETs. *IEEE Transactions on Electron Devices, 59*(2), 292–301.

[51] Krishnamohan, T., Kim, D., Raghunathan, S., & Saraswat, K. (2008, Dec.). Double-Gate Strained-Ge Heterostructure Tunneling FET (TFET) With record high drive currents and« 60mV/dec subthreshold slope. In *2008 IEEE International Electron Devices Meeting* (pp. 1–3). IEEE.

[52] Nayfeh, O. M., Chleirigh, C. N., Hennessy, J., Gomez, L., Hoyt, J. L., & Antoniadis, D. A. (2008). Design of tunneling field-effect transistors using strained-silicon/strained-germanium type-II staggered heterojunctions. *IEEE Electron Device Letters, 29*(9), 1074–1077.

[53] Kim, S. H., Jacobson, Z. A., & Liu, T. J. K. (2010). Impact of body doping and thickness on the performance of Germanium-source TFETs. *IEEE Transactions on Electron Devices, 57*(7), 1710–1713.

[54] Lee, Y., Nam, H., Park, J. D., & Shin, C. (2015). Study of work-function variation for high-$\kappa $/metal-gate Ge-Source tunnel field-effect transistors. *IEEE Transactions on Electron Devices, 62*(7), 2143–2147.

[55] Vandooren, A., Leonelli, D., Rooyackers, R., Hikavyy, A., Devriendt, K., Demand, M., . . . & Huyghebaert, C. (2013). Analysis of trap-assisted tunneling in vertical Si homo-junction and SiGe hetero-junction tunnel-FETs. *Solid-State Electronics, 83*, 50–55.

[56] Hanna, A. N., & Hussain, M. M. (2015). Si/Ge hetero-structure nanotube tunnel field effect transistor. *Journal of Applied Physics, 117*(1), 014310.

[57] Ferhati, H., & Djeffal, F. (2018). Boosting the optical performance and commutation speed of phototransistor using SiGe/Si/Ge tunneling structure. *Materials Research Express, 5*(6), 065902.

[58] Wang, W., Wang, P. F., Zhang, C. M., Lin, X., Liu, X. Y., Sun, Q. Q., . . . & Zhang, D. W. (2013). Design of U-shape channel tunnel FETs with SiGe source regions. *IEEE Transactions on Electron Devices, 61*(1), 193–197.

[59] Bhuwalka, K. K., Born, M., Schindler, M., Schmidt, M., Sulima, T., & Eisele, I. (2006). P-channel tunnel field-effect transistors down to sub-50 nm channel lengths. *Japanese Journal of Applied Physics, 45*(4S), 3106.

[60] Jo, K. W., Kim, W. K., Takenaka, M., & Takagi, S. (2019). Impact of SiGe layer thickness in starting substrates on strained Ge-on-insulator pMOSFETs fabricated by Ge condensation method. *Applied Physics Letters, 114*(6), 062101.

3 Ultra Low Power III-V Tunnel Field-Effect Transistors

J. Ajayan and D. Nirmal

CONTENTS

3.1 TFETs ON GaAs MATERIAL SYSTEMS

III-V TFETs are capable of operating with a power supply (V_{DD}) of 0.15 V to 0.3 V due to their extremely low subthreshold swing enabled by the tunneling charge injection mechanism. III-V TFETs can be classified into two main categories depending on the materials used for their construction. They are homojunction (HJ) TFETs and heterojunction TFETs. Homojunction TFETs use the same materials in the source, channel, and drain regions. On the other hand, heterojunction TFETs use dissimilar materials at the source, channel, and drain regions. In most of the heterojunction TFETs, the drain and channel are made of the same materials and the source is made with a different material. There are three different types of band alignments in heterostructures (Figure 3.1):

1. Straddling gap or Type-I heterostructure
2. Staggered gap or Type-II heterostructure
3. Broken gap or Type-III heterostructure

A type-II staggered bandgap heterostrucure is preferred in III-V TFETs to achieve high performance. In staggered TFETs, usually a low bandgap semiconductor will be used as the source and a wider bandgap semiconductor will be used as channel/drain to effectively reduce the ambipolar current and improve I_{ON} [1–11]. The use of lower bandgap material at the source increases the tunneling rate that improves

DOI: 10.1201/9781003200987-3

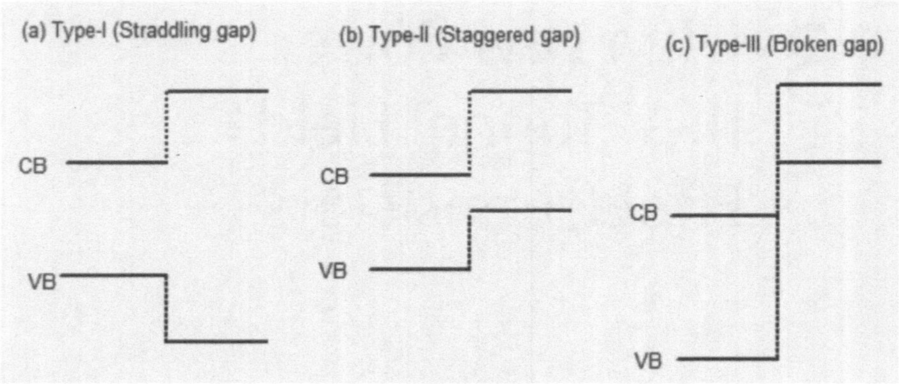

FIGURE 3.1 Band Alignments in Heterostructures

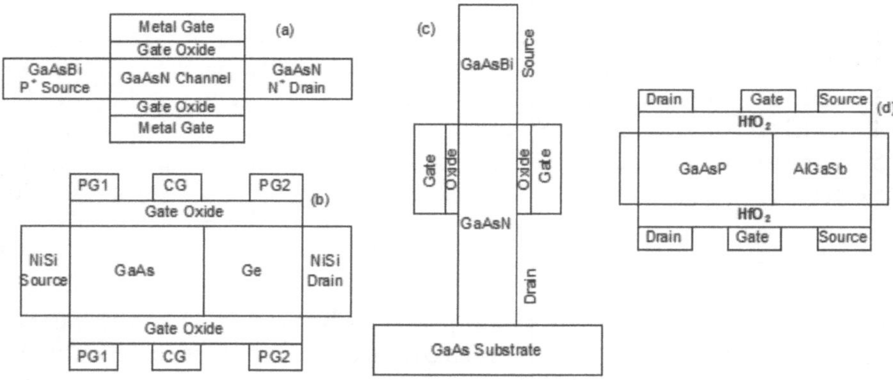

FIGURE 3.2 (a) GaAsBi/GaAsN Heterojunction TFET [11], (b) Polarity Control GaAs-Ge TFET [12], (c) GaAsBi/GaAsN Type-II Staggered TFET [13], and (d) GaAsP/AlGaSb CP-TFET [14]

I_{ON}. On the other hand, by using a wide bandgap semiconductor at channel/drain of TFETs, the ambipolar current and off current (I_{OFF}) can be suppressed significantly. The band-to-band tunneling (BTBT) takes place at source/channel and channel/drain interfaces in TFETs. This BTBT process is the key charge transport mechanism in TFETs. Usually, metals with low work function will be employed at the source and drain end of the gate metal to significantly improve I_{ON} by improving the BTBT rate of charges at the source/channel interface. A wide potential barrier at the channel/drain interface results in the decrease of ambipolar current and I_{OFF}. One disadvantage of using a low work function metal gate at the drain end is that it leads to the increase of C_{gd} (gate-drain parasitic capacitance). Moreover, it also results in the increase of electric field density at the drain end. These two factors limit the

RF performance of TFETs. The use of hetero gate dielectrics in III-V TFETs can significantly minimize the HCEs. The smaller effective electron tunneling mass in Gallium Arsenide (GaAs), GaAsP, AlGaSb, GaAsBi, and GaAsN can be exploited fully to improve the I_{ON} of TFETs. The different architectures of GaAs material system-based TFETs are shown in Figure 3.2.

The bandgap energy of GaAs is 1.42 eV. In GaAsN/GaAsBi heterojunction TFET [Figure 3.2 (a) and Figure 3.2 (c)], the channel and drain regions are made of GaAsN material and the source region is made of GaAsSb material. Adding nitrogen (N) and bismuth (Bi) in GaAs material helps to significantly reduce the bandgap of GaAs. Figure 3.3 (a) depicts the change in drain current as a function of gate over drive voltage and acceptor concentration (N_a) of $GaAs_{0.9}Bi_{0.1}/GaAs_{0.9}N_{0.1}$ TFET. The optimized N_a concentration is $5 \times 10^{19}/cm^3$. Figure 3.3 (b) illustrates the change in I_{ON} as a function of valence band density of states (N_V) and acceptor concentration in source (N_a). A large N_V is required for achieving higher I_{ON}. Figure 3.3 (c) depicts the impact of concentrations of Bi and N on the $GaAs_{1-x}Bi_x/GaAs_{1-y}N_y$ TFETs. The $GaAs_{1-x}Bi_x/GaAs_{1-y}N_y$ TFETs with large Bi and N concentration exhibits higher drive current due to the reduction of bandgap of channel layer. Polarity controlled Ge/GaAs heterojunction TFET [Figure 3.2 (b)] has two gates, namely, a polarity gate (PG) and a control gate (CG). Polarity gate PG1 helps to create a n^+ drain region and PG2 helps to create a p^+ source region. Tunneling barrier modulation can be controlled with the help of CG gate. The drain current of the Ge/GaAs heterojunction TFET increases with an increase in control gate voltage. The influence of different gate dielectric materials and temperature on the drain current of Ge/GaAs heterojunction TFET is shown in Figure 3.4 (a) and Figure 3.4 (b), respectively. The use of high-k gate dielectric materials in Ge/GaAs Type-II TFET improves I_{ON} and also leads to the increase of I_{OFF}. The IOFF of Ge/GaAs TFET also increases with rise in temperature.

Figure 3.5 (a) illustrates the variation of transconductance (g_m) wth V_{GS} for H-TFET (hetero-material TFET), H-DW TFET (hetero-material dual work function TFET), and H-HGD DW TFET (hetero-material hetero gate dielectric dual work function TFET). H-DW TFET and H-HGD DW TFET provide higher g_m compared with H-TFET. However, these devices also exhibit higher drain conductance (g_{ds}) [Figure 3.5 (b)]. But H-HGD DW TFET shows superior RF performance compared to H-TFET and H-DW TFET [Figure 3.5 (c)]. GaAsP/AlGaSb charged plasma (CP) TFET is found to be effective in minimizing ambipolar current and thereby improving I_{ON} [Figure 3.5 (d)]. The CP-TFET can be made using AlGaSb at the source and GaAsP at the drain and channel regions. Intrinsic gain (A_i) is a key parameter that can be used for measuring the analog performance of TFETs. A_i of TFETs can be computed as

$$A_i = \frac{g_m}{g_{ds}} \tag{3.1}$$

A higher A_i is required to achieve good analog performance. Cutoff frequency (f_T) and maximum oscillation frequency (f_{max}) are the two key parameters that are used to measure the RF performance of TFETs. Drain length, source length, gate length, gate to source, and gate to drain spacer lengths and oxide thickness are the

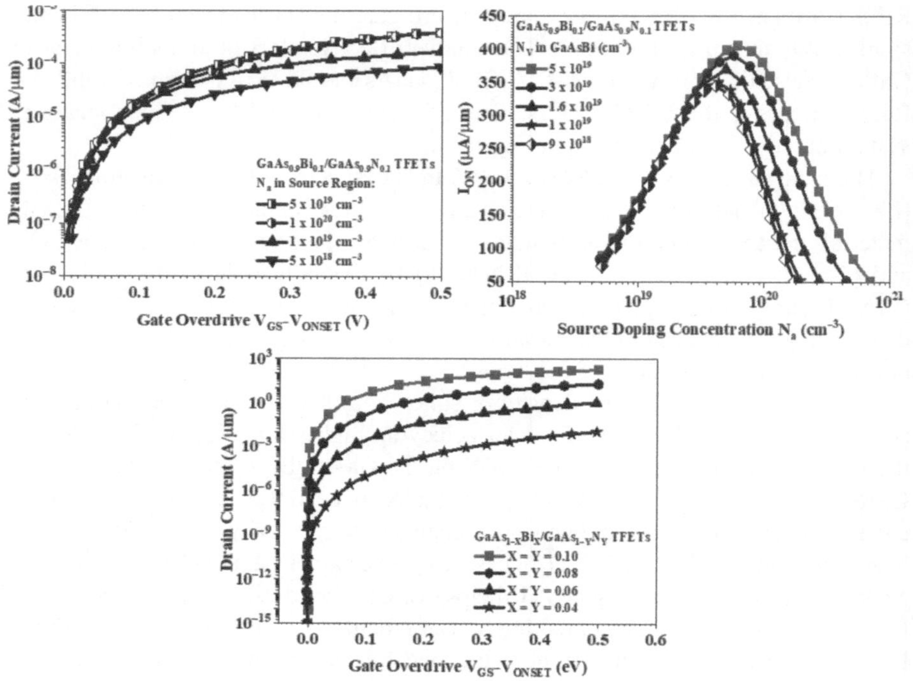

FIGURE 3.3 $GaAs_{0.9}Bi_{0.1}/GaAs_{0.9}N_{0.1}$ Type-II Heterojunction Staggered TFETs: (a) Impact of Gate Overdrive Voltage and Source Doping Concentration (N_a) on Transfer Curves, (b) Impact of N_V and N_a on the I_{ON}, and (c) the Influence of N and Bi Concentrations on the Drive Current [11]

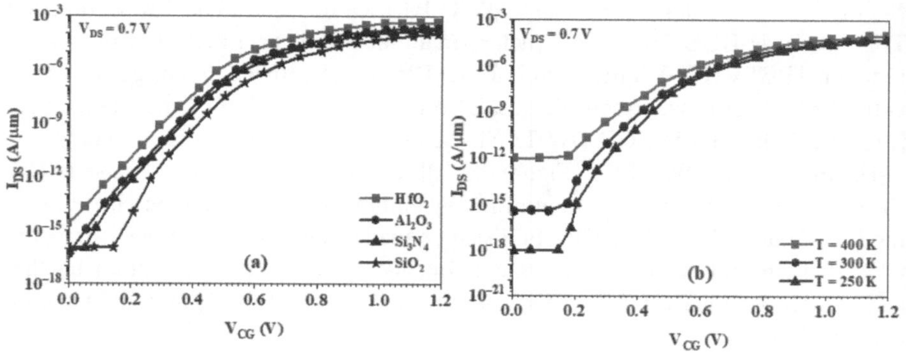

FIGURE 3.4 The Influence of (a) Different Gate Dielectric Materials and (b) Temperature on the Drain Current of Ge/GaAs Heterojunction TFET [12]

key geometrical parameters that significantly affect the analog/RF performance of III-V TFETs. C_{gs} and C_{gd} are the critical factors that limit the f_T and f_{max} of III-V TFETs [16].

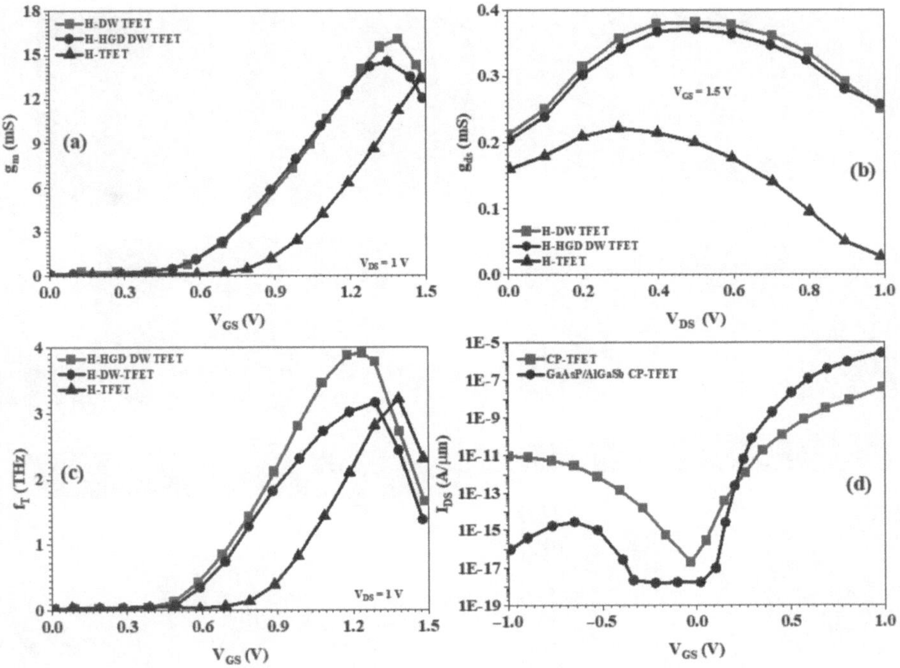

FIGURE 3.5 H-HGD-DW TFET, H-DW TFET, and H-TFETs: (a) g_m Vs V_{GS} Characteristics, (b) g_{ds} Vs V_{DS} Characteristics [15], (c) f_T versus V_{GS} Characteristics [15], and (d) I_{DS} Vs V_{GS} Characteristics for CP-TFET and GaAsP/GaAsSb CP-TFET [14]

3.2 TFETs ON InGaAs MATERIAL SYSTEMS

$In_{1-x}Ga_xAs$ semiconductor has zinc blend structure and $In_{0.53}Ga_{0.47}As$ has a bandgap of 0.74 eV, electron effective mass (m_e) of 0.041m_0, hole effective mass of 0.45m_0, electron affinity of 4.5 eV, lattice constant of 5.87 A^0, electron mobility (μ_e) of 12000 cm^2/Vs, and hole mobility of (μ_h) of 300 cm^2/Vs. In 2009, S. Mookerjee et al. [17, 18] reported the first ever $In_{0.53}Ga_{0.47}As$ TFET that features a 10 nm Al_2O_3 gate dielectric, and it exhibited an I_{SAT} of 20 μA/μm @ V_{GS} = 2 V and SS of greater than 150 mV/dec. TFETs with a gate modulated Zener tunnel junction at the source region is highly preferable for high-speed digital logic integrated circuit applications due to their low SS for a range of gate voltages near OFF state. In 2010, S. Mookerjee et al. [19] reported the successful fabrication of a side wall gated vertical $In_{0.53}Ga_{0.47}As$ TFET. MBE (molecular beam epitaxy) can be used to grow the epitaxial layer structures of vertical TFETs. Ti/Pt/Au metal layers were used for realizing contacts of $In_{0.53}Ga_{0.47}As$ vertical TFETs. Al_2O_3, HfO_2, and ZrO_2 are widely used as gate oxide materials in $In_{0.53}Ga_{0.47}As$ TFETs. Ti/Pt/Au, Cr/Au, AuGe/Ni/Au, Ti/Au, Ti/Pd/Au, Pd/Au, and Mo can be used to create source/drain contacts. TaN, Pd, Pd/Au, Ti/W, etc. can be used to form gate electrodes in $In_{0.53}Ga_{0.47}As$ TFETs. Some of the popular Indium Gallium Arsenide (InGaAs)-based TFET architectures are shown in Figure 3.6. The I_D-V_{DS} curves of side wall gated vertical $In_{0.53}Ga_{0.47}As$ TFET under

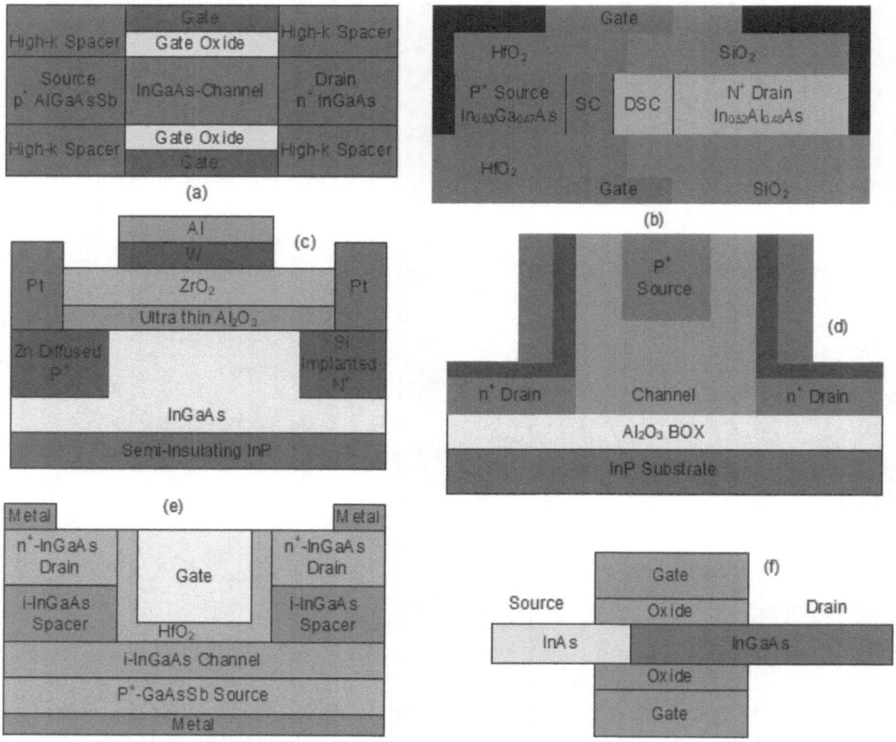

FIGURE 3.6 (a) Hetero-Tunnel-Junction N-Channel Nanowire (NW) TFET [20], (b) HDL-TFET [21], (c) Gate Stack Engineered InGaAs Planar TFET [22], (d) TTFET Structure [23], (e) U-Gate TFET [24], and (f) InAs/InGaAs Double-Gate Heterojunction Enhanced-TFET [25]

different temperatures are shown in Figure 3.7. Under forward bias condition, the side wall gated vertical $In_{0.53}Ga_{0.47}As$ exhibits gate modulated NDR (negative differential resistance), which is evident from Figure 3.7 (a). The SS of the side wall gated vertical $In_{0.53}Ga_{0.47}As$ TFET increases with increase in EOT (equivalent oxide thickness).

Moreover, the I_{SAT} of the side wall gated vertical $In_{0.53}Ga_{0.47}As$ TFET decreases with an increase in EOT. In 2012, Dheeraj Mohata et al. [26] experimentally have done the performance comparison of $In_{0.7}Ga_{0.3}As$ homojunction (homj) TFET, $GaAs_{0.4}Sb_{0.6}/In_{0.65}Ga_{0.35}As$ moderate-stagger Type-II heterojunction (Mod.Hetej) TFET, and $GaAs_{0.35}Sb_{0.65}/In_{0.7}Ga_{0.3}As$ high-stagger Type-II heterojunction (High. Hetj) TFET, which is depicted in Figure 3.7 (b). For the same dimensions, high stagger heterojunction TFET offers higher I_{ON}, lower SS, and higher I_{ON}/I_{OFF} ratio. Guangle Zhou et al. [27] investigated the effect of thin undercut thickness of InGaAs layer and observed that thinner InGaAs undercuts are preferable for achieving higher I_{ON} and I_{ON}/I_{OFF} ratio. A thinner InGaAs undercut effectively minimizes the ambipolar current [Figure 3.7 (c)]. In 2013, Tao Yu et al. [28] reported a $GaAs_{0.5}Sb_{0.5}/In_{0.53}Ga_{0.47}As$ quantum well type-II staggered TFET on InP wafer featuring HfO_2

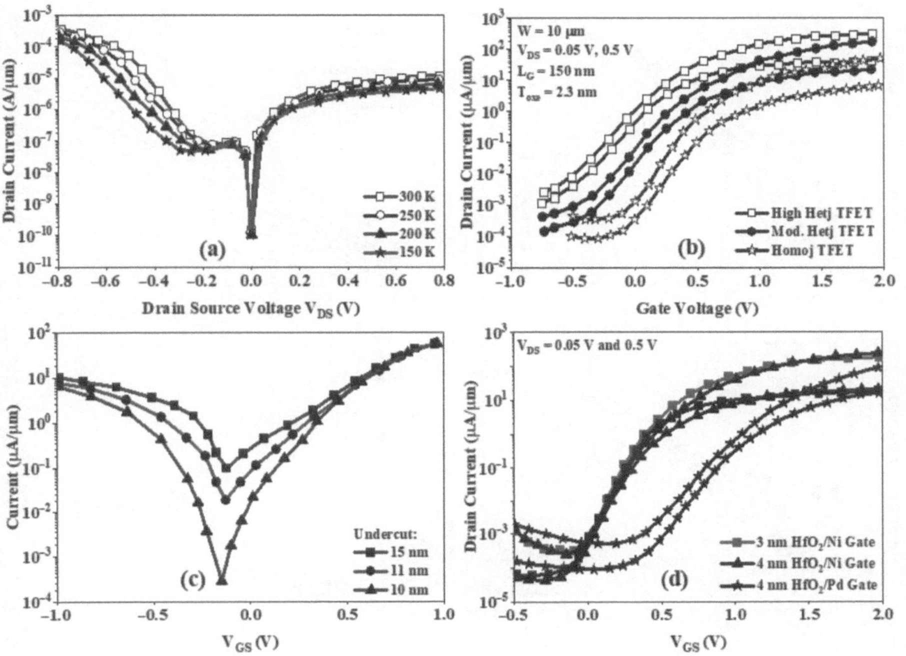

FIGURE 3.7 (a) Temperature Dependence of Output Curves of $In_{0.53}Ga_{0.47}As$ Vertical TFET [19]; (b) Transfer Characteristics of $In_{0.7}Ga_{0.3}As$ Homojunction TFET (homj TFET), $In_{0.65}Ga_{0.35}As/GaAs_{0.4}Sb_{0.6}$ Moderate Stagger Heterojunction TFET (Mod. Hetj TFET), and $In_{0.7}Ga_{0.3}As/GaAs_{0.35}Sb_{0.65}$ High Stagger Heterojunction TFET (High Hetj TFET) [26]; (c) the Impact of InGaAs Thickness Variation in the Undercut Access Region on Transfer Characteristics of Vertical TFETs [27]; and (d) Transfer Characteristics of $GaAs_{0.4}Sb_{0.6}/In_{0.65}Ga_{0.35}As$ vertical TFET with Different Gate Stacks [28]

gate oxide. The I_{ON} of the quantum well TFET is found to be increasing with increase in the gate area. In 2015, Bijesh Rajamohanan et al. [29] experimentally demonstrated the switching performance of $GaAs_{0.4}Sb_{0.6}/In_{0.65}Ga_{0.35}As$ quantum well TFET, which is depicted in Figure 3.7 (d). The main advantage of staggered Type-II TFET is that its effective barrier height can be tuned by changing the material compositions. Figure 3.7 (d) depicts the I_D-V_{DS} curves of $GaAs_{0.4}Sb_{0.6}/In_{0.65}Ga_{0.35}As$ quantum well TFETs with three different gate stacks such as 4 nm HfO_2/Pd, 4 nm HfO_2/Ni, and 3 nm HfO_2/Ni. $GaAs_{0.4}Sb_{0.6}/In_{0.65}Ga_{0.35}As$ quantum well TFETs with HfO_2/Ni gate stack offers higher I_{ON} and I_{ON}/I_{OFF} ratio. In 2014, Yan Zhu et al. [30] intensively studied the effects of temperature on the reliability of $GaAs_{0.35}Sb_{0.65}/In_{0.7}Ga_{0.3}As$ type-II staggered vertical TFET on InP wafer. The effect of temperature on the ambipolar conduction of $GaAs_{0.35}Sb_{0.65}/In_{0.7}Ga_{0.3}As$ type-II staggered vertical TFET is depicted in Figure 3.8.

The ambipolar current is found to be increasing exponentially with a rise in temperature. A higher temperature leads to the increase of SS and decrease of I_{ON}/I_{OFF} ratio of $GaAs_{0.35}Sb_{0.65}/In_{0.7}Ga_{0.3}As$ type-II staggered vertical TFET due to the

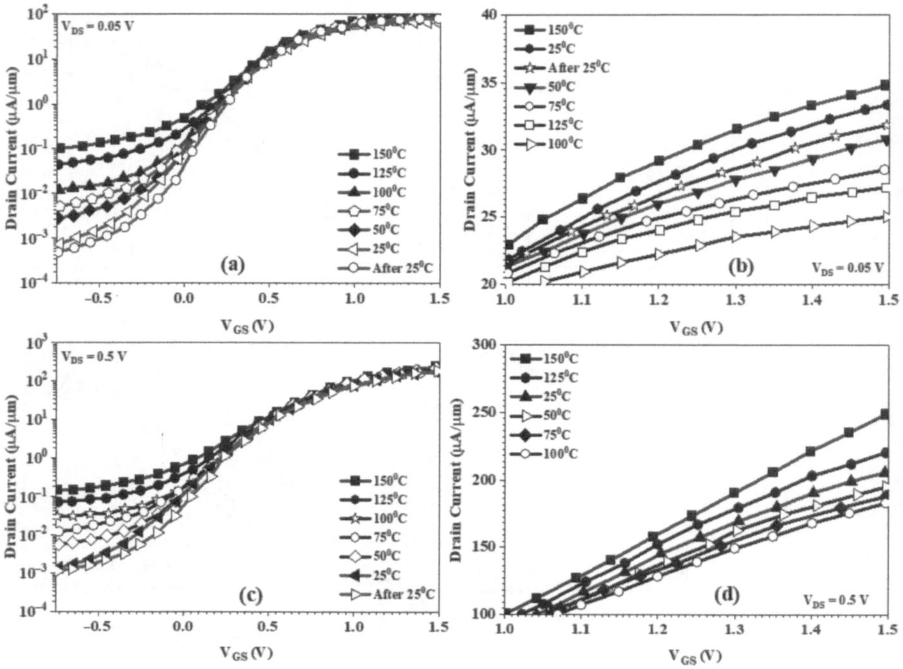

FIGURE 3.8 Ambipolar Current Variations with Temperature on Transfer Characteristics of Staggered Bandgap Type-II Heterojunction-$GaAs_{0.35}Sb_{0.65}/In_{0.7}Ga_{0.3}As$ Vertical TFET on InP Wafer [30]

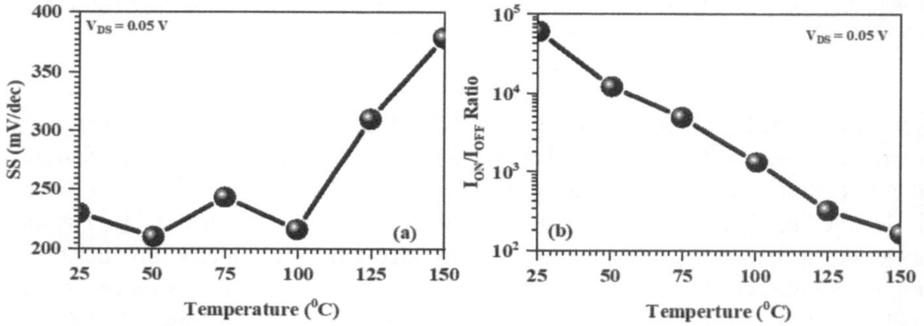

FIGURE 3.9 $GaAs_{0.35}Sb_{0.65}/In_{0.7}Ga_{0.3}As$ Type-II Staggered Bandgap Vertical TFET on InP Wafer (a) SS Vs Temperature and (b) Temperature Vs I_{ON}/I_{OFF} ratio Plot [30]

increase of ambipolar current (Figure 3.9). In 2015, Avik Chakraborty et al. [20] investigated the RF/analog performance of AlGaAsSb/InGaAs type-II staggered heterojunction nanowire TFET. Lower effective tunneling width, lower bandgap, and lower effective tunneling mass are the major advantages of III-V materials like InGaAs, which helps to achieve higher I_{ON}, lower I_{OFF}, and higher I_{ON}/I_{OFF} ratio

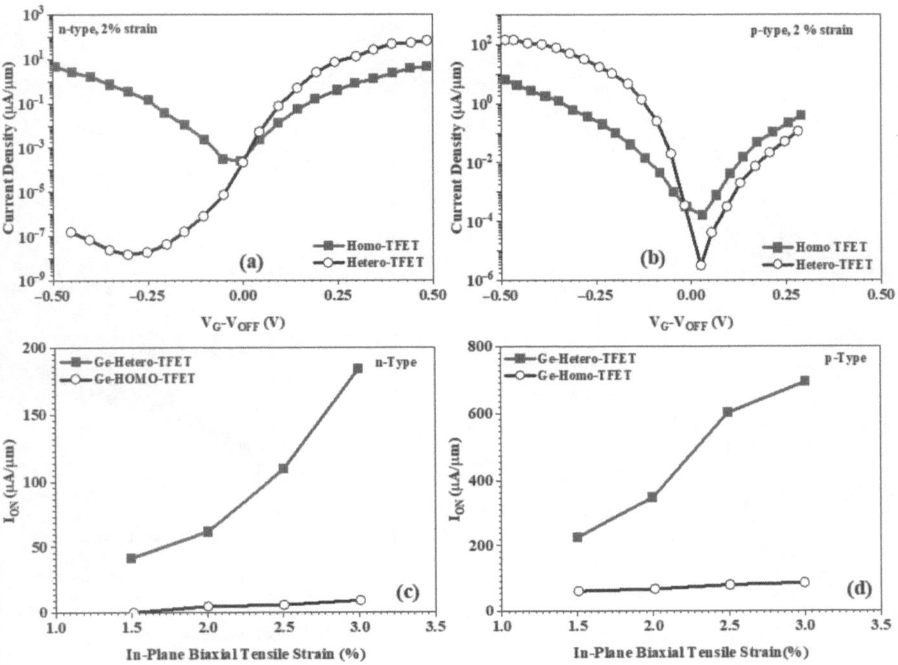

FIGURE 3.10 (a) Current Density Vs V_G-V_{OFF} of N-channel Heterojunction and Homojunction TFETs with 2% Tensile Strain, (b) Current Density Vs V_G-V_{OFF} of P-channel Heterojunction and Homojunction TFETs with 2% Tensile Strain, (c) Biaxial Tensile Strain Vs I_{ON} (N-channel Homo and Heterojunction TFETs), and (d) Biaxial Tensile Strain Vs I_{ON} (P-channel Homo and Heterojunction TFETs) [51]

[31–49]. The nanowire structure provides good gate control over the channel that helps to achieve extremely low SS. The performance of InGaAs TFETs can be improved by introducing a germanium layer on top of the InGaAs channel. The introduction of Ge epi layer creates a tensile strain in the channel that significantly improves the RF/analog characteristics of InGaAs TFETs [50]. The transfer curves of n-channel and p-channel InGaAs homojunction TFET and Ge-InGaAs heterojunction TFET are depicted in Figure 3.10 (a) and Figure 3.10 (b), respectively. The effect of tensile strain on the n and p channel Ge-InGaAs Type-II TFETs is illustrated in Figure 3.10 (c) and Figure 3.10 (d), respectively. For n and p channel TFETs, Ge-InGaAs heterostructure offers higher I_{ON} and lower I_{OFF} due to the suppression of ambipolar current. Also, the I_{ON} of Ge-InGaAs TFETs can be improved by increasing the tensile strain between Ge and InGaAs layers. In 2015, Genquan Han et al. [25] reported a double gate (DG) n-channel Indium Arsenide (InAs)/InGaAs type-II staggered heterojunction TFET.

The effect of distance between InAs/InGaAs heterojunction and source/channel tunneling junction ($L_{T\text{-}H}$) on the performance of n-channel DG-Type-II staggered InAs/InGaAs heterojunction TFET is illustrated in Figure 3.11. A large $L_{T\text{-}H}$ helps to achieve a

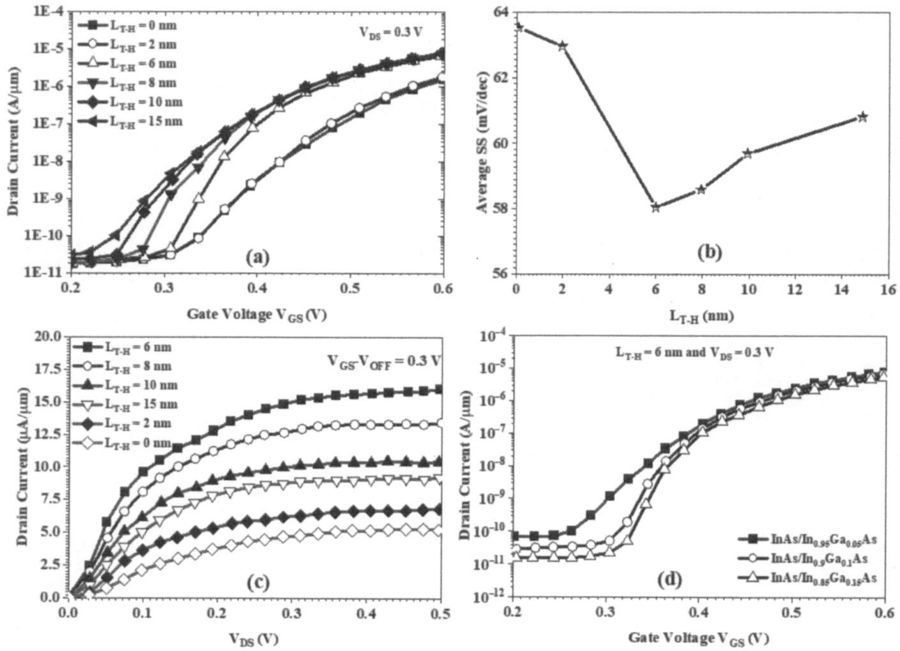

FIGURE 3.11 InAs/InGaAs DG Heterojunction TFET (a) Influence of $L_{T\text{-}H}$ on the I_D-V_{GS} Curves, (b) $L_{T\text{-}H}$ Vs SS Plot, (c) Influence of $L_{T\text{-}H}$ on the I_D-V_{DS} Curves, and (d) Effects of Ga Concentration in the Channel on I_D-V_{GS} Curves [25]

low threshold voltage and high drain current. A $L_{T\text{-}H}$ of 6 nm is considered the optimized value for achieving for achieving lowest SS and highest I_{ON} in InAs/InGaAs TFETs. A higher indium concentration in the channel results in the lowering of its bandgap that leads to the increase of both I_{ON} and I_{OFF} [52–56]. The use of graded channel can significantly improve the I_{ON} and I_{ON}/I_{OFF} ratio by effectively minimizing the I_{OFF}. A lower bandgap is required for improving I_{ON} whereas a higher bandgap is preferable for reducing the I_{OFF}. Therefore, the graded channel can simultaneously improve I_{ON} without the increase of I_{OFF} [57]. In 2017, Pao-Chuan Shih et al. [24] reported a novel GaAsSb/InGaAs type-II staggered heterojunction vertical U-Gate TFET that exhibited a high I_{ON} of 520 µA/µm, I_{ON}/I_{OFF} ratio of over 10^7 at V_{GS}—V_{TH} = 0.3 V. U-Gate vertical TFET uses a raised drain/source region to effectively reduce the tunneling leakage current by a wide energy gap spacer material. A large spacer thickness is preferable for reducing I_{OFF} of U-Gate vertical TFET [Figure 3.12 (a)]. However, the use of a thick spacer layer slightly reduces the I_{ON} of the U-Gate vertical TFET [Figure 3.12 (b)]. In summary, a large spacer thickness in U-Gate vertical TFET can significantly improve the I_{ON}/I_{OFF} ratio [Figure 3.12 (c)]. A lower channel thickness is essential for achieving higher I_{ON} in U-Gate vertical TFETs [Figure 3.12 (d)]. $In_{0.7}Ga_{0.3}As$ channel material provides higher I_{DS}/W and g_m/W compared to $In_{0.53}Ga_{0.47}As$ channel material [Figure 3.13 (a) and Figure 3.13 (b)]. This is mainly due to the reduction of bandgap with an increase in indium content in the channel [58].

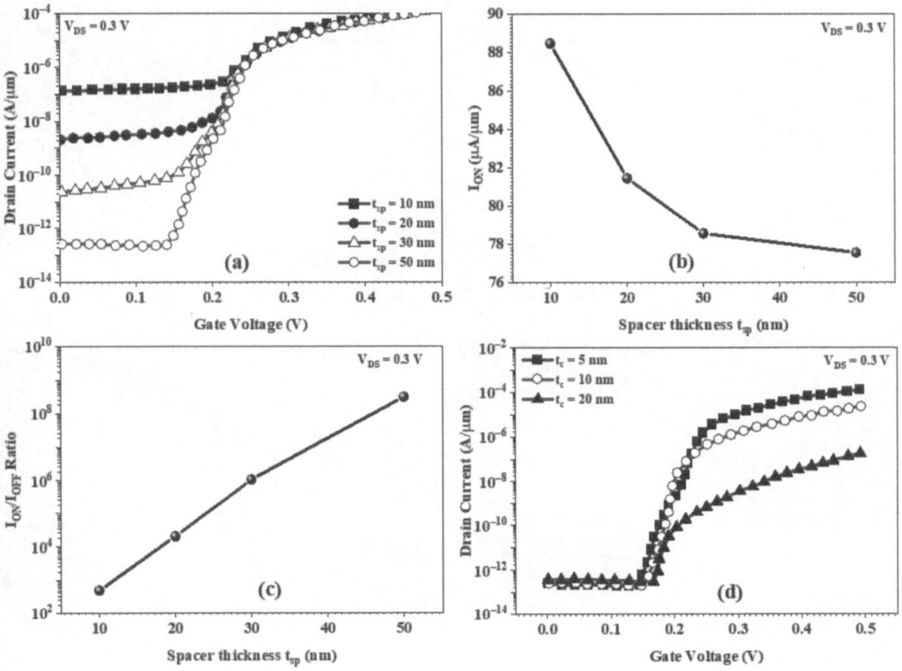

FIGURE 3.12 U-gate GaAsSb/InGaAs Heterojunction TFET (a) Spacer Thickness Effects on Transfer Curves, (b) Spacer thickness Vs I_{ON} Plot, (c) Spacer Thickness Vs I_{ON}/I_{OFF} Ratio Plot, and (d) Effect of Channel Thickness Variation on Transfer Curves [24]

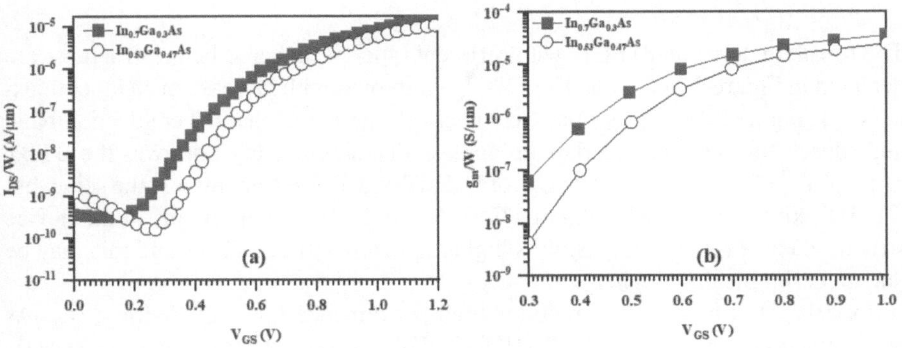

FIGURE 3.13 The Influence of Ga Content in the Channel on the Analog Performance of Homojunction InGaAs TFET [58]

In 2017, Prabhat Kumar Dubey et al. [23] compared the performance of L-shaped InGaAs TFET with T-shaped InGaAs TFET and found that T-shaped TFETs are superior in performance. TTFETs offer high drive current [Figure 3.14 (a)] and lower SS [Figure 3.14 (b)] compared to LTFETs due to the enhanced line tunneling rate.

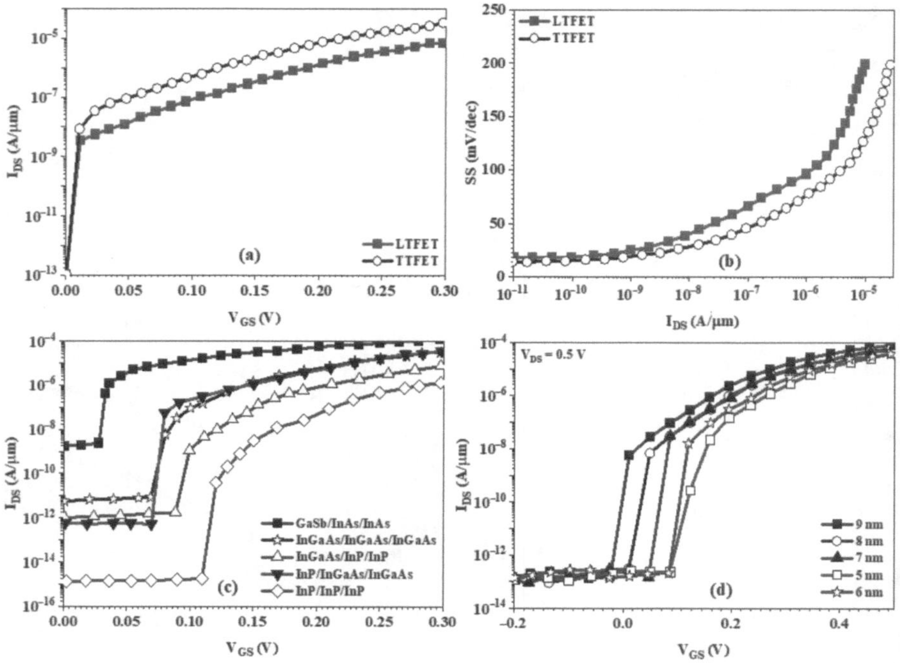

FIGURE 3.14 (a) The Transfer Curves of TTFET and LTFET, (b) SS versus I_{DS} Characteristics of LTFET and TTFET, (c) Transfer Characteristics of TTFETs with Different Materials Systems, and (d) Influence of Channel Thickness on the Transfer Curves of InP\ $In_{0.53}Ga_{0.47}As$\ $In_{0.53}Ga_{0.47}As$ TTFET [23]

The transfer curves of TTFETs with different type-II staggered heterostructures are depicted in Figure 3.14 (c). For TTFETs, I_{ON} improves with increase in channel thickness (t_c) [Figure 3.14 (d)]. For large channel thickness, the channel conduction band and source valence band overlap an increase that significantly improves the BTBT rate and I_{ON} of the TTFETs. The C_{GS} of LTFET and TTFET are almost the same, but TTFET exhibits higher C_{GD} than LTFET due to double drain regions in the device structure. However, TTFET exhibits higher f_T compared to LTFET and this may be due to its significantly higher g_m.

In 2019, Hu Liu et al. [21] reported high-performance $In_{0.52}Al_{0.48}As/In_{0.53}Ga_{0.47}As$ heterojunction dopingless TFET (HDL-TFET), whose electrical performance is depicted in Figure 3.15. The I_{OFF} and I_{ON} of HDL-TFET is sensitive to source side channel length (L_{SC}), and it is found that both I_{OFF} and I_{ON} increase with an increase in L_{SC}. However, a large L_{SC} is suitable for achieving higher I_{ON}/I_{OFF} ratio in HDL-TFETs. But a L_{SC} of around 5 nm is found to be suitable for achieving low SS in InGaAs HDL TFETs. In 2019, Zhirui Yan et al. [59] reported a $In_{0.53}Ga_{0.47}As/GaAs_{0.5}Sb_{0.5}$ type-II staggered Z-Gate heterojunction TFET. The gate bias dependence of f_T of Z-TFET, HGD-ZTFET (heterogate dielectric Z-TFET), HZ-TFET (heterogate Z-TFET), and HGD-HZ-TFET (heterogate dielectric-heterojunction ZTFET) are

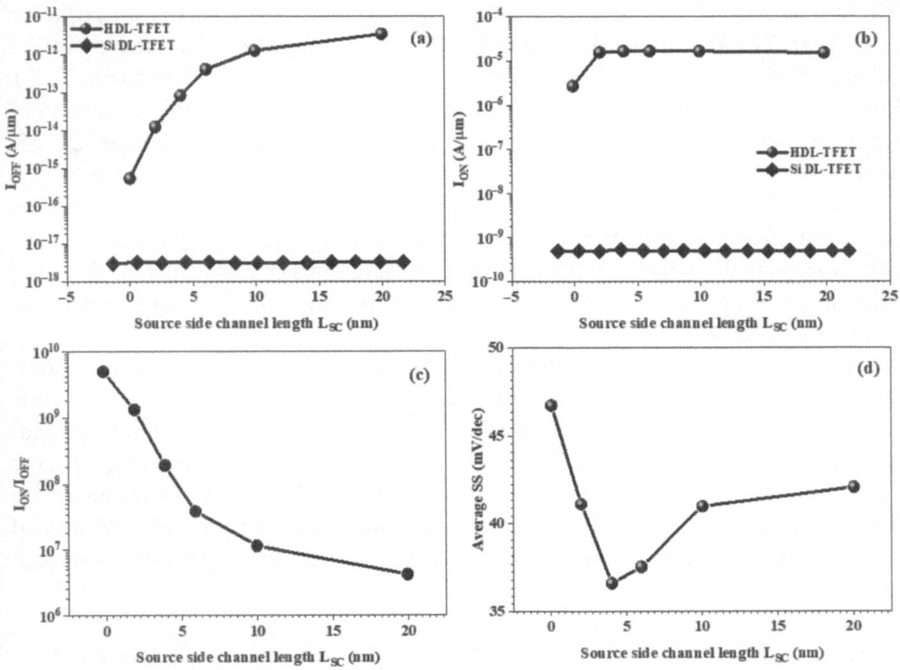

FIGURE 3.15 I_{OFF} Vs L_{SC} Plot for Si DL-TFET and HDL-TFET, (b) I_{ON} Vs L_{SC} Plot for Si DL-TFET and HDL-TFET, (c) I_{ON}/I_{OFF} ratio Vs L_{SC} Plot for HDL-TFET, and (d) SS Vs L_{SC} Plot for HDL-TFET [21]

shown in Figure 3.15 (a). HGD-HZ TFET exhibits higher f_T compared with Z-TFET, HGD-TFET, and HZ-TFET due to its higher transconductance and lower total gate capacitance. The Ga and Sb concentration significantly affects the DC performance of GaAsSb/InGaAs ZTFETs, and it is observed that both I_{ON} and I_{OFF} increase with an increase in Sb concentration. On the other hand, a lower Ga content in InGaAs channel is essential for achieving higher I_{ON} [60, 61].

3.3 TFETs ON InAs MATERIAL SYSTEMS

Like InGaAs, InAs semiconductor also has zinc blend structure and has a bandgap of 0.36 eV, electron effective mass (m_e) of $0.023m_0$, hole effective mass of $0.41m_0$, electron affinity of 4.9 eV, lattice constant of 6.05 A^0, electron mobility (μ_e) of 40000 cm^2/Vs, and hole mobility of (μ_h) of 500 cm^2/Vs. In 2012, Rui Li et al. [65] reported an InAs/AlGaSb TFET on GaSb wafer that exhibited an I_{ON} of 78 μA/μm @ V_{DS} = 0.5 V. In 2013, Anil W. Dey et al. [66] reported an InAs/GaSb nanowire TFET that exhibited an I_{ON} of 310 μA/μm @ V_{DS} = 0.5 V and g_m of 250 mS/mm @ V_{DS} = 0.3 V. MBE process can be used to grow the epitaxial layer heterostructure of TFET. Al_2O_3, HfO_2, ZrO_2, Al_2O_3/HfO_2, etc. can be used as gate dielectrics in InAs-based TFETs [67]. Ti/Au is widely used as source and drain contacts in InAs

TFETs. Dual metal gate (DMG) technology is found to be effective in reducing SS of InAs TFETs without degrading I_{ON} [68]. δ-doping offers high drive current and low IOFF compared to uniform doping in InAs/AlGaSb broken bandgap TFET [Figure 3.16 (a)]. In 2015, Erik Lind et al. [69] reported the successful fabrication of lateral and vertical GaSb/InAs nanowire TFETs on silicon substrate using MOCVD process. InAs, GaSb, and $In_{0.53}Ga_{0.47}As$ materials are having lower bandgaps [63]. GaSb/InAs TFETs without underlap between source and gate regions offer higher I_{ON} compared with GaSb/InAs TFETs with underlaps. The drive current degrades with increase in underlap between gate and source regions [Figure 3.16 (b)]. A higher underlap between gate and source regions degrades I_{ON} and SS characteristics of GaSb/InAs TFETs.

The I_D-V_{GS} curves of n-type and p-type InAs/GaSb TFET with and without underlap are illustrated in Figure 3.16 (c) and Figure 3.16 (d), respectively. The electrostatic control of GaSb/InAs TFETs can be improved by providing GAA (gate-all-around) technology [73]. A lower nanowire diameter is required to achieve high I_{ON} and low I_{OFF} in GaSb/InAs GAA nanowire TFET. The SS of GaSb/InAs GAA nanowire TFET increases with a rise in temperature. In 2016, Ching-Yi Hsu et al. [70] studied the role of InAs channel thickness (t_{InAs}) on the I_D-V_{GS} curves of GaSb/InAs vertical

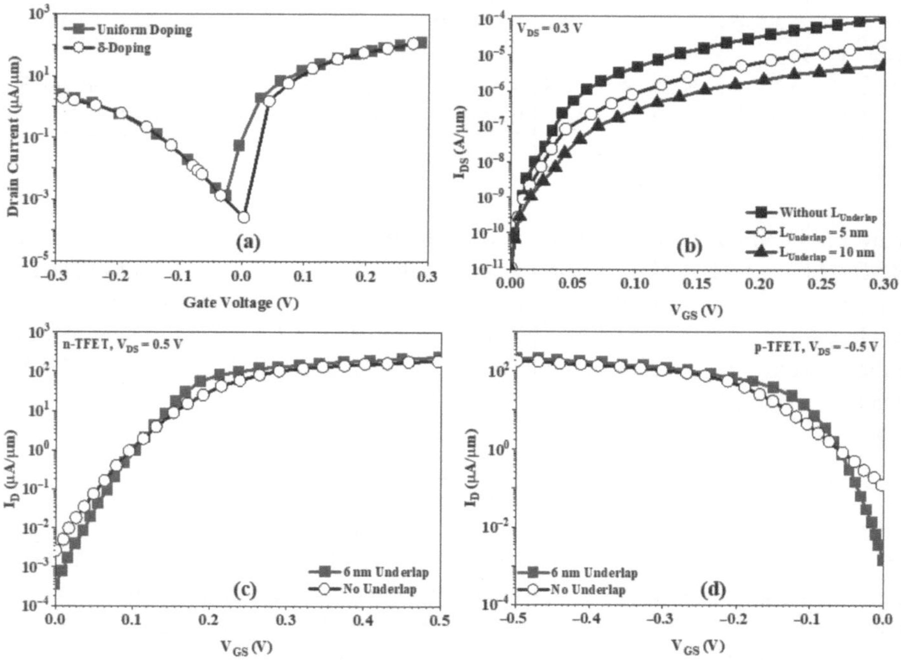

FIGURE 3.16 (a) Doping Effects on Transfer Curves of AlGaSb/InAs TFET [62], (b) Source-Channel Underlap Effects in the Transfer Characteristics of GaSb/InAs TFET [63], (c) Source-Channel Underlap Effects on the I_D-V_{GS} Curves of N-type GaSb/InAs Vertical TFET [64], and (d) Source-Channel Underlap Effects on the I_D-V_{GS} Curves of P-type GaSb/InAs Vertical TFET [64]

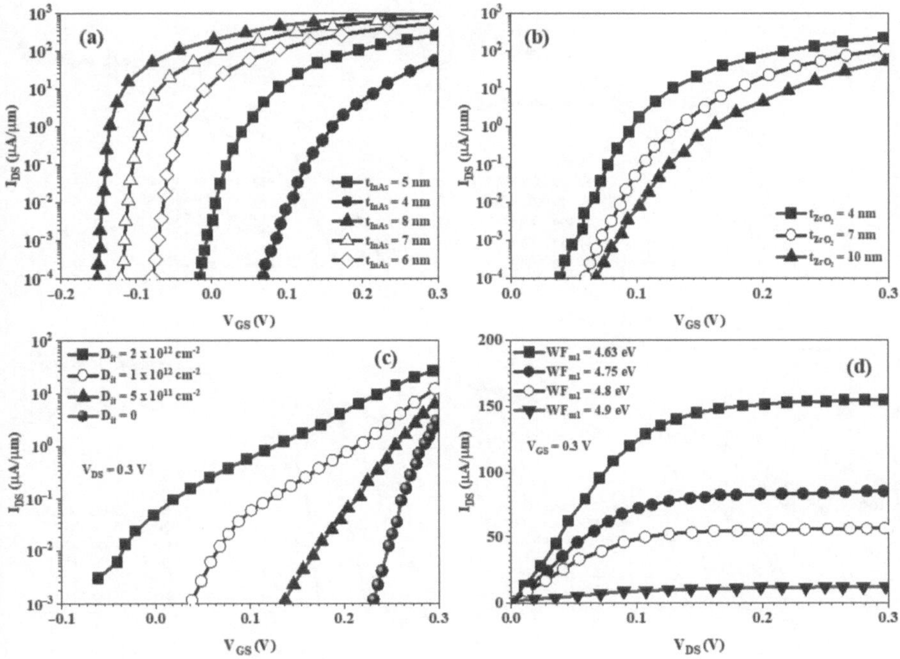

FIGURE 3.17 (a) t_{InAs} Effects on the I_{DS}-V_{GS} Curves of GaSb\InAs Vertical TFET [70], (b) Gate Dielectric Thickness Effects on the I_{DS}-V_{GS} Curves of GaSb\InAs Vertical TFET [70], (c) Surface States (D_{it}) Effects on the I_{DS}-V_{GS} Curves of Vertical GaSb\InAs TFET [71], and (d) WFm1 Effects on the I_{DS}-V_{DS} Curves of DMG Vertical GaSb\InAs TFETs [71]

TFETs and observed that GaSb/InAs TFETs with large t_{InAs} exhibit higher I_{ON} and lower R_{ON} (ON resistance) because of their low tunneling barrier [Figure 3.17 (a)]. Channel doping concentration also plays a key role in determining the I_{ON} of GaSb/InAs TFETs, and it is observed that a large channel doping concentration is required to improve I_{ON} of GaSb/InAs TFETs. Also, the thickness of gate dielectric can significantly affect the drive current of GaSb/InAs vertical TFETs. Usually, a lower dielectric thickness is preferable for improving drive current in GaSb/InAs TFETs [Figure 3.17 (b)]. The surface states are another important factor that affects the switching performance of GaSb/InAs TFETs [Figure 3.17 (c)]. The I_{OFF} and SS of GaSb/InAs TFETs degrade severely with an increase in D_{it} (interface traps), which in turn leads to the degradation of switching performance of the device. The DMG technology can effectively minimize the SS of GaSb/InAs vertical TFETs. Usually, metals with lower work function are used as metal gate-1 and metals with higher work function are used as metal gate-2 in DMG TFETs to achieve low SS and low threshold voltage. The use of metal gate with lower work function as metal gate-1 significantly improves the drive current [Figure 3.17 (d)].

In 2016, S. M. Biswal et al. [72] studied the effect of gate length (L) scaling on the RF/analog performance of InAs nanowire TFET and observed that reduction of gate length improves RF performance and degrade analog performance (Figure 3.18).

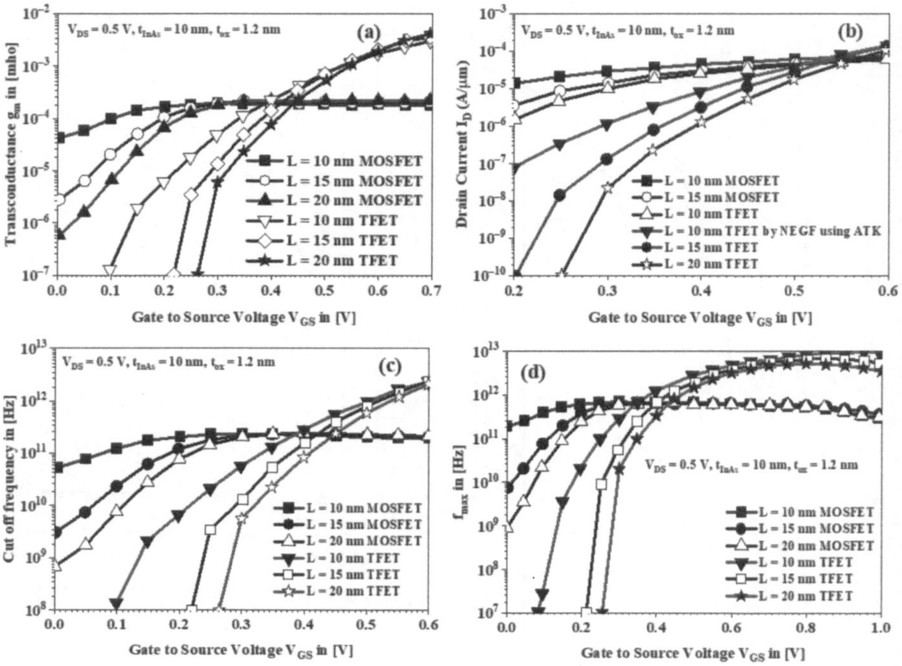

FIGURE 3.18 InAs MOSFETs and InAs TFETs [72] (a) g_m Vs V_{GS} Curves for Different L, (b) I_D Vs V_{GS} Curves for Different L, (c) f_T Vs V_{GS} Curves for Different L, and (d) f_{max} Vs V_{GS} Curves for Different L

Transconductance (g_m), f_T (Equation (3.2)), f_{max}, and gain bandwidth product (GBW) (Equation (3.4)) are the key factors that determine the analog/RF performance of a device.

$$f_T = \frac{g_m}{2\pi C_{gg}} \tag{3.2}$$

$$C_{gg} = C_{gs} + C_{gd} \tag{3.3}$$

$$GBW = \frac{g_m}{2\pi C_{gd}} \tag{3.4}$$

Good linearity is required for RF applications. Output resistance (R_o) and transconductance (g_m) are the two key parameters that determine the linearity of the devices. For better linearity, g_{m3} should be minimum.

$$g_{m3} = \frac{\partial^3 I_{DS}}{\partial V_{GS}^3} \tag{3.5}$$

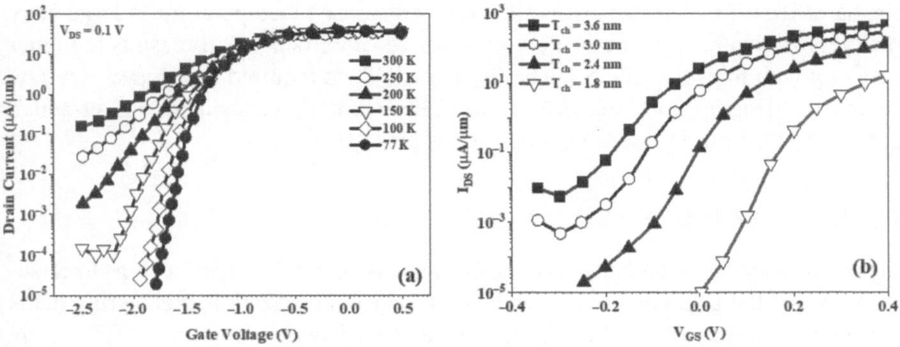

FIGURE 3.19 (a) Temperature Effects on Transfer Characteristics of $Al_{0.5}Ga_{0.5}Sb$\InAs Vertical TFET [74] and (b) Channel Thickness Effects on the Transfer Curves of N-GaSb\InAs TFET [75]

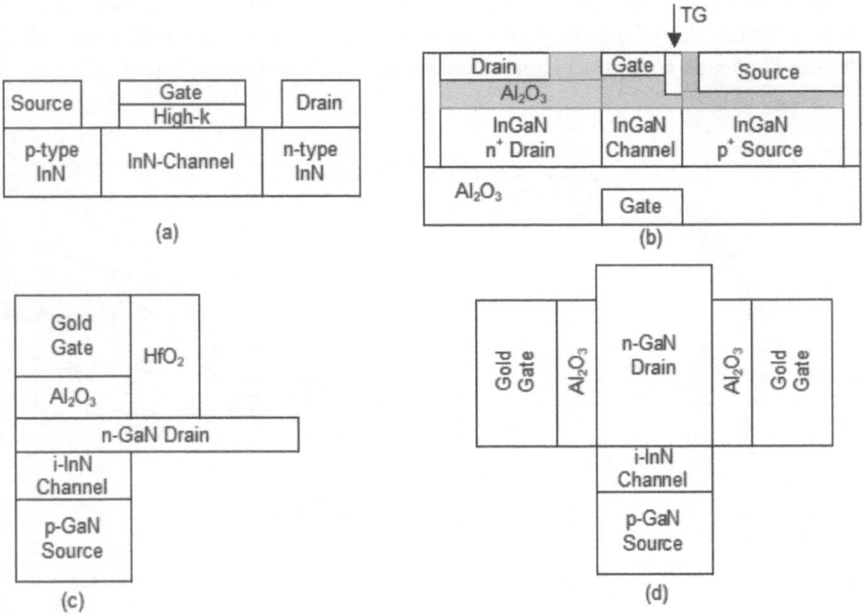

FIGURE 3.20 III-Nitride Material-Based TFET Structures (a) InN Homojunction TFET [82], (b) GE DL TFET [83], (c) In-line Geometry TFET Design [84], and (d) Side Wall Gated Geometry TFET Design [84]

InAs-based TFETs are considered as an attractive technology for future ultra low power and high-performance digital integrated circuits [76, 77]. In 2017, Vinay Kumar Chinni et al. [74] reported a vertical V-shaped AlGaSb/InAs TFET on GaAs wafer that exhibited an I_{ON} of 433 μA/μm @ V_{DS} = 0.5 V. The vertical V-shaped AlGaSb/InAs TFET exhibits higher I_{OFF} at higher temperatures [Figure 3.19 (a)].

The SS of III-V TFETs changes significantly with a rise in temperature. A large body thickness or channel thickness (T_{ch}) in InAs-based TFETs normally results in higher I_{ON} along with high I_{OFF} and SS. That is, a small T_{ch} is required to achieve very low SS and I_{OFF} [Figure 3.19 (b)]. InAs-based TFETs are also considered an attractive technology for future biosensing applications [78–81].

3.4 TFETs ON InGaN MATERIAL SYSTEMS

The use of wide bandgap materials like GaN and AlN in III-V TFETs helps to effectively reduce the I_{OFF}. GaN has a bandgap of 3.4 eV. Polarization engineering helps to achieve interband tunneling in III-Nitride-based heterostructure TFETs. GaN wurtzite crystals lacks inversion symmetry and therefore have polarization dipole along the c-axis. The piezoelectric nature and lattice mismatch in III-Nitrides also contribute polarization discontinuity. In inline III-Nitride TFET design, the gate electrode is usually placed in parallel with the tunnel junction. On the other hand, in side wall double gate III-Nitride TFET design, the gate electrode is usually placed perpendicular to the tunnel junction. III-Nitride materials-based TFET structures are shown in Figure 3.20. The transfer curves and output curves of III-Nitride TFETs with inline and side wall gated geometry designs are illustrated in Figure 3.21 (a) and

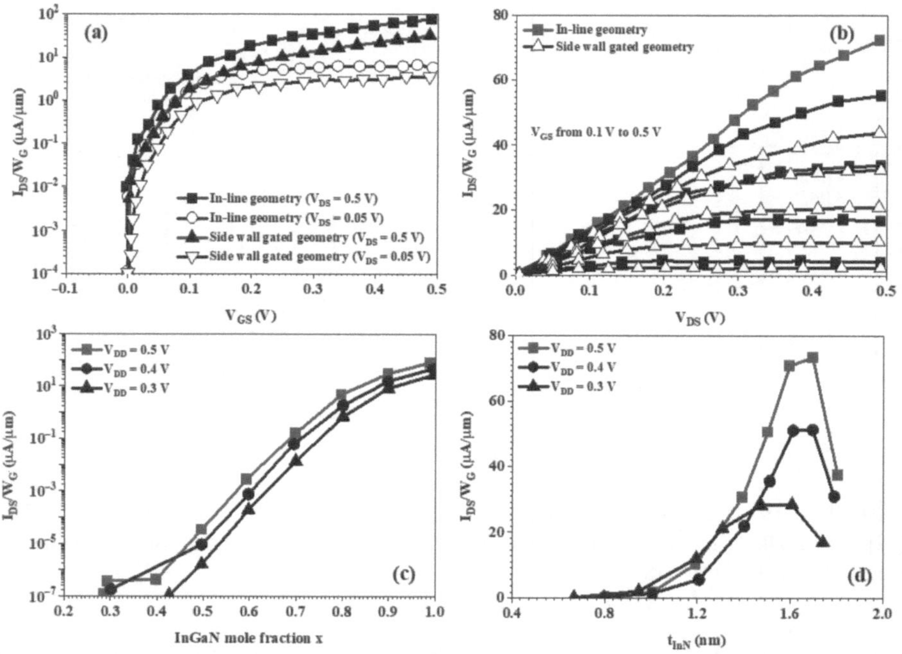

FIGURE 3.21 (a) I_{DS}-V_{GS} Plot of GaN TFETs with In-line Geometry and Side Wall Gated Geometry, (b) I_{DS}-V_{DS} Plot of GaN TFETs with In-line Geometry and Side Wall Gated Geometry, (c) Indium Mole Fraction Vs I_{DS} Plot of GaN Channel TFET, and (d) t_{InN} Vs I_{DS} Plot of GaN TFET [84]

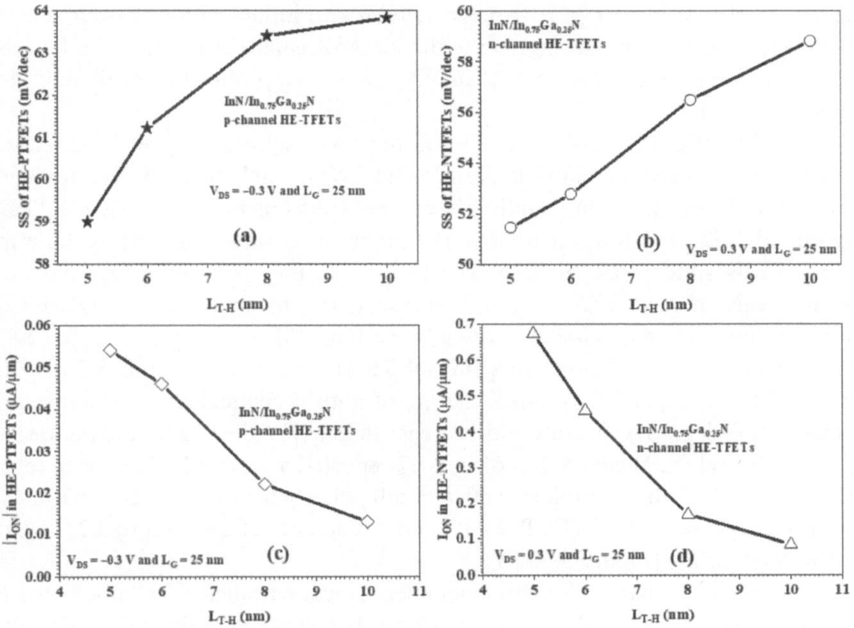

FIGURE 3.22 DG InN/InGaN TFET [85]: (a) SS Vs $L_{T\text{-}H}$ Plot for p-TFET, (b) SS Vs $L_{T\text{-}H}$ Plot for n-TFET, (c) I_{ON} Vs $L_{T\text{-}H}$ Plot for p-TFET, and (d) I_{ON} Vs $L_{T\text{-}H}$ Plot for n-TFET

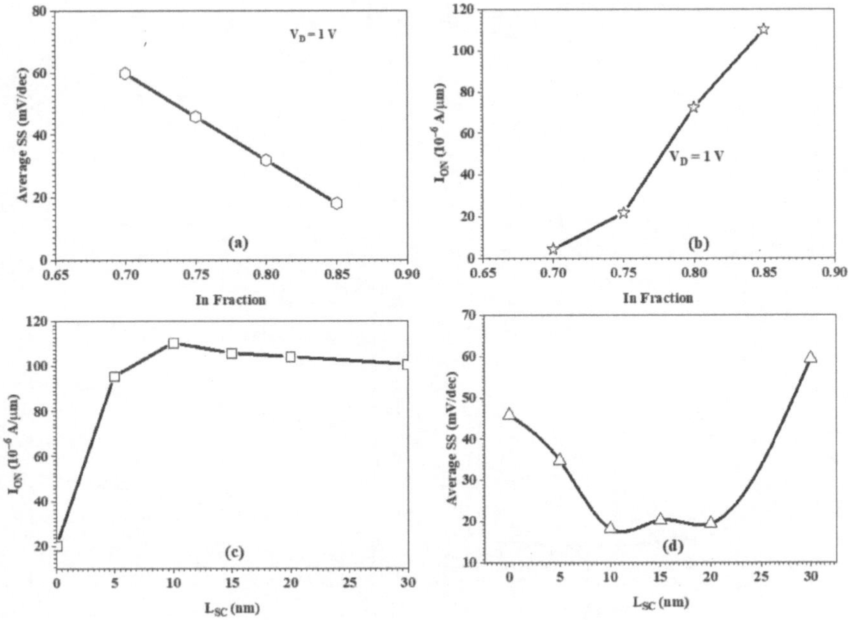

FIGURE 3.23 InGaN GD TFET [86]: (a) Indium Mole Fraction Vs Average SS Plot, (b) Indium Mole Fraction Vs I_{ON} Plot (c), L_{SC} Vs I_{ON} Plot, and (d) L_{SC} Vs Average SS Plot

Figure 3.21 (b), respectively. III-Nitride TFETs with inline geometry offer superior drive current performance compared with side wall gated III-Nitride TFETs. Also, III-Nitride TFETs with larger power supply values (V_{DD}) offer higher drive current [Figure 3.21 (c)].

The drive current of III-Nitride TFETs improves significantly with increase in indium content in Indium Gallium Nitride (InGaN) or InN channels. The thickness of InN (t_{InN}) layer also significantly affects the drive current of III-Nitride TFETs [Figure 3.21 (d)]. When t_{InN} increases, the tunneling distance as well as the effective band offset increases. In 2016, Yue Peng et al. [85] reported a heterojunction enhanced InN/InGaN TFET. $L_{T\text{-}H}$ [85] represents the distance between InN/InGaN heterojunction and channel-source tunnel junction. When $L_{T\text{-}H}$ increases, the SS of both p-channel and n-channel InN/InGaN TFETs increases [Figure 3.22 (a) and Figure 3.22 (b), respectively]. Also, the I_{ON} of both p-channel and n-channel InN/InGaN TFETs degrades severely with increase in $L_{T\text{-}H}$ [Figure 3.22 (c) and Figure 3.22 (d), respectively]. X. Duan et al. [86] in 2017 reported a graded InGaN drain region TFET (GD-TFET) that exhibited reduced ambipolar behavior. A large indium content in InGaN layer of GD-TFET results in the reduction of SS [Figure 3.23 (a)] and improvement of I_{ON} [Figure 3.23 (b)].

L_{SC} (source side channel length) is another critical parameter that affects the I_{ON} and SS of GD-TFETs. An L_{SC} of 10 to 15 nm is considered suitable for achieving

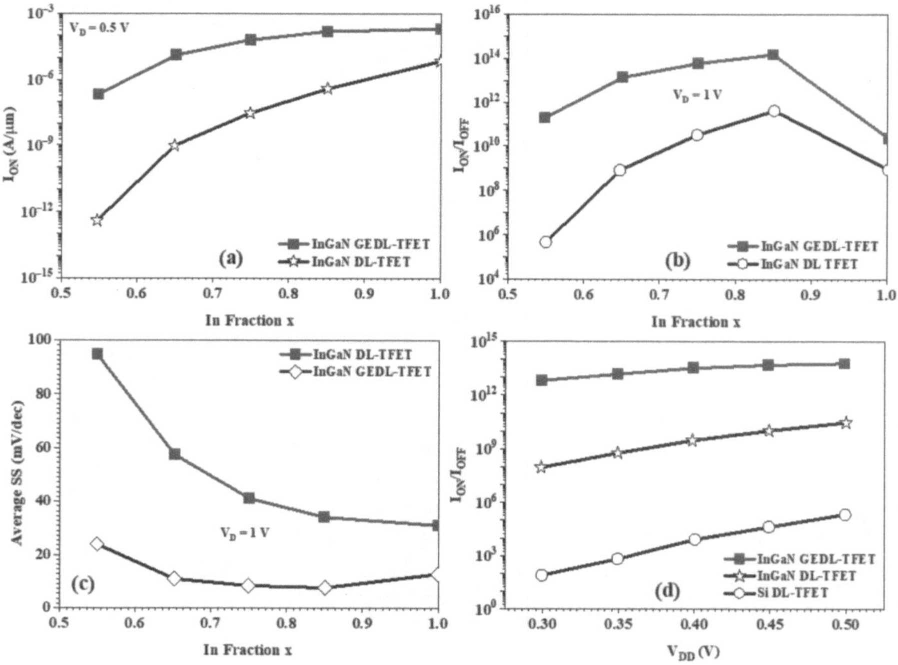

FIGURE 3.24 InGaN GE DL TFET [83]: (a) I_{ON} Vs Indium Mole Fraction, (b) I_{ON}/I_{OFF} Ratio Vs Indium Mole Fraction, (c) Average SS Indium Mole Fraction, and (d) I_{ON}/I_{OFF} Ratio Vs V_{DD} Scaling

high I_{ON} as well as low SS in GD-TFETs [Figure 3.23 (c) and Figure 3.23 (d)]. In 2018, X. Duan et al. [83] reported a gate engineered (GE) high-performance dopingless (DL) InGaN TFET that exhibited an I_{ON} of 80.2 μA/μm and f_T of 119 GHz. This GE-DL InGaN TFET is highly suitable for low-power digital logic and RF applications. Charge plasma technique can be used to form drain and source regions in DL-TFETs. The advantage of using charge plasma technique is that it can eliminate the random dopant fluctuations. Moreover, the avoidance of physical doping process in DL-TFETs can effectively reduce the bulk-trap assisted tunneling, which in turn improves the SS characteristics. InGaN GE DL TFET provides better I_{ON} [Figure 3.24 (a)], higher I_{ON}/I_{OFF} ratio [Figure 3.24 (b) and Figure 3.24 (d)], and lower SS [Figure 3.24 (c)] compared with InGaN DL TFETs. Work function of the tunneling gate (ϕ_{TG}) also plays a key role on the performance of GE-DL TFET (Figure 3.25). It is observed that a low ϕ_{TG} is required to achieve higher I_{ON} [Figure 3.25 (a)] and lower SS [Figure 3.25 (b)]. This is due to the fact that a lower ϕ_{TG} helps to achieve a lower tunneling distance and higher transconductance [Figure 3.25 (c)]. Also, a lower ϕ_{TG} can help to improve f_T of the GE-DL TFETs [Figure 3.25 (d)]. In 2019, Ashwani Kumar et al. [87] reported a p-channel GaN TFET with high I_{ON}/I_{OFF} ratio of 10^{14}. A lower channel width is required in p-channel GaN TFET to achieve low SS [Figure 3.26 (a)]. But use of low channel width significantly reduces the I_{ON} of the device [Figure 3.26 (b)]. A lower gate to drain length (L_{GD}) and lower gate length (L_G)

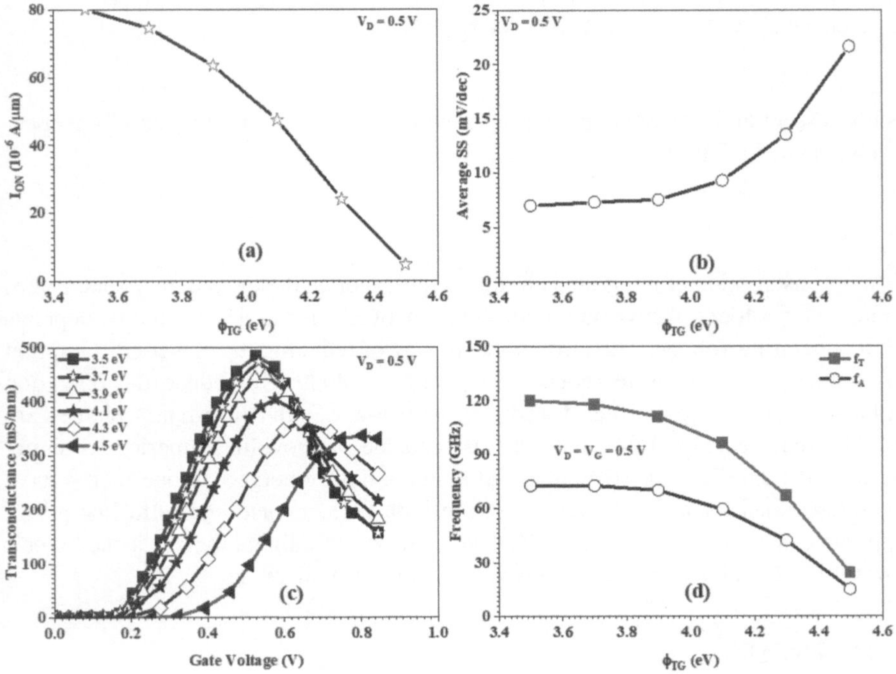

FIGURE 3.25 InGaN GE DL TFET [83]: (a) I_{ON} Vs ϕ_{TG} Plot, (b) Average SS Vs ϕ_{TG} Plot, (c) g_m Vs ϕ_{TG} Plot, and (d) f_T and f_A Vs ϕ_{TG} Plot

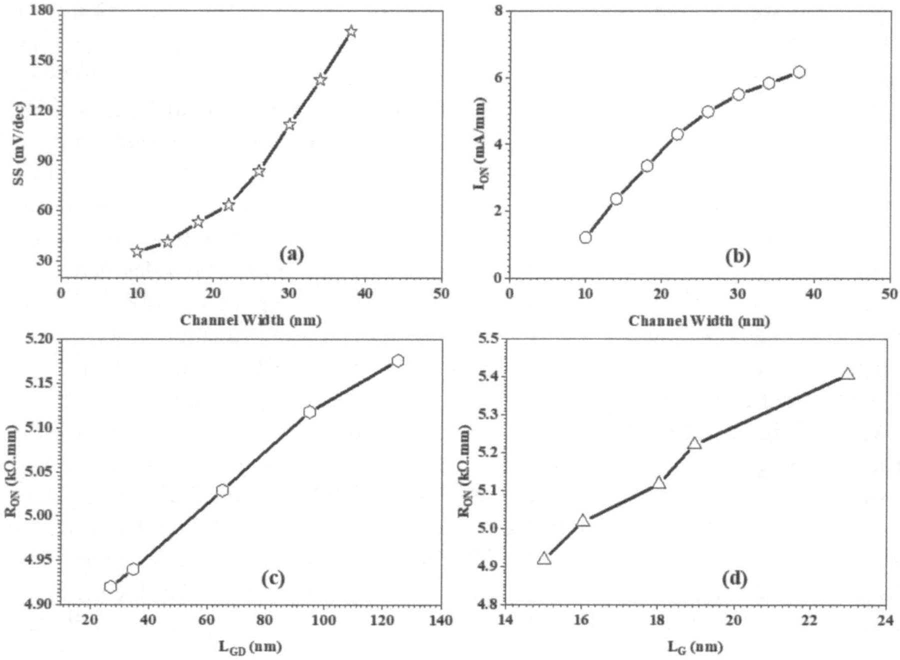

FIGURE 3.26 P-type GaN TFET [87]: (a) Channel Width Vs SS Plot, (b) Channel Width Vs I_{ON} Plot, (c) L_{GD} Vs R_{ON} Plot, and (d) L_G Vs R_{ON} Plot

is also essential for achieving low R_{ON} [Figure 3.26 (c) and Figure 3.26 (d)], respectively, in GaN p-TFETs.

3.5 CONCLUSION

Power consumption of integrated circuits is the major challenge the microelectronics industry faces today. Active power consumption of electronic circuits mainly depends on its operating voltage. The most effective way of reducing the power consumption of electronic circuits is to reduce its operating voltage. To reduce the operating voltage of electronic circuits, transistors with low SS and threshold voltages are the key requirements. III-V materials are capable of providing superior switching performance. TFETs on III-V material systems have emerged as one of the most potential transistor technologies for future high-performance and ultra low power applications such as SRAM, DRAM, flash memory, and digital logic ICs due to their low threshold voltage and below 60 mV/dec subthreshold swing.

REFERENCES

[1] Lemtur, Alemienla, Dheeraj Sharma, Jyoti Patel, Priyanka Suman, Chithraja Rajan: Two-stage Op-amp and Integrator Realisation through GaAsP/AlGaSb Nanowire CP-TFET, *Micro & Nano Letters* 14 (9) (2019) 980.

[2] Wang, Hongjuan, Genquan Han, Yan Liu, Shengdong Hu, Chunfu Zhang, Jincheng Zhang, Yue Hao: Theoretical Investigation of Performance Enhancement in GeSn/SiGeSn Type-II Staggered Heterojunction Tunneling FET, *IEEE Transactions on Electron Devices* 63 (1) (2016) 303.

[3] Visciarelli, Michele, Elena Gnani, Antonio Gnudi, Susanna Reggiani, Giorgio Baccarani: Impact of Strain on Tunneling Current and Threshold Voltage in III-V Nanowire TFETs, *IEEE Electron Device Letters* 37 (2016) 560.

[4] Memisevic, E., M. Hellenbrand, J. Svensson, L.-E. Wernersson: Impact of Band-Tails on the Subthreshold Swing of III-V Tunnel Field-Effect Transistor, *IEEE Electron Device Letters* 38 (12) (2017) 1661.

[5] Liu, Jheng-Sin, Michael B. Clavel, Mantu K. Hudait: An Energy-Efficient Tensile-Strained Ge/InGaAs TFET 7T SRAM Cell Architecture for Ultralow-Voltage Applications, *IEEE Transactions on Electron Devices*, 64 (5) (2017) 2193.

[6] Li, Mingda (Oscar), David Esseni, Joseph J. Nahas, Debdeep Jena, Huili Grace Xing: Two-Dimensional Heterojunction Interlayer Tunneling Field Effect Transistors (Thin-TFETs), *IEEE Journal of the Electron Devices Society* 3 (3) (2015) 200.

[7] Afzalian, A., G. Doornbos, T.-M. Shen, M. Passlack, J. Wu: A High-Performance InAs/GaSb Core-Shell Nanowire Line-Tunneling TFET: An Atomistic Mode-Space NEGF Study, *IEEE Journal of the Electron Devices Society* 7 (2018) 88.

[8] Grillet, Corentin, Alessandro Cresti, Marco G. Pala: Vertical GaSb/AlSb/InAs Heterojunction Tunnel-FETs: A Full Quantum Study, *IEEE Transactions on Electron Devices*, 65 (2018) 3038.

[9] Baravelli, Emanuele, Elena Gnani, Roberto Grassi, Antonio Gnudi, Susanna Reggiani, Giorgio Baccarani: Optimization of n- and p-type TFETs Integrated on the Same InAs/AlxGa1–xSb Technology Platform, *IEEE Transactions on Electron Devices*, 61 (2014) 178.

[10] Knoch, Joachim, Joerg Appenzeller: Modeling of High-Performance p-Type III–V Heterojunction Tunnel FETs, *IEEE Electron Device Letters* 31 (2010) 305.

[11] Wang, Yibo, Genquan Han, Yan Liu, Chunfu Zhang, Qian Feng, Jincheng Zhang, Yue Hao: Investigation of GaAsBi/GaAsN Type-II Staggered Heterojunction TFETs with the Analytical Model, *IEEE Transactions on Electron Devices* 64 (4) (2017) 1541.

[12] Nigam, K., P. Kondekar, D. Sharma: DC Characteristics and Analog/RF Performance of Novel Polarity Control GaAs-Ge Based Tunnel Field Effect Transistor, *Superlattices and Microstructures* 92 (2016) 224.

[13] Wang, Y., Y. Liu, G. Han, H. Wang, C. Zhang, J. Zhang, Y. Hao: Theoretical Investigation of GaAsBi/GaAsN Tunneling Field-effect Transistors with Type-II Staggered Tunnelling Junction, *Superlattices and Microstructures* 106 (2017) 139.

[14] Lemtur, A., D. Sharma, P. Suman, J. Patel, D. S. Yadav, N. Sharma: Performance Analysis of Gate All around GaAsP/AlGaSb CP-TFET, *Superlattices and Microstructures* 117 (2018) 364.

[15] Raad, Bhagwan Ram, Kaushal Nigam, Dheeraj Sharma, P. N. Kondekar: Performance Investigation of Bandgap, Gate Material Work Function and Gate Dielectric Engineered TFET with Device Reliability Improvement, *Superlattices and Microstructures* 94 (2016) 138.

[16] Cho, Seongjae, Hyungjin Kim, Heesauk Jhon, In Man Kang, Byung-Gook Park, James S. Harris: Mixed-Mode Simulation of Nanowire Ge/GaAsHeterojunction Tunneling Field-Effect Transistor for Circuit Applications, *IEEE Journal of the Electron Devices Society* 1 (2013) 48.

[17] Mookerjea, S., D. Mohata, R. Krishnan, J. Singh, A. Vallett, A. Ali, T. Mayer, V. Narayanan, D. Schlom, A. Liu, and S. Datta: Experimental Demonstration of 100 nm Channel Length In0.53Ga0.47As-based Vertical Inter-band Tunnel Field Effect Transistors (TFETs) for Ultra Low-power Logic and SRAM Applications, *IEDM Technical Digest* (2009) 949.

[18] Zhao, Han, Y. Chen, Y. Wang, F. Zhou, F. Xue, J. Lee: In0.7Ga0.3As Tunneling Field-Effect Transistors with an Ion of 50 μA/μm and a Subthreshold Swing of 86 mV/dec Using HfO_2 Gate Oxide, *IEEE Electron Device Letters* 31 (12) (2010) 1392.

[19] Mookerjea, Saurabh, Dheeraj Mohata, Theresa Mayer, Vijay Narayanan, Suman Datta: Temperature-Dependent I–V Characteristics of a Vertical In0.53Ga0.47As Tunnel FET, *IEEE Electron Device Letters* 31 (6) (2010) 564.

[20] Avik, Chakraborty, Angsuman Sarkar: Investigation of Analog/RF Performance of Staggered Heterojunctions-based Nanowire Tunneling Field-effect Transistors, *Superlattices and Microstructures* 80 (2015) 125.

[21] Liu, Hu, Lin An Yang, Zhi Jin, Yue Hao: An In0.53Ga0.47As/In0.52Al0.48As Heterojunction Dopingless Tunnel FET With a Heterogate Dielectric for High Performance, *IEEE Transactions on Electron Devices* 66 (7) (2019) 3229.

[22] Ahn, Dae-Hwan, Sang-Hee Yoon, Kimihiko Kato, Taiichirou Fukui, Mitsuru Takenaka, Shinichi Takagi: Effects of ZrO2/Al2O3 Gate-Stack on the Performance of Planar-Type InGaAs TFET, *IEEE Transactions on Electron Devices* 66 (4) (2019) 1862.

[23] Dubey, Prabhat Kumar, Brajesh Kumar Kaushik: T-Shaped III-V Heterojunction Tunneling Field-Effect Transistor, *IEEE Transactions on Electron Devices* 64 (8) (2017) 3120.

[24] Shih, Pao-Chuan, Wei-Chih Hou, Jiun-Yun Li: A U-Gate InGaAs/GaAsSb Heterojunction TFET of Tunneling Normal to the Gate with Separate Control Over ON- and OFF-State Current, *IEEE Electron Device Letters* 38 (12) (2017) 1751.

[25] Han, Genquan, Bin Zhao, Yan Liu, Hongjuan Wang, Mingshan Liu, Chunfu Zhang, Shengdong Huc, Yue Hao: Investigation of Performance Enhancement in InAs/InGaAs Heterojunction-enhanced N-channel Tunneling Field-effect Transistor, *Superlattices and Microstructures* 88 (2015) 90.

[26] Mohata, Dheeraj, Student Member, IEEE, Bijesh Rajamohanan, Theresa Mayer, Mantu Hudait, Joel Fastenau, Dmitri Lubyshev, Amy W. K. Liu, Suman Datta: Barrier-Engineered Arsenide-Antimonide Heterojunction Tunnel FETs With Enhanced Drive Current, *IEEE Electron Device Letters* 33 (11) (2012) 1568.

[27] Zhou, Guangle, Yeqing Lu, Rui Li, Qin Zhang, Qingmin Liu, Tim Vasen, Haijun Zhu, Jenn-Ming Kuo, Tom Kosel, Mark Wistey, Patrick Fay, Alan Seabaugh, Huili Xing: InGaAs/InP Tunnel FETs With a Subthreshold Swing of 93 mV/dec and I_{ON}/I_{OFF} Ratio Near 10^6, *IEEE Electron Device Letters* 33 (6) (2012) 782.

[28] Yu, Tao, James T. Teherani, Dimitri A. Antoniadis, Judy L. Hoyt: In0.53Ga0.47As/GaAs0.5Sb0.5 Quantum-Well Tunnel-FETs With Tunable Backward Diode Characteristics, *IEEE Electron Device Letters* 34 (12) (2013) 1503.

[29] Rajamohanan, Bijesh, Rahul Pandey, Varistha Chobpattana, Canute Vaz, David Gundlach, Kin P. Cheung, John Suehle, Susanne Stemmer, Suman Datta: 0.5 V Supply Voltage Operation of In0.65Ga0.35As/GaAs0.4Sb0.6 Tunnel FET, *IEEE Electron Device Letters* 36 (2015) 20.

[30] Zhu, Yan, Dheeraj K. Mohata, Suman Datta, Mantu K. Hudait: Reliability Studies on High-Temperature Operation of Mixed As/Sb Staggered Gap Tunnel FET Material and Devices, *IEEE Transactions on Device and Materials Reliability* 14 (2014) 245.

[31] Ajayan, J., T. Ravichandran, P. Mohankumar, P. Prajoon, J. C. Pravin, D. Nirmal: Investigation of DC and RF Performance of Novel MOSHEMT on Silicon Substrate for Future Submillimetre Wave Applications, *Semiconductors* 52 (16) (2018) 1991–1997.

[32] Ajayan, J., D. Nirmal, K. Dheena, P. Mohankumar, L. Arivazhagan, A. S. Augustine Fletcher, T. D. Subash, M. Saravanan: Investigation of Impact of Gate Underlap/overlap on the Analog/RF Performance of Composite Channel Double Gate MOSFETs, *Journal of Vacuum Science & Technology B* 37 (6) (2019) 06221.

[33] Ajayan, J., D. Nirmal, P. Mohankumar, L. Arivazhagan, M. Saravanan, S. Saravanan: LG = 20 nm High Performance GaAs Substrate Based Metamorphic Metal Oxide Semiconductor

High Electron Mobility Transistor for Next Generation High Speed Low Power Applications, *Journal of Nanoelectronics and Optoelectronics* 14 (8) (2019) 1133–1142.

[34] Ajayan, J., D. Nirmal: 20-nm T-gate Composite Channel Enhancement-mode metamorphic HEMT on GaAs Substrates for Future THz Applications, *Journal of Computational Electronics* 15 (2016) 1291–1296.

[35] Ajayan, J., D. Nirmal: 20 nm High Performance Enhancement Mode InP HEMT with Heavily Doped S/D Regions for Future THz Application, *Superlattices and Microstructures* 100 (2016) 526–534.

[36] Ajayan, J., T. Ravichandran, P. Mohankumar, P. Prajoon, J. C. Pravin, D. Nirmal: Investigation of DC-RF and Breakdown Behaviour in Lg = 20 nm Novel Asymmetric GaAs MHEMTs for Future Submillimetre Wave Applications, *AEU-International Journal of Electronics and Communications* 84 (2018) 387–393.

[37] Ajayan, J., D. Nirmal: A Review of InP/InAlAs/InGaAs Based Transistors for High Frequency Applications, *Superlattices and Microstructures* 86 (2015) 1–19.

[38] Ajayan, J., T. D. Subash, D. Kurian: 20 nm High Performance Novel MOSHEMT on InP Substrate for Future High Speed Low Power Applications, *Superlattices and Microstructures* 109 (2017) 183–193.

[39] Ajayan, J., T. Ravichandran, P. Prajoon, J. C. Pravin, D. Nirmal: Investigation of Breakdown Performance in Lg = 20 nm Novel Asymmetric InP HEMTs for Future High-speed High-power Applications, *Journal of Computational Electronics* 17 (1) (2018) 265–272.

[40] Ajayan, J., D. Nirmal, P. Prajoon, J. C. Pravin: Analysis of Nanometer-Scale InGaAs/InAs/InGaAs composite Channel MOSFETs Using High-K Dielectrics for High Speed Applications, *AEU-International Journal of Electronics and Communications* 79 (2017) 151–157.

[41] Ajayan, J., D. Nirmal: 22 nm In0:75Ga0: 25As Channel-based HEMTs on InP/GaAs Substrates for Future THz Applications, *Journal of Semiconductors* 38 (2017) 27–32.

[42] Ajayan, J., D. Nirmal, T. Ravichandran, P. Mohankumar, P. Prajoon, L. Arivazhagan, K. S. Chandan: InP High Electron Mobility Transistors for Submillimetre Wave and Terahertz Frequency Applications: A Review, *International Journal of Electronics and Communications* 94 (2018) 199–214.

[43] Ajayan, J., D. Nirmal: 20-nm Enhancement-mode Metamorphic GaAs HEMT with Highly Doped InGaAs Source/drain Regions for High-frequency Applications, *International Journal of Electronics* 104 (2017) 504–512.

[44] Ajayan, J., D. Nirmal, P. Mohankumar, K. Dheena, F. Augustine, L. Arivazhagan, B. Santhosh Kumar: GaAs Metamorphic High Electron Mobility Transistors for Future Deep Space-biomedical-millitary and Communication System Applications: A Review, *Microelectronics Journal* 92 (2019) 104604.

[45] Charles Pravin, J., D. Nirmal, P. Prajoon, A. Ajayan: Implementation of Nanoscale Circuits Using Dual Metal Gate Engineered Nanowire MOSFET with High-k Dielectrics for Low Power Applications, *Physica E: Low-dimensional Systems and Nanostructures* 83 (2016) 95–100.

[46] Charles Pravin, J., D. Nirmal, P. Prajoon, N. Mohan Kumar, J. Ajayan: Investigation of 6T SRAM Memory Circuit Using High-k Dielectrics Based Nano Scale Junctionless Transistor, *Superlattices and Microstructures* 104 (2017) 470–476.

[47] Ajayan, J., D. Nirmal, P. Mohankumar, L. Arivazhagan: Investigation of Impact of Passivation Materials on the DC/RF Performances of InP-HEMTs for Terahertz Sensing and Imaging, *Silicon* 12 (2020) 1225–1230.

[48] Ajayan, J., T. Ravichandran, P. Mohankumar, P. Prajoon, J. Charles Pravin, D. Nirmal: Investigation of RF and DC Performance of E-Mode $In_{0.80}Ga_{0.20}As/InAs/In_{0.80}Ga_{0.20}As$ Channel based DG-HEMTs for Future Submillimetre Wave and THz Applications, *IETE Journal of Research* (2018), DOI: 10.1080/03772063.2018.1553641.

[49] Ajayan, J., D. Nirmal, Ribu Mathew, Dheena Kurian, P. Mohankumar, L. Arivazhagan, D. Ajitha: A Critical Review of Design and Fabrication Challenges in InP HEMTs for Future Terahertz Frequency Applications, *Materials Science in Semiconductor Processing* 128 (2021) 105753.

[50] Clavel, Michael, Patrick Goley, Nikhil Jain Yan Zhu, Mantu K. Hudait: Strain-Engineered Biaxial Tensile Epitaxial Germanium for High-Performance Ge/InGaAs Tunnel Field-Effect Transistors, *IEEE Journal of the Electron Devices Society* 3 (2015) 184.

[51] Liu, Jheng-Sin, Michael B. Clavel, Mantu K. Hudait: Performance Evaluation of Novel Strain-Engineered Ge-InGaAs Heterojunction Tunnel Field-Effect Transistors, *IEEE Transactions on Electron Devices* 62 (10) (2015) 3223.

[52] Sant, Saurabh, Andreas Schenk: Methods to Enhance the Performance of InGaAs/InP Heterojunction Tunnel FETs, *IEEE Transactions on Electron Devices* 63 (2016) 2169.

[53] Franco Jacopo, AliReza Alian, Anne Vandooren, Anne S. Verhulst, D. Linten, N. Collaert, A. Thean: Intrinsic Robustness of TFET Subthreshold Swing to Interface and Oxide Traps: A Comparative PBTI Study of InGaAs TFETs and MOSFETs, *IEEE Electron Device Letters* 37 (2016) 1055.

[54] Smets, Quentin, Anne S. Verhulst, Salim El Kazzi, David Gundlach, Curt A. Richter, Anda Mocuta, Nadine Collaert, Aaron Voon-Yew Thean, Marc M. Heyns: Calibration of the Effective Tunneling Bandgap in GaAsSb/InGaAs for Improved TFET Performance Prediction, *IEEE Transactions on Electron Devices* 63 (2016) 4248.

[55] Zhao, Xin, Alon Vardi, Jesús A. del Alamo: Sub-thermal Subthreshold Characteristics in Top-down InGaAs/InAs Heterojunction Vertical Nanowire Tunnel FETs, *IEEE Electron Device Letters* 38 (2017) 855.

[56] Smets, Quentin, Anne S. Verhulst, Eddy Simoen, David Gundlach, Curt Richter, Nadine Collaert, Marc M. Heyns: Calibration of Bulk Trap-Assisted Tunneling and Shockley–Read–Hall Currents and Impact on InGaAs Tunnel-FETs, *IEEE Transactions on Electron Devices* 64 (2017) 3622.

[57] Zhu, Jiadi, Yang Zhao, Qianqian Huang, Cheng Chen, Chunlei Wu, Rundong Jia, Ru Huang: Design, Simulation of a Novel Graded-Channel Heterojunction Tunnel FET with High ION/IOFF Ratio and Steep Swing, *IEEE Electron Device Letters* 38 (9) (2017) 1200.

[58] Bordallo, Caio C. M., João Antonio Martino, Paula G. D. Agopian, Alireza Alian, Yves Mols, Rita Rooyackers, Anne Vandooren, Anne S. Verhulst, Eddy Simoen, Cor Claeys, Nadine Collaert: The Influence of Oxide Thickness and Indium Amount on the Analog Parameters of InxGa1-xAs nTFETs, *IEEE Transactions on Electron Devices* 64 (2017) 3595.

[59] Yan, Zhirui, Cong Li, Jiamin Guo, Yiqi Zhuang: A GaAs0.5Sb0.5/In0.53Ga0.47As Heterojunction Z-gate TFET with Hetero-gate-dielectric, *Superlattices and Microstructures* 129 (2019) 282.

[60] Zhao, H., Y. Chen, Y. Wang, F. Zhou, F. Xue, J. Lee: InGaAs Tunneling Field-Effect-Transistors with Atomic-Layer-Deposited Gate Oxides, *IEEE Transactions on Electron Devices* 58 (9) (2011) 2990.

[61] Zhou, Guangle, Yeqing Lu, Rui Li, Qin Zhang, Wan Sik Hwang, Qingmin Liu, Tim Vasen, Chen Chen, Haijun Zhu, Jenn-Ming Kuo, Siyuranga Koswatta, Tom Kosel, Mark Wistey, Patrick Fay, Alan Seabaugh, Huili Xing: Vertical InGaAs/InP Tunnel FETs With Tunneling Normal to the Gate, *IEEE Electron Device Letters* 32 (11) (2011) 1516.

[62] Lu, Yeqing, Guangle Zhou, Rui Li, Qingmin Liu, Qin Zhang, Timothy Vasen, Soo Doo Chae, Thomas Kosel, Mark Wistey, Huili Xing, Alan Seabaugh, Patrick Fay: Performance of AlGaSb/InAs TFETs With Gate Electric Field and Tunneling Direction Aligned, *IEEE Electron Device Letters* 33 (2012) 655.

[63] Hsu, Chih-Wei, Ming-Long Fan, Vita Pi-Ho Hu, Pin Su: Investigation and Simulation of Work-Function Variation for III–V Broken-Gap Heterojunction Tunnel FET, *IEEE Journal of the Electron Devices Society* 3 (2015) 194.

[64] Sharma, Ankit, Arun Goud Akkala, Jaydeep P. Kulkarni, Kaushik Roy: Source-Underlapped GaSb–InAs TFETs With Applications to Gain Cell Embedded DRAMs, *IEEE Transactions on Electron Devices* 63 (2016) 2563.

[65] Li, Rui, Yeqing Lu, Guangle Zhou, Qingmin Liu, Soo Doo Chae, Tim Vasen, Wan Sik Hwang, Qin Zhang, Patrick Fay, Tom Kosel, Mark Wistey, Huili Xing, Alan Seabaugh: AlGaSb/InAs Tunnel Field-Effect Transistor With On-Current of 78 μA/μm at 0.5 V, *IEEE Electron Device Letters* 33 (2012) 363.

[66] Dey, Anil W., B. Mattias Borg, Bahram Ganjipour, Martin Ek, Kimberly A. Dick, Erik Lind, Claes Thelander, Lars-Erik Wernersson: High-Current GaSb/InAs(Sb) Nanowire Tunnel Field-Effect Transistors, *IEEE Electron Device Letters* 34 (2013) 211.

[67] Jiang, Zhengping, Yeqing Lu, Yaohua Tan, Yu He, Michael Povolotskyi, Tillmann Kubis, Alan C. Seabaugh, Patrick Fay, Gerhard Klimeck: Quantum Transport in AlGaSb/InAs TFETs With Gate Field In-Line With Tunneling Direction, *IEEE Transactions on Electron Devices* 62 (2015) 2445.

[68] Beneventi, Giovanni Betti, Elena Gnani, Antonio Gnudi, Susanna Reggiani, Giorgio Baccarani: Dual-Metal-Gate InAs Tunnel FET With Enhanced Turn-On Steepness and High ON-Current, *IEEE Transactions on Electron Devices* 61 (2014) 776.

[69] Lind, Erik, Elvedin Memisevic, Anil W. Dey, Lars-Erik Wernersson: III-V Heterostructure Nanowire Tunnel FETs, *IEEE Journal of the Electron Devices Society* 3 (2015) 96.

[70] Hsu, Ching-Yi, Yuping Zeng, Chen-Yen Chang, Chenming Hu, Edward Yi Chang: Study of Inherent Gate Coupling Nonuniformity of InAs/GaSb Vertical TFETs, *IEEE Transactions on Electron Devices* 63 (2016) 4267.

[71] Hsu, Ching-Yi, Chun-Yen Chang, Edward Yi Chang, Chenming Hu: Suppressing Non-Uniform Tunneling in InAs/GaSb TFET With Dual-Metal Gate, *IEEE Journal of the Electron Devices Society* 4 (2015) 60.

[72] Biswal, S. M., B. Baral, D. De, A. Sarkar: Study of Effect of Gate-length Downscaling on the Analog/RF Performance and Linearity Investigation of InAs-based Nanowire Tunnel FET, *Superlattices and Microstructures* 91 (2016) 319.

[73] Memišević, Elvedin, Johannes Svensson, Markus Hellenbrand, Erik Lind, Lars-Erik Wernersson: Scaling of Vertical InAs-GaSb Nanowire Tunneling Field-Effect Transistors on Si, *IEEE Electron Device Letters* 37 (2016) 549.

[74] Chinni, Vinay Kumar, Mohammed Zaknoune, Christophe Coinon, Laurence Morgenroth, David Troadec, Xavier Wallart, Ludovic Desplanque: V-Shaped InAs/Al0.5Ga0.5Sb Vertical Tunnel FET on GaAs (001) Substrate With ION = 433 μA.μm−1 at VDS = 0.5 V, *IEEE Journal of the Electron Devices Society* 5 (2017) 53.

[75] Huang, Jun Z., Pengyu Long, Michael Povolotskyi, Gerhard Klimeck, Mark J. W. Rodwell: Scalable GaSb/InAs Tunnel FETs With Nonuniform Body Thickness, *IEEE Transactions on Electron Devices* 64 (2017) 96.

[76] Strangio, Sebastiano, Pierpaolo Palestri, Marco Lanuzza, Felice Crupi, David Esseni, Luca Selmi: Assessment of InAs/AlGaSb Tunnel-FET Virtual Technology Platform for Low-Power Digital Circuits, *IEEE Transactions on Electron Devices* 63 (2016) 2749.

[77] Long, Pengyu, Evan Wilson, Jun Z. Huang, Gerhard Klimeck, Mark J. W. Rodwell, Michael Povolotskyi: Design and Simulation of GaSb/InAs 2D Transmission-Enhanced Tunneling FETs, *IEEE Electron Device Letters* 37 (2016) 107.

[78] Ajay, Rakhi Narang, Manoj Saxena, Mridula Gupta: Model of GaSb-InAs p-i-n Gate All Around (GAA) BioTunnel FET (BTFET), *IEEE Sensors Journal* 19 (7) (2019) 2605.

[79] Visciarelli, Michele, Elena Gnani, Antonio Gnudi, Susanna Reggiani, Giorgio Baccarani: Impact of Traps and Strain on Optimized n- and p-Type TFETs Integrated on the Same InAs/AlGaSb Technology Platform, *IEEE Transactions on Electron Devices* 64 (2017) 3108.

[80] Huang, Jun Z., Pengyu Long, Michael Povolotskyi, Hesameddin Ilatikhameneh, Tarek A. Ameen, Rajib Rahman, Mark J. W. Rodwell, Gerhard Klimeck: A Multiscale Modeling of Triple-Heterojunction Tunneling FETs, *IEEE Transactions on Electron Devices* 64 (2017) 2728.

[81] Memisevic, Elvedin, Johannes Svensson, Erik Lind, Lars-Erik Wernersson: InAs/InGaAsSb/GaSb Nanowire Tunnel Field-Effect Transistors, *IEEE Transactions on Electron Devices* 64 (2017) 4746.

[82] Ghosh, Krishnendu, Uttam Singisetti: RF Performance and Avalanche Breakdown Analysis of InN Tunnel FETs, *IEEE Transactions on Electron Devices* 61 (2014) 3405.

[83] Duan, Xiaoling, Jincheng Zhang, Shulong Wang, Yao Li, Shengrui Xu, Yue Hao: A High-Performance Gate Engineered InGaN Dopingless Tunnel FET, *IEEE Transactions on Electron Devices* 65 (2018) 1223.

[84] Li, Wenjun, Saima Sharmin, Hesameddin Ilatikhameneh, Rajib Rahman, Yeqing Lu, Jingshan Wang, Xiaodong Yan, Alan Seabaugh, Gerhard Klimeck, Debdeep Jena, Patrick Fay: Polarization-Engineered III-Nitride Heterojunction Tunnel Field-Effect Transistors, *IEEE Journal on Exploratory Solid-State Computational Devices and Circuits* 1 (2015) 28.

[85] Peng, Yue, Genquan Han, Hongjuan Wang, Chunfu Zhang, Yan Liu, Yibo Wang, Shenglei Zhao, Jincheng Zhang, Yue Hao: InN/InGaN Complementary Heterojunction-enhanced Tunneling Field-Effect Transistor with Enhanced Subthreshold Swing and Tunneling Current, *Superlattices and Microstructures* 93 (2016) 144.

[86] Xiaoling Duan, X., Jincheng Zhang, Shulong Wang, Rudai Quan, Yue Hao: Effect of Graded InGaN Drain Region and 'In' Fraction in InGaN Channel Onperformances of InGaN Tunnel Field-effect Transistor, *Superlattices and Microstructures* 112 (2017) 671.

[87] Kumar, Ashwani, Maria Merlyne De Souza: A p-Channel GaN Heterostructure Tunnel FET With High ON/OFF Current Ratio, *IEEE Transactions on Electron Devices* 66 (2019) 2916.

4 Performance Analysis of Carbon Nanotube and Graphene Tunnel Field-Effect Transistors

K. Ramkumar, Singh Rohitkumar Shailendra, and V. N. Ramakrishnan

CONTENTS

4.1 INTRODUCTION

Since the introduction of integrated circuits (ICs) in 1952 [1] and the acknowledgment of the first IC at Texas Instrument in 1958 [2], the last five decades witnessed an exponential development of silicon (Si)-based microelectronics industry. Rapid advancement in this industry is accomplished mainly due to consistent scaling or scaling down of all electronic components (active and passive) integrated on the ICs. IC scaling down technique supported the scaling of traditional complementary metal oxide semiconductor (CMOS) device, and metallic interconnects are used for the connection

DOI: 10.1201/9781003200987-4

of device terminal with a power supply voltage [3]. Scaling down in IC technology makes less testing necessities at the system level and achieves significant cost-saving and faster switching. It leads to compact, low-power, and highly reliable designs. As an outcome, it gives faster and enhanced ICs for high-definition digital television, digital receiver, high-speed microprocessor, communication, DSP, traffic control, business transactions, weather monitoring system, space guidance, medical treatment system, and many other commercial, industrial, and scientific enterprises [4]. Further, as per Moore's prediction, the transistor count on the ICs is expected to grow exponentially with time [5]. This expectation ended up being valid as given in Figure 4.1 [6, 7]. It represents the increment in a number of transistors per processor chip as a function of time. As it can be observed, the transistor count on the ICs doubles every 18 months. To meet the IC density anticipated by Moore's law, technology scaling has been sought aggressively until today since the 1970s. The length of the channel in a metal oxide semiconductor field-effect transistors (MOSFETs) is scaled down by a factor of 0.7 every two years. Since 2006, at 65 nm technology node, the gate length of the MOSFET has touched at sub-micron/nano range. Today, researchers expect 10 nm and 7 nm technology node as a feature size in the near future [7, 8].

One of the major issues is increased leakage current that occurs due to various quantum mechanical tunneling, including band-to-band tunneling, source to drain tunneling, and direct gate oxide tunneling. Different issues are extensive process variations, the impact of crystal misalignments, the arbitrariness of discrete doping, and the addition of interface scattering so that the mean free path of electrons becomes comparable to component dimensions [9, 10]. These device-level impacts

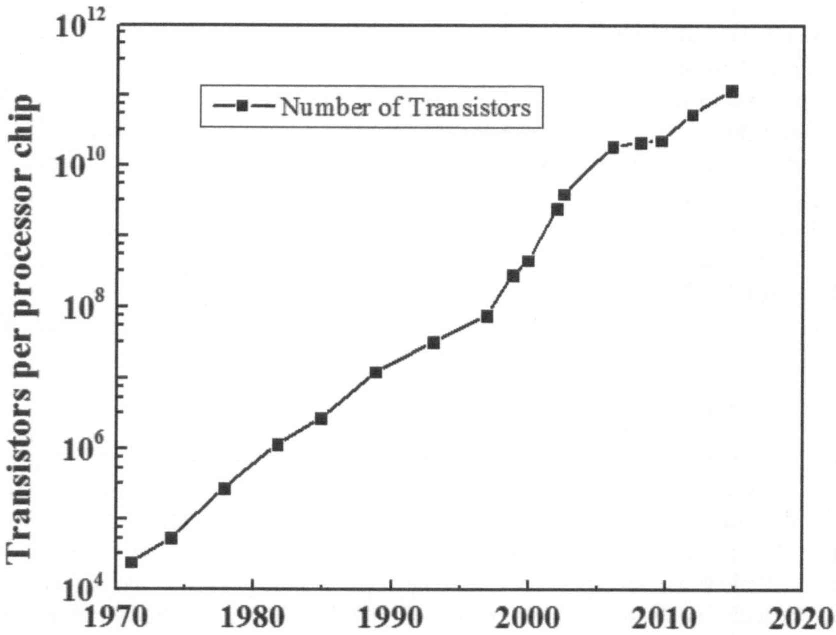

FIGURE 4.1 Evolution of MOSFET

cause the current-voltage (I-V) characteristics to be different from well-tempered traditional Si-MOSFET. Thus, researchers have significant concerns with regard to further improving device performance by scaling down the element size of MOSFET. In addition, circuit-level effects corresponding to short-channel effects, increment within the resistance of metallic on-chip interconnects, and power dissipation can surely reduce the suitableness of MOSFET for advance applications.

Device engineers have developed multi-gate FET and FinFET devices [11, 12] to reduce the short-channel effects. In these devices, better gate controllability is obtained, which ends up in more control on the channel and substantial reduction in the leakage current. Researchers have even begun the investigation of the latest devices and channel material in the sub-10 nm technology node, which could be the attainable alternatives to traditional Si-CMOS. Based on ITRS [13], a number of the rising devices that have the capabilities to exchange traditional Si-technology in post Si era are nanowire field-effect transistor (NWFET) [14], III-V compound semiconductor field-effect transistor [15–17], graphene field-effect transistor (G-FET) [18–20] and carbon nanotube field-effect transistor (CNTFET) [21–24].

The examination on tunnel FETs (TFETs) is as yet in the exploratory stage, and scientists are exploring diverse material to acknowledge TFETs, including different allotropes of carbon (C) atom. The bandgaps of graphene and CNT are adjustable and adequately little for use in a TFET. The supreme electrostatic control on the channel potential, extraordinary carrier transport property of graphene and CNT, and ability to function at ultra low voltages make carbon-based TFETs particularly interesting for future low power circuits. Also, because of the underlying similitude of a MOSFET and a TFET, progressions in carbon-based FETs are straightforwardly helping research on carbon-based TFETs. In this chapter, we will initially discuss CNT/graphene material properties, followed by TFETs based on the aforementioned and some of the device optimization techniques.

4.2 CARBON NANOTUBES (CNTs)

In 1991, S. Ijima found carbon nanotube at NEC Central Research Laboratories at Tsukuba in Japan [25]. CNTs are different forms of carbon having a tube-like structure. An atom thick sheet of carbon with a planar structure called graphene is shown in Figure 4.2 (a). CNTs are mainly two types: single-walled carbon nanotube (SWNT) and multi-walled carbon nanotube (MWNT).

If the rolling sheet of carbon has only one cylinder structure, it is called SWNT as shown in Figure 4.2 (b), and if it has more than one cylinder structure, it is called MWNT as shown in Figure 4.2 (c). An SWNT is made by rolling up a sheet of graphene along a chiral vector $C_h = n_1a+n_2b$, where n_1 and n_2 are positive integers that will give the chirality of the tube, and a and b are the lattice unit vectors as shown in Figure 4.3. Depending upon the chirality value of n_1 and n_2, the behavior of SWNT can be either metallic or semiconducting as shown in Table 4.1. If the difference of chirality value n_1 and n_2 means n_1–n_2 is a multiple of 3, the behavior of SWNT is metallic or else it is semiconducting. Similarly, SWNT is classified into three groups depending on the value of chiral vector n_1 and n_2: (i) armchair type CNT when the value of $n_1 = n_2 = n$, (ii) zigzag type CNT when $n_1 = 0$ or $n_2 = 0$, and (iii) chiral type

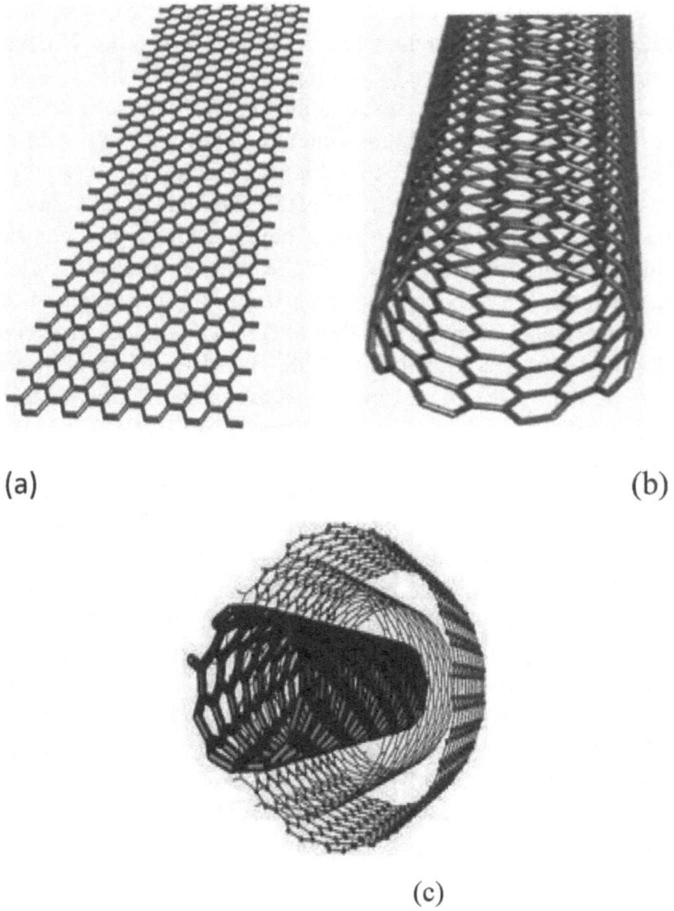

FIGURE 4.2 Structure of (a) Graphene, (b) SWNTs, and (c) MWNTs

CNT when the difference of chiral vector n_1 and n_2 is nonzero. All armchair types of CNT behave as conductor or metallic. On the other hand, zigzag and chiral types of CNTs show conductor or metallic behavior when the difference of two chiral vectors (n_1–n_2) is a multiple of 3, otherwise the behavior of CNTs are semiconducting, which are used in CNTFET [26].

In Figure 4.4 (a), we can see that structure of CNT with chiral vector (10, 10) where the value of $n_1 = n_2 = n$, which shows armchair type of CNT as shown in Table 4.1, whereas in Figure 4.4 (b), we can see the structure of CNT with chiral vector (10, 0) where the value of $n_2 = 0$, which shows the zigzag type of CNT as shown in Table 4.1. While in Figure 4.4 (c) we can see the structure of CNT with chiral vector (10, 6) where the value of $n_1 – n_2 \neq 3$ integer, which shows the chiral type of CNT as shown in Table 4.1 and Figure 4.4 (d), we can see that structure of CNT with chiral vector (10, 7) where the value of $n_1 – n_2 = 3$ integer, which shows quasi semiconducting type of CNT as shown in Table 4.1.

TABLE 4.1
Type of CNT with the Value of Chiral Vector (n_1, n_2)

S. no.	Types of CNT	n_1 and n_2 Values	Properties
1	Armchair	$n_1 = n_2$	Metallic
2	Zigzag	$n_2 = 0$	Semiconducting
3	Chiral	$n_1–n_2 \neq 3 \times$ integer	Wide bandgap
4	Quasi semiconducting	$n_1–n_2 = 3 \times$ integer	Narrow bandgap

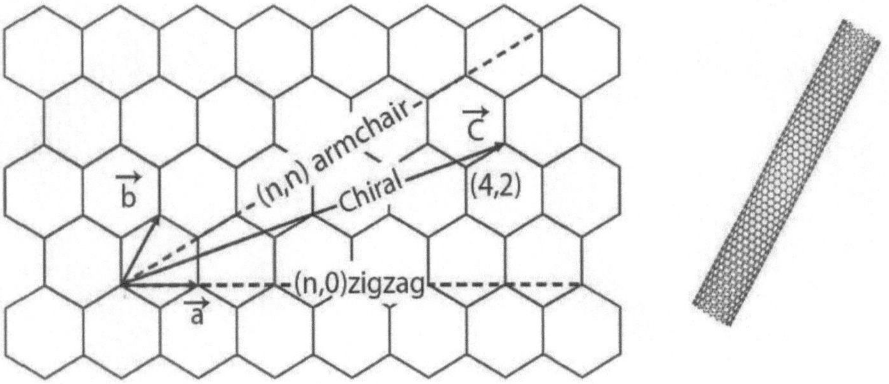

FIGURE 4.3 Planar Sheet of Graphene and Rolled Lattice Sheet of CNT

The density of states (DOS) without narrow bandgap effects can be derived for different CNTs by using the tight-binding approach. Figure 4.5 (a) has chiral vector (10, 10), which shows the metallic characteristics of CNT because of overlapping of energy band diagram; Figure 4.5 (b) has chiral vector (10, 0), which shows semiconducting characteristics because of non-overlapping of energy band diagram; Figure 4.5 (c) has chiral vector (10, 6), which shows properties of large bandgap; and Figure 4.5 (d) has chiral vector (10, 7), which shows the properties of the small bandgap. In general, the DOS in CNT is smaller than metals by four orders of magnitude and results in poor conductivity. However, the 1-D nature of nanotubes results in long mean free path of the order of 1–10 μm for carriers, in contrast to that of metals, which is around 40 nm for copper. Additional states have to be available for the scattered carrier, to where the charge state can be scattered. However, additional states for small-angle scattering events are energetically far away due to the circumferential quantization. The only allowed states are in the direction along the CNT. The phase space for scattering is therefore strongly restricted, even at the room temperature. At low bias voltage, the only energetically allowed state to be scattered is just backwards in the opposite direction. The density of state of CNT with different chiral vector

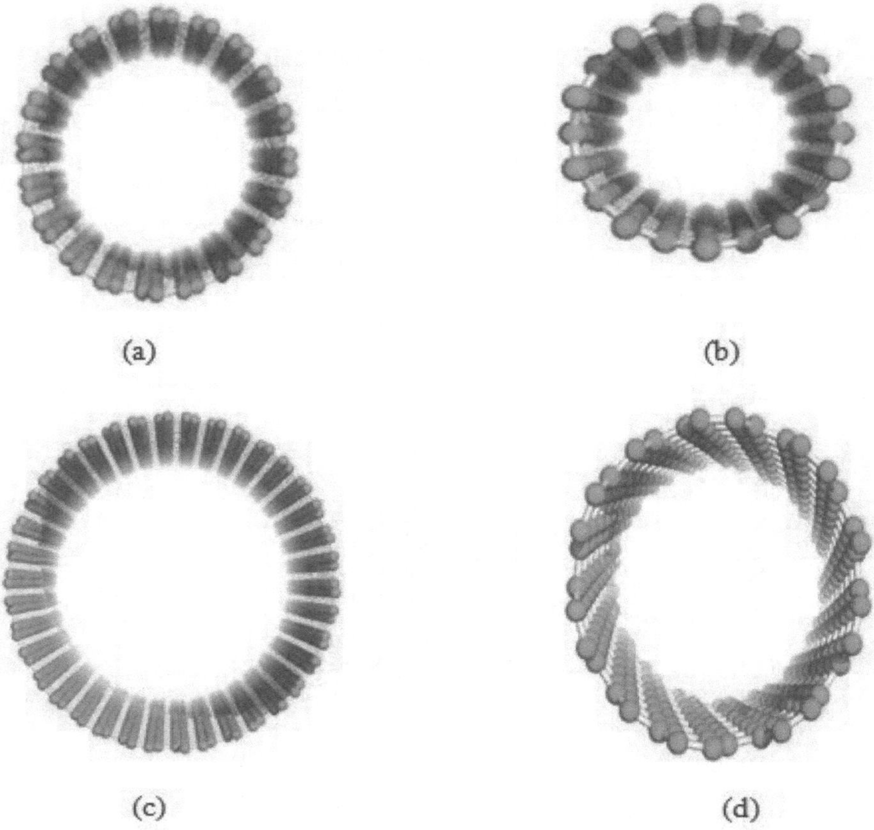

FIGURE 4.4 Structure of CNT with Chiral Vector: (a) (10, 10), (b) (10, 0), (c) (10, 6), and (d) (10, 7)

is shown in Figures 4.5 (a), (b), (c), and (d). From these figures, we see that various chiral vectors have a different type of properties as shown in Table 4.1.

The properties that make CNTs especially engaging for applications in FETs are given next [27, 28]:

1. High electric current can be sustained in the CNTs due to its inert nature.
2. The gate-all-around architecture of CNTs exhibits excellent electrostatic control on the channel potential. High-k materials can be used as a gate oxide to enhance gate control further.
3. Even in long channel devices, quasi-ballistic transport can be experienced due to the long mean free paths in carbon nanotube.

Therefore, CNTs are in effect broadly examined for use as a channel material in FETs for future CMOS applications. CNT FETs have already demonstrated quasi-ballistic transport and ON-state current (ION) greater than silicon-based MOSFETs in experiments.

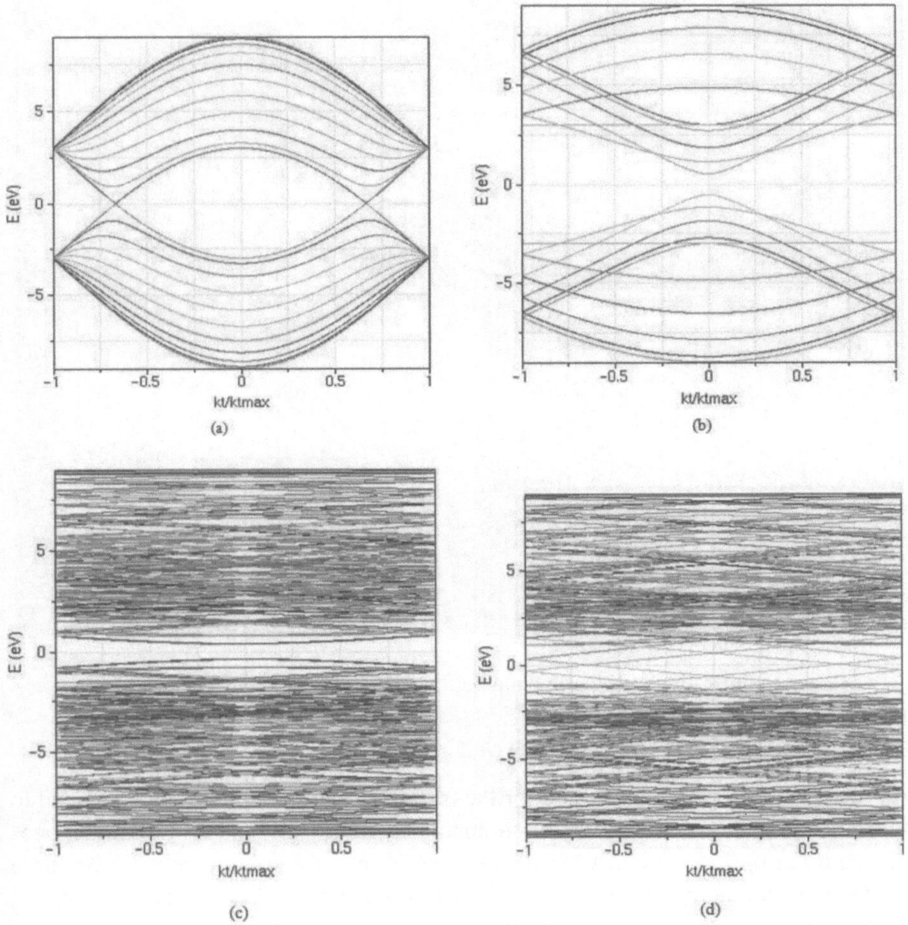

FIGURE 4.5 Sub-Band Structure of CNT with Chiral Vector: (a) (10, 10), (b) (10, 0), (c) (10, 6), and (d) (10, 7)

4.2.1 Graphene

Graphene, a two-dimensional, one-layer thickness of sp^2 hybridized carbon particles, has pulled in gigantic consideration and exploration interest, attributable to its excellent physical, thermal, and mechanical properties. Different types of graphene-based materials, like graphene oxide, reduced graphene oxide, and exfoliated graphite, have been dependably created in enormous scope. The processibility and functionalization along with reliable properties help graphene-derived materials be an ideal possibility when considering different useful materials [29]. Critically, graphene and its subsidiaries have been investigated in a wide scope of utilizations, like photonic and electronic devices, clean energy, and sensors.

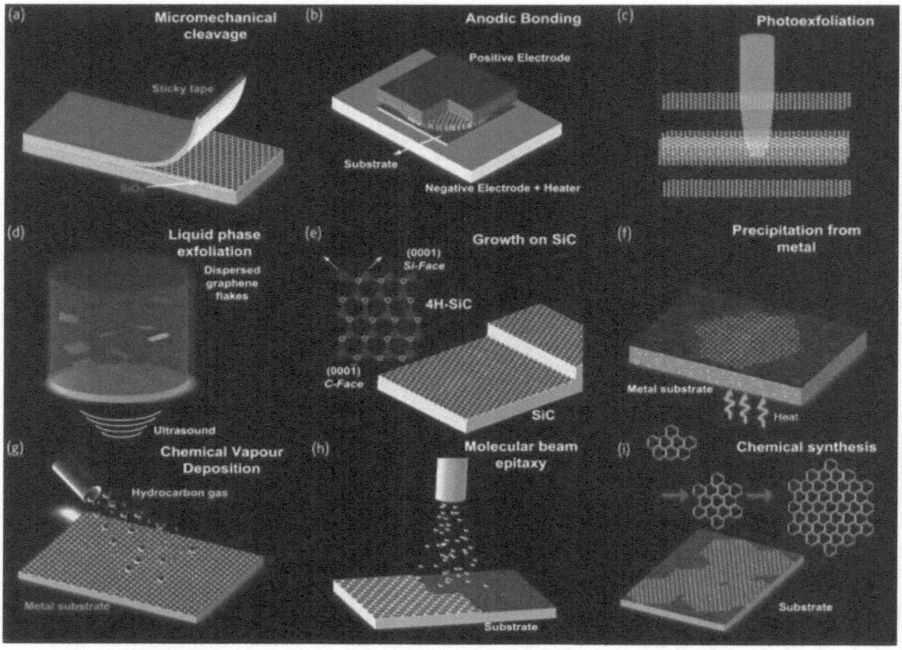

FIGURE 4.6 Illustration of Experimental Setups for Producing Graphene [31]

4.2.2 Incurring an Atomic Sheet of Carbon

A few strategies have been created for the isolation of graphene, from mechanical to substance to epitaxial ones. We will momentarily outline the most well-known ones by summing up the substance pertinent to us from zeroing in principally on three procedures and momentarily presenting other arising ones. Among these strategies, the "Scotch-tape" technique brings about the greatest single-layer graphene as far as the best carrier transport, while different techniques give a large-scale manufacturing course to graphene manufacture, and the nature of graphene has improved altogether by tweaking the cycle boundaries as of late [30].

4.2.3 Properties of Graphene

The following properties of graphene can be utilized in TFETs to get better electrical characteristics [32]:

1. A narrow bandgap can be created and controlled by diverse techniques, which results in high I_{ON} in graphene-based TFETs.
2. The highest electron mobility of all material found yet with theoretical limit of 200,000 cm^2/V-s (> 100* faster than Si).
3. Excellent electrical conductivity (13* better than copper).
4. An atomic layer of thickness allows better gate control and diminished short-channel effects.

5. Defect-free, single-layer graphene is the strongest material ever tested (300* harder than steel).
6. High thermal conductivity (2* better than diamond).
7. Stretchability: graphene can be stretched up to 25%—highly relevant for flexible electronics.
8. Impermeable.
9. Large surface area (2,630 m^2/g).
10. Transparent.

4.3 CARRIER TRANSPORT IN CARBON NANOTUBE AND GRAPHENE

The enormous current carrying capacity, low intrinsic resistivity, and quantized conductance in a CNT have been tentatively exhibited and theoretically settled [33, 34]. The carrier transport mechanism in a CNT can be either diffusive or ballistic. A ballistic transport happens when the electrons transmitted with no defect, impurity, or phonon scattering. At the point when the length of a CNT is less than the mean free path length in the material, ballistic transport is conceivable. The mean free path length in a CNT can be very long (in excess of 200 or 300 nanometers). Subsequently, ballistic transport can be seen at room temperature in a CNT in any event, for lengths near 100 nm. The ballistic transport is significant in CNT semiconductors since the conduction component assumes a significant part in deciding the electrical characteristics like I_{ON} and subthreshold swing of the device. Furthermore, if the transport component is simply ballistic in a semiconductor, at that point there is no energy dissipation in the channel.

The TFET device simulations considering both phonon scattering and ballistic mechanism show that the subthreshold swing is lesser than 60 mV/decade at 300 K. Still, the existence of phonon scattering prompts degradation of the subthreshold swing since phonon-absorption-assisted transport is considered to be a critical part in the OFF-state [35, 36]. Thus, the subthreshold swing shows stronger dependence on the temperature within the sight of phonon scattering. In the ON-state, the drive current of the TFET is determined by the properties of the tunneling barrier and the impact of phonon scattering in the channel is somewhat restricted.

In brief, the carrier transport theory in 2-D graphene sheets is given as: at high carrier density, the conductivity of graphene relies upon the density of carrier, the dielectric constant of the substrate, and the properties of the impurity potential, which would all be dealt with utilizing the Boltzmann transport formalism. The disorder of carrier causes the local random fluctuations at low carrier density. As a result, the carrier transport at the Dirac point is profoundly inhomogeneous. The dependence of carrier transport on the carrier density in mono-, bi-, and tri-layer graphene has been given as follows: Figure 4.7 shows the mobility vs carrier density at different temperatures (from 4.2 K to 350 K) for monolayer, bi-layer, and tri-layer graphene, respectively. In monolayer graphene, the mobility decreases as we increase the carrier density, while the mobility increases in bi-layer/tri-layer graphene. From Figure 4.8, it is observed that in monolayer graphene, the mobility decreases with

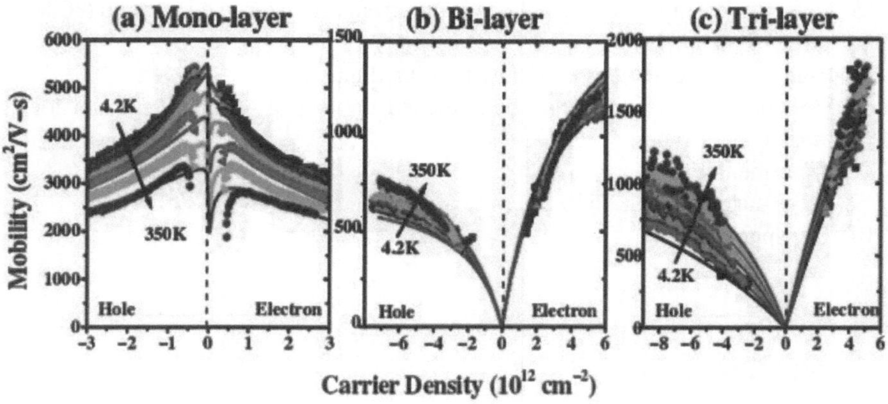

FIGURE 4.7 Mobility vs Carrier Density at Temperatures from 4.2K to 350K in (a) Monolayer, (b) Bi-Layer, and (c) Tri-Layer Graphene

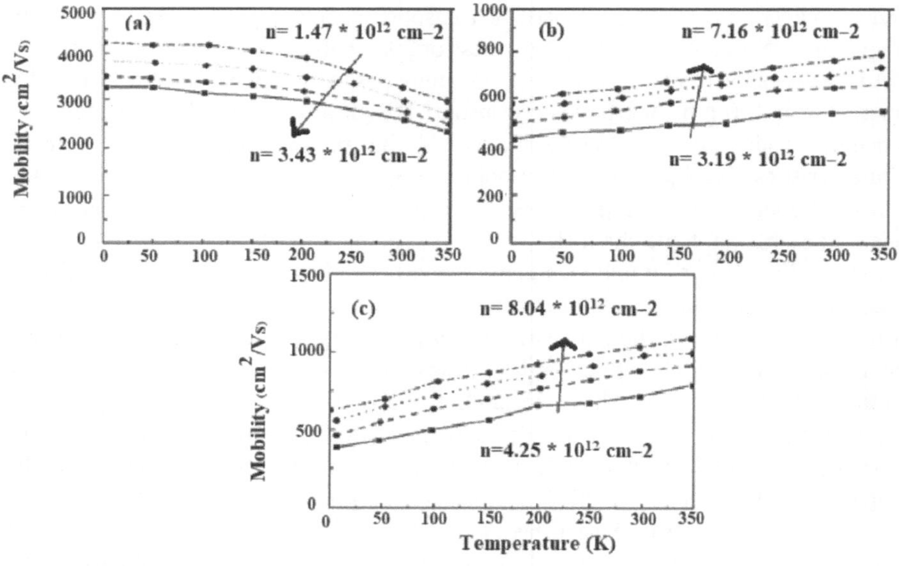

FIGURE 4.8 Mobility vs Temperature at Different Carrier Densities in (a) Monolayer, (b) Bi-Layer, and (c) Tri-Layer Graphene

temperature, especially when the temperature is above ~200K, while the mobility increases with temperature for bi- and tri-layer graphene [37].

4.4 BI-LAYER GRAPHENE TFETs

The bi-layer graphene (BLG) has been the choice of channel material in a TFET due to little and tunable bandgap that can be availed in BLG. Either by applying electric field opposite to the layer of graphene or chemical doping, the required bandgap is acquired [38–40]. The imbalance of charge between the layers of graphene prompts a variable bandgap. Hence, the bandgap of BLG is given as [38, 41].

$$E_g = \frac{\Delta\gamma}{\sqrt{\gamma^2 + \Delta^2}} \tag{4.1}$$

where Eg represents the bandgap of BLG, Δ denotes the energy distinction between the layers of graphene because of charge imbalance, and γ denotes hopping energy of interlayer. By maintaining the potential difference between the two gates (V_{GS} and V_{bias}), electric field is developed in BLG-based TFETs.

The bandgap in BLG is opened by both the vertical electric field due to asymmetric chemical doping in the top-bottom and the application of V_{bias}. The regions of drain and source were formed by changing the work function of the electrodes. Figure 4.9 depicts the BLG TFET, the I_{ON} of 67 μA/μm, I_{OFF} of 20.3 nA/μm, I_{ON}/I_{OFF} ratio of 2.9 ×10^3, and SS of 35 mV/decade were obtained at V_{bias} = 0.9 V and V_{DD} = 0.2 V [38]. As we increase the input voltage, the leakage current and subthreshold swing increase due to ambipolar conduction. By using V_{bias}, the bandgap and threshold voltage in BLG could be changed. However, an increase in V_{bias} to change the threshold voltage also leads to an increase in the bandgap of the material and the reduction in the ambipolar conduction as shown in Figure 4.10. A fascinating part of BLG TFET is

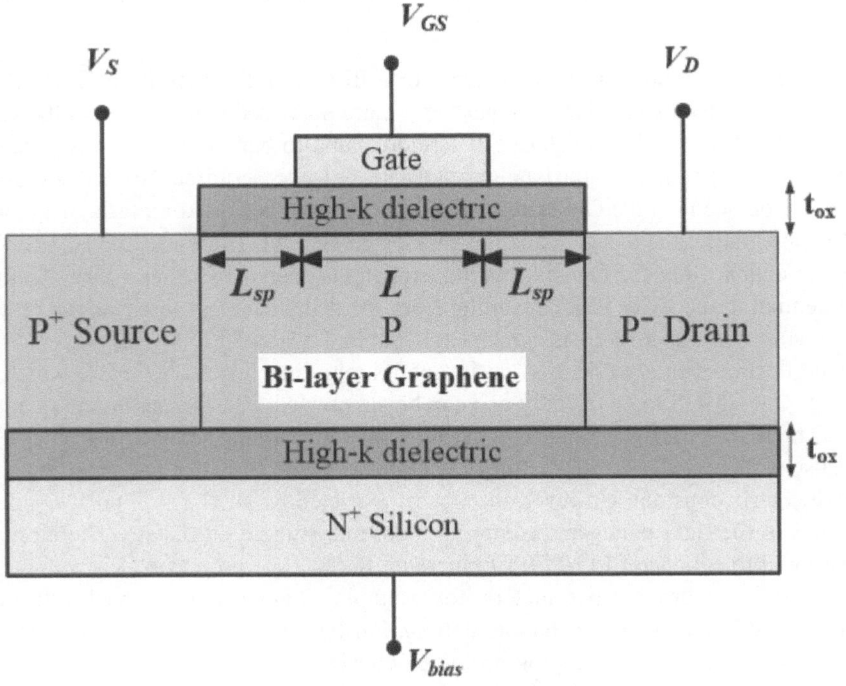

FIGURE 4.9 Schematic Diagram of Bi-Layer Graphene TFETs [38]

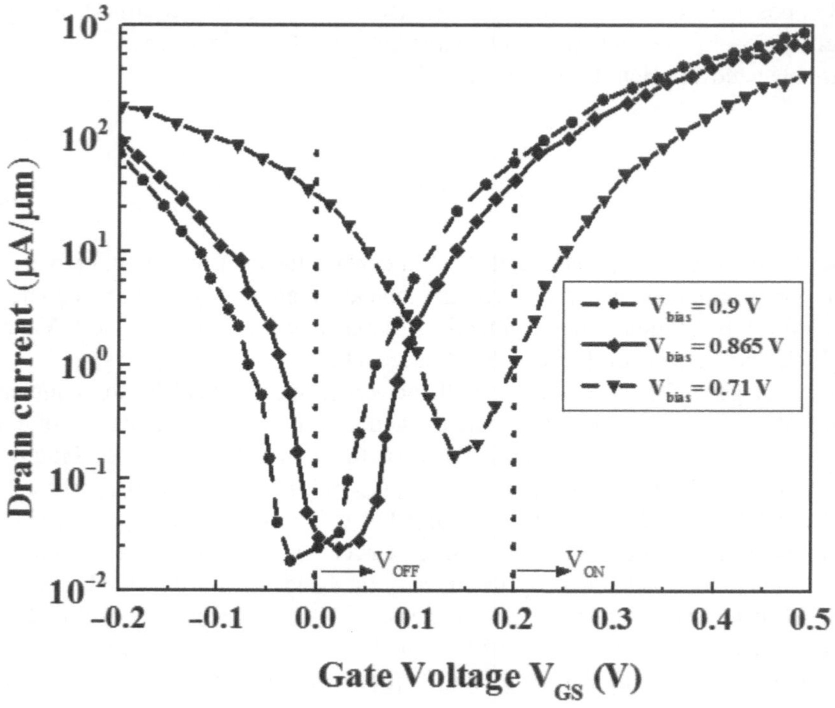

FIGURE 4.10 I_D vs V_{GS} Displaying Shift in the V_{TH} Using V_{bias} to Align V_{OFF} at 0 V [38]

the electrostatic control of the bandgap. For a BLG TFET, when the FET is in the OFF-state, the bandgap is large because of a more noteworthy distinction in the voltage of the top gate and the back gate. Therefore, band-to-band is additionally reduced because of an expansion in the bandgap and a lower I_{OFF} is acquired. Notwithstanding, when the device is in the ON-state, the bandgap lessens because of a little difference in the potential of the top-gate and the back-gate bias [38]. Therefore, the I_{ON} is additionally enhanced in the ON-state as the bandgap is decreased. This is a key feature differentiating the BLG TFET resulting from the dependence of the bandgap on the electrostatics of the device and can be exploited in the later TFETs.

The desire of this work is to design and streamline this BLG-TFET, which is both digital CMOS logic friendly and can be acknowledged utilizing process strategies. The BLG-TFET guarantees higher I_{ON} than conventional semiconductor TFETs because of its tunable band-structure, low effective mass, little bandgap opening, and low dielectric constant. However, the I_{OFF} is restricted by BLG's ambipolar conduct because of the tight bandgap, prompting more unfortunate OFF state. Therefore, it is essential to enhance BLG-TFET to increase the I_{ON}/I_{OFF} proportion via cautiously tuning different process parameters, for example, channel doping, back-gate bias voltage, and induced source/channel doping. Further, we assess the performance of this device with evaluated doping profile by considering diffusion of metal contact-induced source/channel doping in graphene.

4.5 GRAPHENE NANORIBBON TFETs

A p-i-n type device structure, with an intrinsic channel formed between highly doped p+ and n+ regions, forms a graphene nanoribbon TFET (GNR-TFET). The highly doped regions act as source and drain terminals. The device operates under reverse biasing condition. The predominant carrier transport mechanism in a GNR-TFET in ON-state is the interband tunneling from valence band of the source to the conduction band of the channel. It helps the GNR-TFET devices to achieve low subthreshold swing. Besides, transistors like GNR- and CNT-FETs also suffer from ambipolarity conduction in their OFF-states, especially under negative gate biases. The origin for ambipolar conduction is BTBT between the conduction band of the drain and valance band of the source. Ambipolarity is, in any case, a significant disadvantage that should be defeated in these FETs. A couple of exploration bunches have effectively proposed different strategies both in GNR-FETs [42] and CNT-FETs [43–45] to decrease the BTBT in their OFF-states and henceforth beat their ambipolarities. Utilization of uniaxial tensile strain across the channel of a double-gated GNR-FET [42] has appeared to improve the device ON/OFF proportion by four orders of magnitudes while reducing the ON current by up to around two significant degrees.

Device simulation of GNR tunneling FETs shown in Figure 4.11 is done by using 3-D atomistic simulations [46]. It is observed that the relaxation of edge bond has a significant impact on the characteristics of the p-i-n GNR tunneling FETs, which differentiates it from CNT TFETs. A semiconducting armchair edge GNR (AGNR) is used as the channel material [47–49]. The AGNR has an index of n = 13, which results in a width of ~1.6 nm and a bandgap of ~0.86 eV. The AGNR channel is

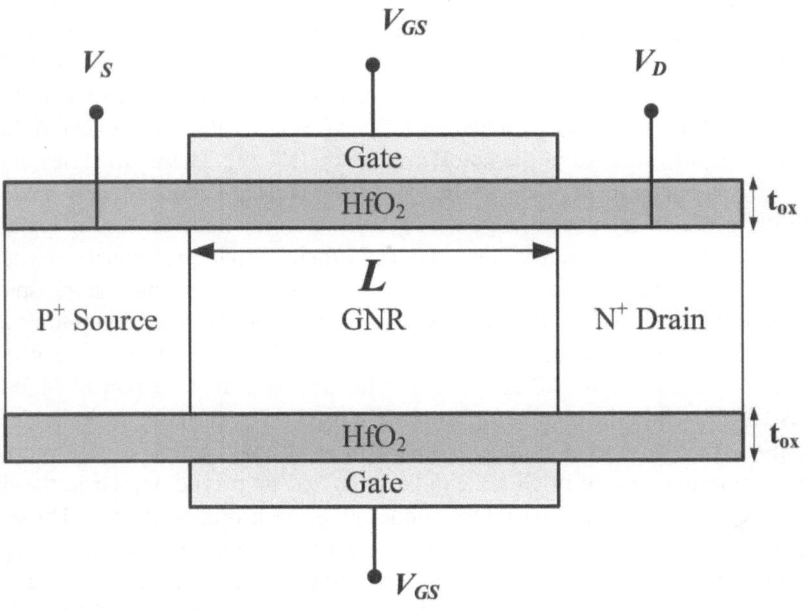

FIGURE 4.11 Schematic Representation of a GNR TFETs from [46]

intrinsic and has the same length as the gate, Lch = 30 nm. The p-type doping density of the semi-infinite source extension is 0.01 dopant/atom, and the drain extension is n-type doped to the same density. The work function of the gate electrode is adjusted to a value so that the minimal leakage current appears at $V_G = 0$, and a variation of the gate work function results in a shift of the threshold voltage. A power supply voltage of V_{DD}= 0.4 V and room temperature operation T = 300 K are assumed. A subthreshold swing of 14 mV/dec and a large ON-OFF ratio are obtained at the ballistic transport limit in the presence of edge bond relaxation. However, the device exhibits ambipolar I-V characteristics that are not preferred for digital electronics applications. By using an asymmetric source-drain doping or a gate underlap design, the ambipolar characteristics can be significantly suppressed.

4.6 ANALOG/RF PERFORMANCE OF GRAPHENE NANORIBBON TFETs

Graphene has been the best choice of material for analog/RF applications because of its excellent electronic properties, like higher carrier mobility and velocity saturation that would permit high transconductance and enhance the device cutoff frequency [50–52]. However, the shortfall of bandgap in graphene prompts quasi-saturation furthermore, negative differential resistance in the output characteristics of long-channel and short-channel devices [53, 54]. This produces higher output conductance compared with transconductance and therefore decreases the intrinsic gain furthermore, the oscillation frequency. The investigations demonstrated that graphene FETs (GFETs) have accomplished the cutoff frequencies of 1.4 THz for a gate length of 45 nm [55], which is higher than the greatest cutoff estimated in conventional Si MOSFETs, III-V HEMTs, yet with a more modest intrinsic gain of ~6 V/V for a gate length of 1 μm [56]. The previously mentioned issues of graphene transistors have empowered the examination on making bandgap by setting graphene on h-BN, graphene nanoribbon, or bi-layer graphene, where the current saturation is accomplished at the expense of decrease in the cutoff frequency [57–59]. These strategies, in any case, added the genuine creation challenges in the present semiconductor handling. Accordingly, the principal challenge with graphene semiconductors for analog applications is to improve the saturation conduct in their output characteristics without essentially degrading its present levels. It has been accounted for that the transport in GFET is generally contributed by the band-to-band tunneling (BTBT) up to channel lengths of 50 nm [60]. Furthermore, Alarcon et al. [61] have as of late observed that the source-to-channel (S-C) BTBT is liable for the saturation current in MOSFET such as GFET, while the channel-to-drain (C-D) BTBT and the thermionic current strength smother saturation conduct. Ganapathi et al. [62] have as of late proposed that the saturation current in GFET can be improved by p-type doping in the drain underlap region to upgrade the BTBT tunneling current predominance. These perceptions recommend that the BTBT is the prevailing current segment in GFET, and furthermore the one that decides the conduct in their output characteristics. Thus, a passage FET (TFET), in which the S-C BTBT current is constrained by gate voltage, can be a reasonable alternative in a graphene device to accomplish better saturation

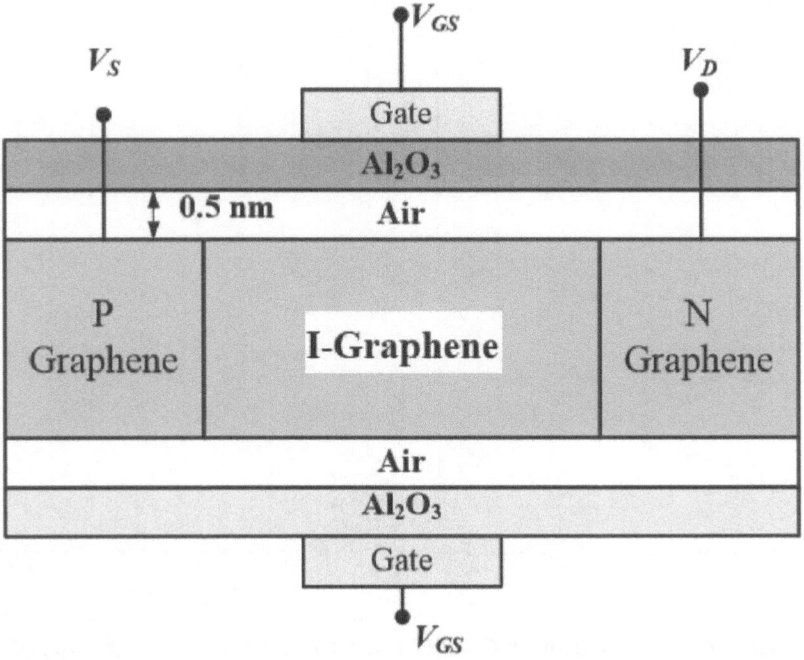

FIGURE 4.12 Schematic Diagram of T-GFET [70]

current without fundamentally degrading its current levels. In addition, a graphene TFET (T-GFET) can be a seriously encouraging contender for great door electrostatic control since a graphene gadget has more modest quantum capacitance; furthermore, TFET math offers a significantly more modest quantum capacitance [63, 64]. Despite the fact that TFETs are especially appropriate for advanced applications due to more extreme subthreshold swing [65], a portion of the TFET models have appeared promising RF execution measurements, and they can be thought of as a solid possibility for simple/RF applications [66–69].

The structure of graphene TFET is shown in Figure 4.12 as a length of $L_S = L_G = L_D = 25$ nm. These proposed structures are reasonable for computerized applications because of their high ON/OFF proportion, yet they are surely not helpful for analog/RF applications in light of exceptionally restricted current immersion area. Along these lines, this work thinks about an organizer p-i-n T-GFET with MOSFET-like activity for analog/RF applications. This paper is predominantly centered around the extraction of the characteristic analog/RF execution measurements of T-GFET, specifically the characteristic increase and cutoff recurrence.

A T-GFET has a barely higher (20%) intrinsic gain compared with C-GFET. It has shown that the fitting choice of device boundaries in the T-GFET can fundamentally stifle the C-D tunneling current and improve the saturation current. Specifically, an 8-nm overlap with the reasonable doping profile prompts a 2.2× improvement in the output resistance, a 3.6× improvement in gm, and an 8× improvement in intrinsic gain, particularly at larger V_{DS} as shown in Figure 4.13. In this manner, the TFET

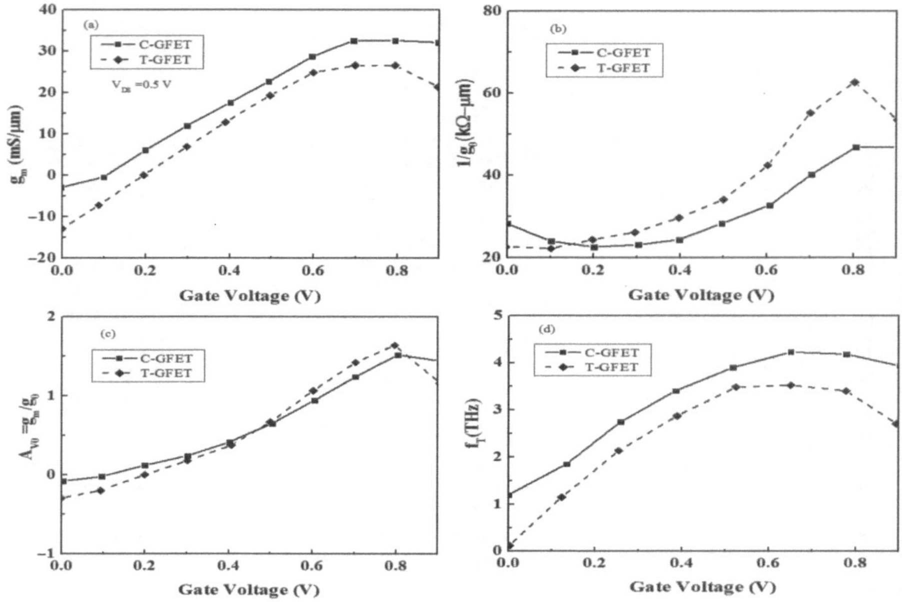

FIGURE 4.13 Comparison between C-GFET and T-GFET Obtained at *VDS* = 0.5 V: (a) Transconductance, (b) Output Resistance, (c) Intrinsic Gain, and (d) Cutoff Frequency [70]

with graphene offers a few advantages, and it is expected to be an alternate choice to C-GFET for analog/RF applications [70].

4.7 ELECTROSTATIC DOPING IN GRAPHENE TFETs

An electrostatically doped bi-layer graphene-based tunnel field-effect transistor (BED-TFET) is discussed [71]. Unlike graphene nanoribbon TFETs in which the edge states weaken the OFF-state execution, BED-TFETs work based on bandgaps prompted by vertical electric fields in the drain, source, and channel region with no chemical doping. The performance of the semiconductor is assessed by self-consistent quantum transport simulations. This device has a few favorable circumstances: (i) ultra low power (V_{DD}=0.1 V), (ii) high performance ($I_{ON}/I_{OFF}>10^4$), (iii) steep subthreshold swing (SS<10 mV/dec), and (iv) electrically configurable between N-TFET and P-TFET post manufacture. It was difficult to realize a tunnel FET (TFET) with high ON current and a low subthreshold swing at the same time, particularly with a low voltage (V_{DD}~0.1 V). The high current can be accomplished by bringing the transmission probability through the source-channel tunneling barrier near solidarity, which can be acknowledged by limiting the effective mass of the channel material and the screening length [72, 73] across the tunneling barrier. As to prerequisite of small effective mass, bi-layer graphene (BLG) is right around an ideal applicant. Notwithstanding, regardless of its small effective mass, noteworthy mobility, and initial guarantee for high-performance devices [74], the absence of an intrinsic bandgap

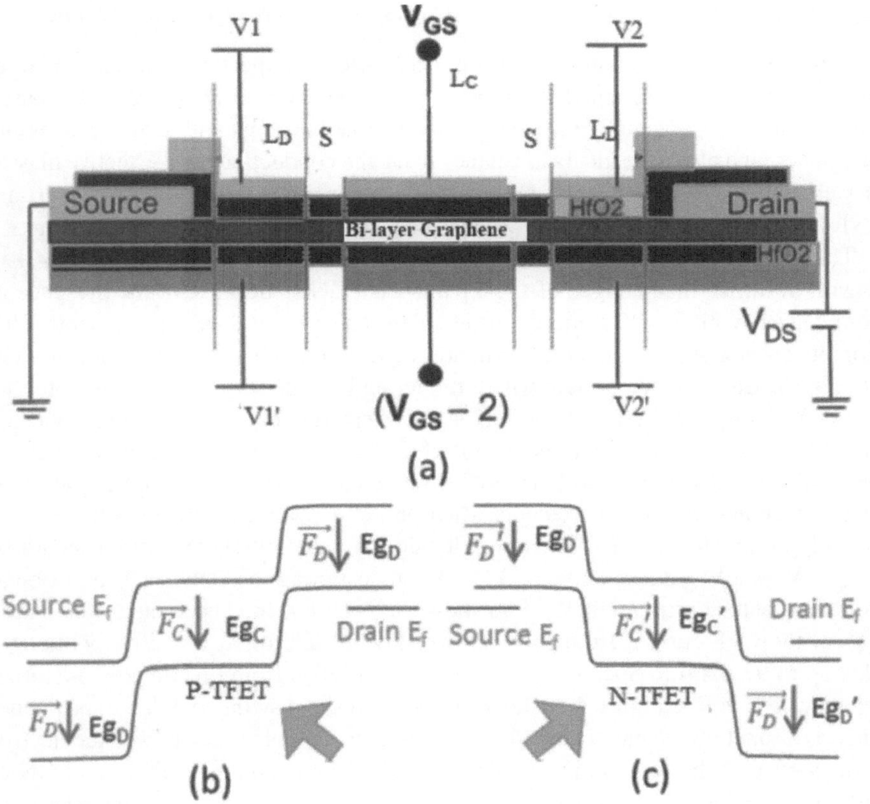

FIGURE 4.14 (a) Schematic of p-i-n BLG TFET: The Energy Band Profile in the OFF State (b) N-TFET, and (c) P-TFET [71]

prevents graphene FETs from turning off. Then again, a tunable bandgap greater than 200 meV can be made in BLG by an electric field [75, 76]. Here, we discussed BED-TFET as a superior steep SS device that empowers V_{DD} to downsize beneath 0.1V. Appropriately, an excellent energy-delay product is acquired in this device.

The bandgap size of every region is too tunable by potential difference (ΔV) between the top and base gates in that region. The prompted bandgaps are indicated by Eg_C furthermore, Eg_D as shown in Figure 4.14. Likewise, a counterfeit heterostructure can be made, however, if electric fields of various regions are unique. One of the primary favorable circumstances of the BED-TFET is its actual low energy-delay product. The BED-TFET has the least energy-postpone item among the contemplated gadgets. This is because of the steep 1V and high I_{ON} obtained in the BED-TFET indeed, even with a low V_{DD} of 0.1 V. This shows the significance of low bandgap materials for low V_{DD} steep gadgets. It is proved that with the appropriate design, the BED-TFET can accomplish I_{ON}/I_{OFF} ratios of more than 10^4, ON current of 45 μA/μm, and a subthreshold swing around 10 mV/dec, all at a low overdrive voltage of V_{DD} = 0.1 V at 300 K.

4.8 LINE EDGE ROUGHNESS IN GRAPHENE NANORIBBON TFETs

In fact, TFETs based on graphene nanoribbon (GNR) are expected to have much larger ON currents, lower OFF currents, as well as steeper SS than Si or even III-V-based semiconductors, are actually uncritical, fully compatible along with planar processing techniques, and also have identical valance band and conduction band effective masses as well as band width-tunable. Nevertheless, managing the breadth of sub-10-nm GNRs precisely and keeping clear of line side roughness seems quite challenging.

The incorporation of line edge roughness as shown in Figure 4.15 [77] will in general diminish the bandgap of the 5.1 nm-wide GNRs because of the presence of atomic layers with $N = 20$ atoms deposited between layers with $N = 21$ elements. This prompts an upgrade of tunneling from source to drain. This impact is more articulated in the transistor OFF-state where tunneling leakage through the gate potential barrier is encouraged than in the ON-state where tunneling relies more upon the strength of the electric field at the p-I interface than on a little decrease of the channel bandgap. Subsequently, the mean OFF-state current, which is the normal of 50 distinct samples, quickly increments when probability (P) increments to arrive at a worth I_{OFF} =15 μA/μm at P = 20%. Simultaneously, the normal ON current separated at $V_{gs} = V_{OFF} + V_{DD}$ doesn't go past 375 μA/μm, lessening the ON/OFF proportion to an unsatisfactory estimation of 25 as shown in Figure 4.16. The standard deviation σ_{OFF} of the OFF current from its mean worth is 10 μA/μm at P = 20%, while σ_{ON} adds up to 97 μA/μm. Such enormous current varieties from one device to another are undesired in ICs. Figure 4.17 shows that subthreshold swing and the transconductance as a function of channel length. In short-channel devices due to weaker electrostatic control of the gate, a higher subthreshold swing is obtained. In case of GNRs with perfect edges, the transconductance remains constant with the channel due to ballistic transport of carriers. However, because of the carrier scattering at rough edges, the transconductance of such GNRs decreases with the length of the channel.

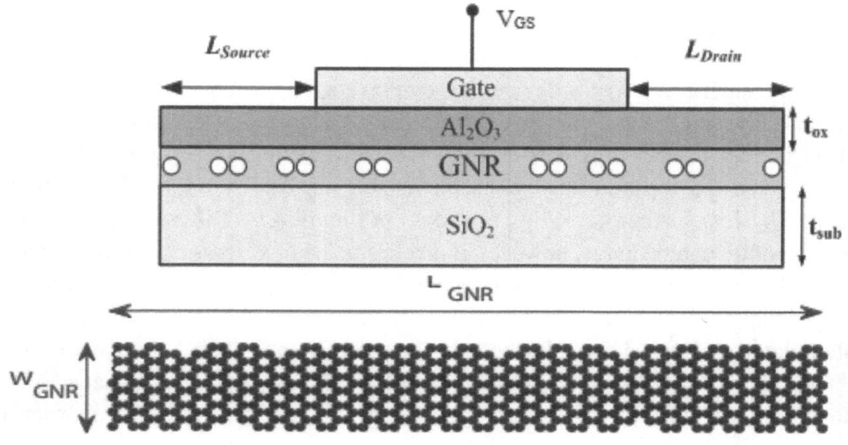

FIGURE 4.15 (Top) Schematic of a p-i-n GNR Tunneling Transistor Where the GNR Is Deposited on a SiO_2 Insulator Layer and (Bottom) Armchair GNR with LER [77]

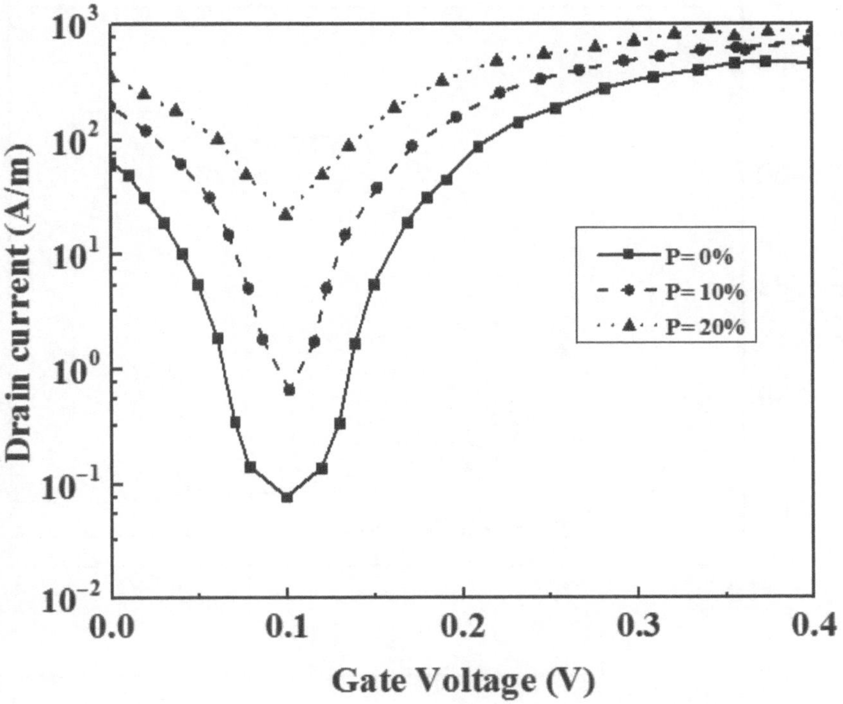

FIGURE 4.16 I_D vs V_{GS} Characteristics of GNR-Based Tunneling Transistors with LER at $V_{DD} = 0.1$ V (Probability P Limited between 0 and 20%) [77]

4.9 CARBON NANOTUBE TFETs

The properties of CNTs that are invaluable for conventional FETs are advantageous for TFETs too. In any case, a portion of the properties of CNTs that are particularly appropriate for carrying out TFETs are as per the following [79, 80]:

1. CNT is a direct bandgap material with a small effective carrier mass. These outcomes in higher band-to-band tunneling (BTBT) in a CNT TFET contrasted with a TFET dependent on an indirect bandgap material like silicon.
2. The bandgap of CNTs is moderate and it very well may be changed over a wide range dependent on their breadth and chirality.
3. The one-dimensional carrier transport in a CNT permits tight control of the gate voltage over the bands in the CNT channel. Consequently, the valence band in the channel of a CNT TFET can be raised over the conduction band in the source with the application of a little gate voltage.
4. The small diameter of a CNT increases BTBT resulting in a larger ON-state current.

Thus, CNT TFETs have pulled in a lot of consideration. The soonest experimental show of a TFET displaying subthreshold swing lesser than 60 mV/decade was for a

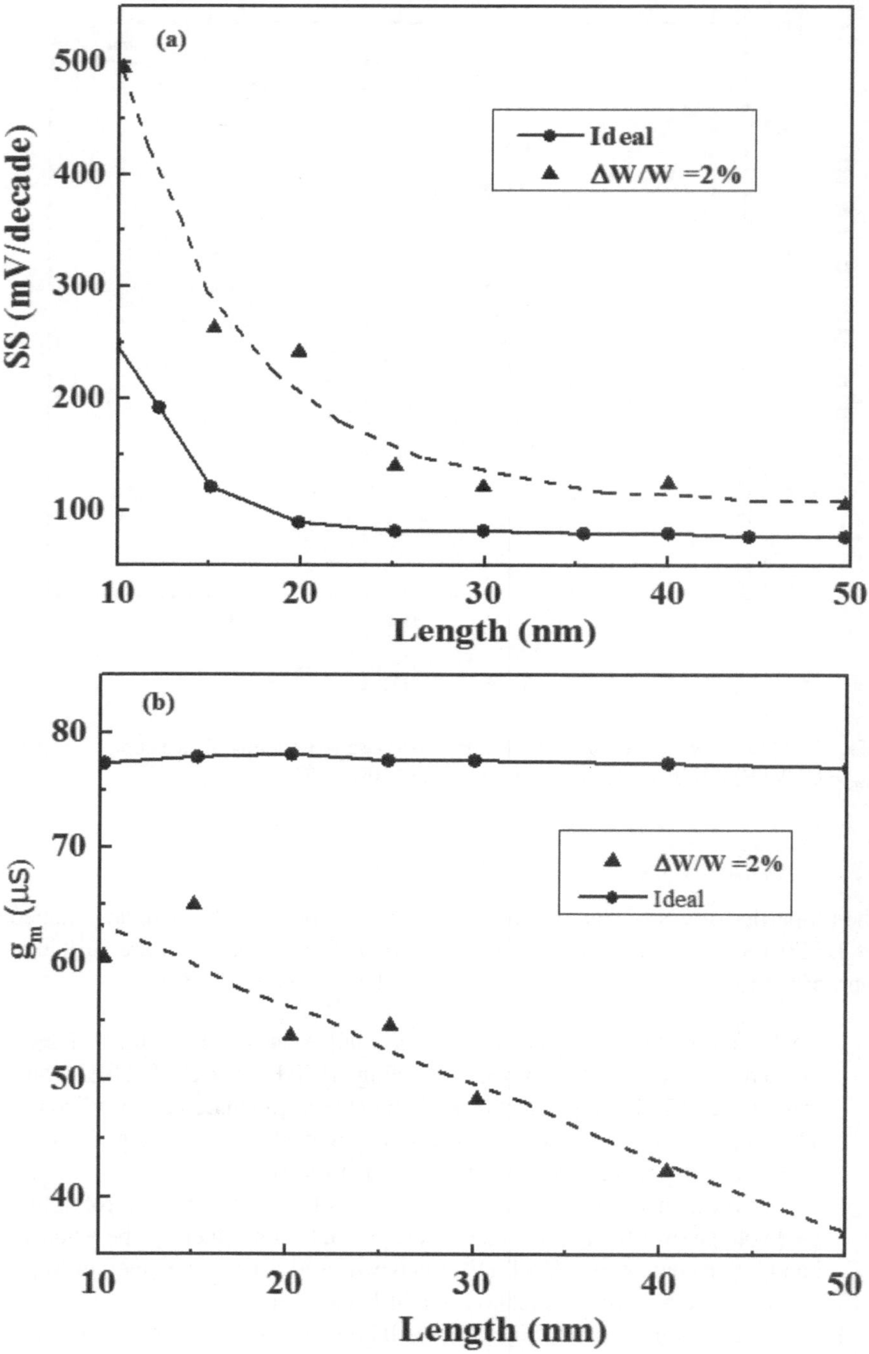

FIGURE 4.17 Difference between the Averages of (a) Subthreshold Swing and (b) the Transconductance of Rough GNR-FETs and Those of a GNR-FET with Perfect Edges as Functions of the Channel Length [78]

CNT TFET in 2004 [81]. In this segment, we will take a look at a basic execution of CNT TFET introduced in [82] and examine the distinct properties of CNT TFETs. The device channel is made up of (13,0) zigzag CNT with a diameter of 1 nm is shown in the figure [83]. The GAA-CNT-TFET consists of t_{ox} = 2 nm HfO_2. The source and drain have a linear doping concentration of 0.8 dopants/nm. A significant highlight note is that the source and channel areas are not covered with high-κ gate oxide, since high-κ gate oxide over source expands the fringe field, which is found to bring about a lower tunneling current [83]. On the other hand, expanding the gate drain fringing can smother the ambipolar conduction; however, it could degrade the device performance at high frequencies because of larger fringe capacitance [84]. The results in [83] show that, with V_{DD} = 0.3 V, the minimum achievable OFF-state current is $I_{OFF} \approx 5 \times 10^{-6}$ μA at 300K. However, under these conditions, the ON-state current is minuscule, $I_{ON} \approx 1$ μA.

The two double gate CNT TFET structures are with chirality of (13,0) having a length of 15 nm as shown in Figure 4.18 [85]. In one design, we put the gate dielectric just under the gate and in another, the gate dielectric is reached out from source to drain. CNT in this chirality has a bandgap of 0.75 eV and a diameter of 1 nm. While the source and drain regions were doped with p and n, respectively, having doping density 5×10^{-3}, which means source and drain are doped with 5 dopant atoms among 1,000 carbon particles. As the gate dielectric we utilized the HfO_2 of the thickness of 2 nm with k = 16 whose equivalent oxide thickness (EOT) is 0.5 nm. The dielectric constant of gate dielectric is varied for double gate CNT MOSFET and CNT TFET. The drive current performance of CNT MOSFET improves with

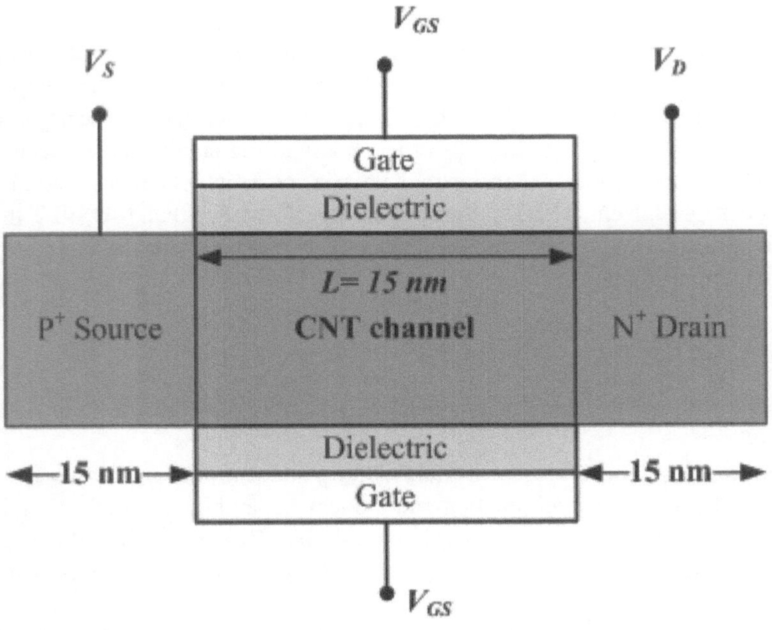

FIGURE 4.18 Schematic Representation of CNT TFET [85]

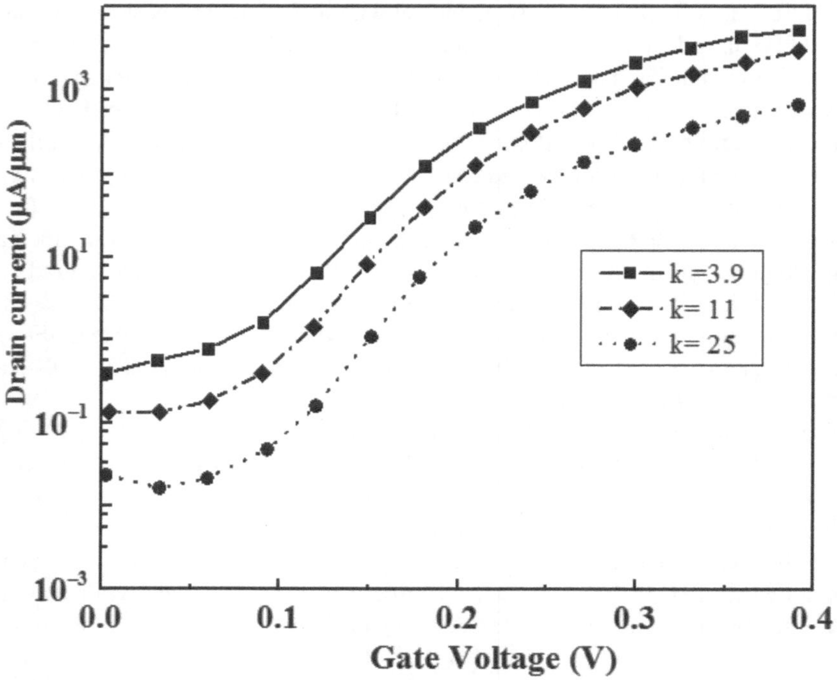

FIGURE 4.19 I_D-V_{GS} Characteristics of CNT TFET for Different Gate Dielectric Materials [85]

an increase in oxide dielectric constants, while in CNT TFET, the drive current decreases as shown in Figure 4.19. The OFF state flow diminishes with expanding dielectric consistent for the CNT TFET, in spite of the fact that it is nearly constant for the CNT MOSFET. The I_{ON}/I_{OFF} proportion and the subthreshold execution of both gadgets improve with expansion in dielectric constants. The investigation shows that besides ON current, TFET has much better performance than MOSFET in other execution boundaries and is by all accounts a likely contender for fast and low power switching applications.

4.10 GAA CNT TFETs STRUCTURAL ENGINEERING

To enhance the performance of graphene TFETs, the following techniques can be employed:

Drain doping: The decrease in drain doping concentration increases tunneling barrier on the drain-channel junction and reduces the ambipolar conduction. It also increases the series resistance, which is undesirable. Hence, a lightly doped spacer region in addition to the heavily doped extended region can be used in the drain for better operation [84].

Source doping: The increase in the source doping concentration decreases the tunneling barrier width at the source-channel junction; as a consequence, tunneling current increases [84]. The maximum source doping in CNT TFETs should satisfy the proportionality of dopant impurity ions and number of carbon atoms per unit length. Ion implantation techniques, which are commonly used to achieve the required doping profile, are unsuitable for CNT TFETs, because the substitution of carbon atoms with impurity atoms in CNT degrades the nanotube properties [27]. As an outcome, getting desired doping concentration is troublesome in CNT-based devices.

Diameter and chirality: The bandgap of CNTs is a function of diameter and chirality. For example, the bandgap of (10, 0), (13, 0) and (16, 0) are 1.053, 0.817, and 0.667eV, respectively [35]. The device simulation of CNT TFETs shows that the tunneling current increases with a decrease in bandgap. However, the ambipolar conduction increases with a decrease in bandgap of the material. Hence, additional care must be taken when lower bandgap materials are used in CNT-TFETs.

Strain: The application of axial tensile strain in CNT TFETs can decrease the I_{OFF} by a considerable amount and improve the subthreshold swing due to bandgap modulation. However, the strain reduces the I_{ON} in TFETs [82, 86]. Hence, the trade-off between I_{ON} and I_{OFF} can be optimized by strain.

Multiple CNTs: The I_{ON} in CNT TFETs can be enhanced by using multi-single-walled CNTs. The diameter and the position of each CNTs can be optimized to improve the I_{ON}/I_{OFF} ratio in multiple CNTs. In addition, from a reliability perspective, employing multiple CNTs is better, since even when some tubes fail, the transistor remains operating [86].

REFERENCES

[1] Roberts, D. 'Silicon integrated circuits: A personal view of the first 25 years', *Electronics and Power, 30*(4), 282–284, 1984.

[2] Kilby, J. 'The integrated circuit's early history', *Proceedings of the IEEE, 88*(1), 109–111, 2000.

[3] Srivastava, N. and Banerjee, K. 'Performance analysis of carbon nanotube interconnects for VLSI applications', in 'Proceedings of the 2005 IEEE/ACM International conference on Computer-aided design', *IEEE Computer Society*, 383–390, 2005.

[4] Kang, S.-M., Leblebici, Y. and Kim, C. 'CMOS digital integrated circuits: Analysis and design', Technical report, McGraw-Hill Higher Education, 2014.

[5] Mollick, E. 'Establishing Moore's law', *IEEE Annals of the History of Computing, 28*(3), 62–75, 2006.

[6] Horowitz, M., Alon, E., Patil, D., Naffziger, S., Kumar, R. and Bernstein, K., 'Scaling, power, and the future of CMOS', in *Electron Devices Meeting, 2005. IEDM Technical Digest. IEEE International*, IEEE, 2005.

[7] Schwierz, F., Pezoldt, J. and Granzner, R. 'Two-dimensional materials and their prospects in transistor electronics', *Nanoscale, 7*(18), 8261–8283, 2015.

[8] Arden, W. M. 'The international technology roadmap for semiconductors perspectives and challenges for the next 15 years', *Current Opinion in Solid State and Materials Science, 6*(5), 371–377, 2002.

[9] De, V. and Borkar, S. 'Technology and design challenges for low power and high performance', in *Proceedings of the 1999 International Symposium on Low Power Electronics and Design*, ACM, pp. 163–168, 1999.

[10] Roy, K., Mukhopadhyay, S. and Mahmoodi-Meimand, H. 'Leakage current mechanisms and leakage reduction techniques in deep-submicrometer CMOS circuits', *Proceedings of the IEEE, 91*(2), 305–327, 2003.

[11] Hisamoto, D., Lee, W.-C., Kedzierski, J., Anderson, E., Takeuchi, H., Asano, K., King, T.-J., Bokor, J. and Hu, C. 'A folded-channel MOSFET for deep-sub-tenth micron era', *IEDM Technical Digest, 38*, 1032–1034, 1998.

[12] Kavalieros, J., Doyle, B., Datta, S., Dewey, G., Doczy, M., Jin, B., Lionberger, D., Metz, M., Rachmady, W., Radosavljevic, M., et al. 'Tri-gate transistor architecture with high-k gate dielectrics, metal gates and strain engineering', in *VLSI Technology, 2006. Digest of Technical Papers. 2006 Symposium on VLSI Technology*, IEEE, pp. 50–51, 2006.

[13] Hoefflinger, B. 'ITRS: The International Technology Roadmap for Semiconductors, in Chips 2020', *Springer*, pp. 161–174, 2011.

[14] Cui, Y., Zhong, Z., Wang, D., Wang, W. U. and Lieber, C. M. 'High performance silicon nanowire field effect transistors', *Nano Letters, 3*(2), 149–152, 2003.

[15] Ashley, T., Barnes, A., Buckle, L., Datta, S., Dean, A., Emery, M., Fearn, M., Hayes, D., Hilton, K., Jefferies, R., et al., 'Novel InSb-based quantum well transistors for ultra-high speed, low power logic applications', in *Solid-State and Integrated Circuits Technology, 2004. Proceedings 7th International Conference, Vol. 3*, IEEE, pp. 2253–2256, 2004.

[16] Datta, S., Ashley, T., Brask, J., Buckle, L., Doczy, M., Emeny, M., Hayes, D., Hilton, K., Jefferies, R., Martin, T., et al., '85 nm gate length enhancement and depletion mode InSb quantum well transistors for ultra high speed and very low power digital logic applications', in *Electron Devices Meeting, 2005. IEDM Technical Digest,* IEEE International, pp. 763–766, 2005.

[17] Kuo, C.-I., Hsu, H.-T., Wu, C.-Y., Chang, E. Y., Miyamoto, Y., Chen, Y.-L. and Biswas, D., 'A 40-nm-gate InAs/in 0.7 Ga 0.3 as composite-channel HEMT with 2200ms/mm and 500-ghz ft', in *Indium Phosphide and Related Materials, 2009. IPRM'09 IEEE International Conference on*, IEEE, pp. 128–131, 2009.

[18] Kim, S., Nah, J., Jo, I., Shahrjerdi, D., Colombo, L., Yao, Z., Tutuc, E. and Banerjee, S. K. 'Realization of a high mobility dual-gated graphene field-effect transistor with Al_2O_3 dielectric', *Applied Physics Letters, 94*(6), 2009.

[19] Lemme, M. C., Echtermeyer, T. J., Baus, M. and Kurz, H. 'A graphene field effect device', *IEEE Electron Device Letters, 28*(4), 282–284, 2007.

[20] Wang, X., Ouyang, Y., Li, X., Wang, H., Guo, J. and Dai, H. 'Room temperature all-semiconducting sub-10-nm graphene nanoribbon field-effect transistors', *Physical Review Letters, 100*(20), 2008.

[21] Deng, J. and Wong, H.-S. P. 'A compact spice model for carbon-nanotube field-effect transistors including non-idealities and its application—part ii: Full device model and circuit performance benchmarking', *IEEE Transactions on Electron Devices, 54*(12), 3195–3205, 2007.

[22] Deng, J., Patil, N., Ryu, K., Badmaev, A., Zhou, C., Mitra, S. and Wong, H. P. 'Carbon nanotube transistor circuits: Circuit-level performance benchmarking and design options for living with imperfections', in *Solid-State Circuits Conference, 2007. ISSCC 2007. Digest of Technical Papers. IEEE International*, IEEE, pp. 70–588, 2007.

[23] Javey, A., Guo, J., Wang, Q., Lundstrom, M. and Dai, H. 'Ballistic carbon nanotube field-effect transistors', *Nature, 424*(6949), 654, 2003.

[24] Kim, Y.-B., 'Challenges for nanoscale MOSFETs and emerging nanoelectronics', *Transactions on Electrical and Electronic Materials, 11*(3), 93–105, 2010.

[25] Iijima, S., 'Helical microtubules of graphitic carbon', *Nature, 354*(6348), 56–58, 1991.

[26] Hanson, G. W., *Fundamentals of Nanoelectronics*, Pearson/Prentice Hall, Upper Saddle River, 2008.

[27] Knoch, J. and Muller, M. R. 'Electrostatic doping-controlling the properties of carbon-based FETs with gates', *IEEE Transactions on Nanotechnology, 13*, pp. 1044–1052, Nov. 2014.

[28] Knoch, J. and Appenzeller, J. 'A novel concept for field-effect transistors: The tunneling carbon nanotube FET', in *Proceedings of the 63rd DRC*, pp. 153–156, IEEE, June 2005.
[29] Huang, X., Yin, Z., Wu, S., Qi, X., He, Q., Zhang, Q., Yan, Q., Boey, F. and Zhang, H. 'Graphene-based materials: Synthesis, characterization, properties, and applications', *Small*, *7*(14), pp. 1876–1902, 2011.
[30] Yung, K. C. and Wu, W. M. 'Introduction to graphene electronics–a new era of digital transistors and devices', *Contemporary Physics*, *54*(5), pp. 233–251, 2013.
[31] Bonaccorso, F., Lombardo, A., Hasan, T., Sun, Z., Colombo, L. and Ferrari, A. C. 'Production and processing of graphene and 2d crystals', *Materials Today*, *15*(12), pp. 564–589, 2012.
[32] Lee, S. K., Kim, H. and Shim, B. S. 'Graphene: An emerging material for biological tissue engineering', *Carbon Letters*, *14*(2), pp. 63–75, 2013.
[33] Poncharal, S. F. S, P., Wang, Z. L. and Heer, W. A. 'Carbon nanotube quantum resistors', *Science*, *280*, pp. 1744–46, June 1998.
[34] Poncharal, P., Berger, C., Yi, Y., Wang, Z. L. and de Heer, W. A. 'Room temperature ballistic conduction in carbon nanotubes', *Journal of Physical Chemistry B*, *106*, pp. 12104–12114, Nov. 2002.
[35] Koswatta, S. O., Lundstrom, M. S. and Nikonov, D. 'Performance comparison between p-i-n tunneling transistors and conventional MOSFETs', *IEEE Transactions on Electron Devices*, *56*, pp. 456–465, Mar. 2009.
[36] Register, L. F., Hasan, M. M. and Banerjee, S. K. 'Stepped broken-gap heterobarrier tunneling field-effect transistor for ultralow power and highspeed', *IEEE Electron Device Letters*, *32*, pp. 743–745, June 2011.
[37] Zhu, W., Perebeinos, V., Freitag, M. and Avouris, P. 'Carrier scattering, mobilities and electrostatic potential in mono-, bi-and tri-layer graphenes', In *APS March Meeting Abstracts*, Vol. 2010, pp. J21–002, March 2010.
[38] Agarwal, T. K., Nourbakhsh, A., Raghavan, P., Radu, I., Gendt, S. D., Heyns, M., Verhelst, M. and Thean, A. 'Bilayer graphene tunneling FET for sub-0.2 V digital CMOS logic applications', *IEEE Electron Device Letters*, *35*, pp. 1308–1310, Dec. 2014.
[39] Fiori, G. and Iannaccone, G. 'Ultralow-voltage bilayer graphene tunnel FET', *IEEE Electron Device Letters*, *30*, pp. 1096–1098, Oct. 2009.
[40] Chen, F., Ilatikhameneh, H., Klimeck, G., Chen, Z. and Rahman, R. 'Configurable electrostatically doped high performance bilayer graphene tunnel FET', *IEEE Journal of the Electron Devices Society*, *4*(3), pp. 124–128, 2016.
[41] Samuels, A. J. and Carey, J. D. 'Molecular doping and band gap opening of bilayer graphene', *ACS Nano*, *7*, pp. 2790–2799, Feb. 2013.
[42] Moslemi, M. R., Sheikhi, M. H., Saghafi, K. and Moravvej-Farshi, M. K. 'Electronic properties of a dual-gated GNR-FET under uniaxial tensile strain', *Microelectronics Reliability*, *52*(11), pp. 2579–2584, 2012.
[43] Yousefi, R., Saghafi, K. and Moravvej-Farshi, M. K. 'Numerical study of lightly doped drain and source carbon nanotube field effect transistors', *IEEE Transactions on Electron Devices*, *57*(4), pp. 765–771, 2010.
[44] Yousefi, R. and Ghoreyshi, S. S. 'Numerical study of Ohmic-Schottky carbon nanotube field effect transistor', *Modern Physics Letters B*, *26*(15), p. 1250096, 2012.
[45] Moghadam, N., Moravvej-Farshi, M. K. and Aziziyan, M. R. 'Design and simulation of MOSCNT with band engineered source and drain regions', *Microelectronics Reliability*, *53*(4), pp. 533–539, 2013.
[46] Zhao, P., Chauhan, J. and Guo, J. 'Computational study of tunneling transistor based on graphene nanoribbon', *Nano Letters*, pp. 684–688, 2009.
[47] Han, M. Y., Özyilmaz, B., Zhang, Y. and Kim, P. 'Energy band-gap engineering of graphene nanoribbons', *Physical Review Letters*, *98*(20), p. 206805, 2007.

[48] Chen, Z., Lin, Y. M., Rooks, M. J. and Avouris, P. 'Graphene nano-ribbon electronics', *Physica E: Low-dimensional Systems and Nanostructures, 40*(2), pp. 228–232, 2007.

[49] Wang, X., Ouyang, Y., Li, X., Wang, H., Guo, J. and Dai, H. 'Room-temperature all-semiconducting sub-10-nm graphene nanoribbon field-effect transistors', *Physical Review Letters, 100*(20), p. 206803, 2008.

[50] Novoselov, K. S., et al. 'Electric field effect in atomically thin carbon films', *Science, 306*(5696), pp. 666–669, 2004.

[51] Schwierz, F. 'Graphene transistors: Status, prospects, and problems', *Proceedings of the IEEE, 101*(7), pp. 1567–1584, July 2013.

[52] Meric, I., et al. 'High-frequency performance of graphene field effect transistors with saturating IV-characteristics', in *Proceedings of the IEEE Int. Electron Devices Meeting (IEDM)*, Dec. 2011, pp. 2.1.1–2.1.4.

[53] Meric, I., Han, M. Y., Young, A. F., Ozyilmaz, B., Kim, P. and Shepard, K. L. 'Current saturation in zero-bandgap, top-gated graphene field-effect transistors', *Nature Nanotechnology, 3*(11), pp. 654–659, Dec. 2008.

[54] Fiori, G. 'Negative differential resistance in mono and bilayer graphene p-n junctions', *IEEE Electron Device Letters, 32*(10), pp. 1334–1336, Oct. 2011.

[55] Liao, L., et al. 'Sub-100 nm channel length graphene transistors', *Nano Letters, 10*(10), pp. 3952–3956, 2010.

[56] Szafranek, B. N., Fiori, G., Schall, D., Neumaier, D. and Kurz, H. 'Current saturation and voltage gain in bilayer graphene field effect transistors', *Nano Letters, 12*(3), pp. 1324–1328, 2012.

[57] Zheng, J., et al. 'Sub-10 nm gate length graphene transistors: Operating at terahertz frequencies with current saturation', *Scientific Reports, 3*, Feb. 2013, Art. ID 1314.

[58] Fiori, G., Neumaier, D., Szafranek, B. N. and Iannaccone, G. 'Bilayer graphene transistors for analog electronics', *IEEE Transactions on Electron Devices, 61*(3), pp. 729–733, Mar. 2014.

[59] Imperiale, I., Bonsignore, S., Gnudi, A., Gnani, E., Reggiani, S. and Baccarani, G. 'Computational study of graphene nanoribbon FETs for RF applications', in *Proceedings of the IEEE Electron Devices Meeting (IEDM)*, Dec. 2010, pp. 32.3.1–32.3.4.

[60] Chauhan, J. and Guo, J. 'Assessment of high-frequency performance limits of graphene field-effect transistors', *Nano Research, 4*(6), pp. 571–579, 2011.

[61] Alarcon, A., et al., 'Pseudosaturation and negative differential conductance in graphene field-effect transistors', *IEEE Transactions on Electron Devices, 60*(3), pp. 985–991, Mar. 2013.

[62] Ganapathi, K., Yoon, Y., Lundstrom, M. and Salahuddin, S. 'Ballistic I—V characteristics of short-channel graphene field-effect transistors: Analysis and optimization for analog and RF applications', *IEEE Transactions on Electron Devices, 60*(3), pp. 958–964, Mar. 2013.

[63] Koswatta, S. O, Lundstrom, M. S. and Nikonov, D. E. 'Performance comparison between p-i-n tunneling transistors and conventional MOSFETs', *IEEE Transactions on Electron Devices, 56*(3), pp. 456–465, Mar. 2009.

[64] Fang, T., Konar, A., Xing, H. and Jena, D. 'Carrier statistics and quantum capacitance of graphene sheets and ribbons', *Applied Physics Letters, 91*(9), p. 092109, 2007.

[65] Ionescu, A. M. and Riel, H. 'Tunnel field-effect transistors as energy efficient electronic switches', *Nature, 479*, pp. 329–337, Nov. 2011.

[66] Liu, H., Datta, S. and Narayanan, V. 'Steep switching tunnel FET: A promise to extend the energy efficient roadmap for post-CMOS digital and analog/RF applications', in *Proceedings of the IEEE International Symposium on Low Power Electronics Design (ISLPED)*, Sept. 2013, pp. 145–150.

[67] Mallik, A. and Chattopadhyay, A. 'Tunnel field-effect transistors for analog/mixed-signal system-on-chip applications', *IEEE Transactions on Electron Devices*, *59*(4), pp. 888–894, Apr. 2012.

[68] Fiori, G., Betti, A., Bruzzone, S. and Iannaccone, G. 'Lateral graphene—hBCN heterostructures as a platform for fully two-dimensional transistors', *ACS Nano*, *6*(3), pp. 2642–2648, 2012.

[69] Zhao, P., Feenstra, R. M., Gu, G. and Jena, D. 'SymFET: A proposed symmetric graphene tunneling field-effect transistor', *IEEE Transaction Electron Devices*, *60*(3), pp. 951–957, Mar. 2013.

[70] Rawat, B. and Paily, R. 'Analysis of graphene tunnel field-effect transistors for analog/RF applications', *IEEE Transaction Electron Devices*, *62*(8), pp. 2663–2669, 2015.

[71] Chen, F. W., Ilatikhameneh, H., Klimeck, G., Chen, Z. and Rahman, R. 'Configurable electrostatically doped high performance bilayer graphene tunnel FET', *IEEE Journal of the Electron Devices Society*, *4*(3), pp. 124–128, 2016.

[72] Ilatikhameneh, H., Ameen, T., Klimeck, G., Appenzeller, J. and Rahman, R. 'Dielectric engineered tunnel field-effect transistor', *Electron Device Letters, IEEE*, *36*(10), pp. 1097–1100, Sept. 2015.

[73] Ilatikhameneh, H., Klimeck, G., Appenzeller, J. and Rahman, R. 'Scaling theory of electrically doped 2D transistors', *IEEE Electron Device Letters*, *36*(7), May 2015.

[74] Schwierz, F. 'Graphene transistors', *Nature Nanotechnology*, *5*(7), pp. 487–496, 2010.

[75] Fiori, G. and Iannaccone, G. 'Ultralow-voltage bilayer graphenetunnel FET', *Electron Device Letters, IEEE*, *30*(10), pp. 1096–1098, Sept. 2009.

[76] Nilsson, J., Neto, A. C., Guinea, F. and Peres, N. 'Electronic properties of graphene multilayers', *Physical Review Letters*, *97*(26), p. 266801, 2006.

[77] Luisier, M. and Klimeck, G. 'Performance analysis of statistical samples of graphene nanoribbon tunneling transistors with line edge roughness', *Applied Physics Letters*, *94*(22), p. 223505, 2009.

[78] Goharrizi, A. Y., Pourfath, M., Fathipour, M. and Kosina, H. 'Device performance of graphene nanoribbon field-effect transistors in the presence of line-edge roughness', *IEEE Transactions on Electron Devices*, *59*(12), pp. 3527–3532, 2012.

[79] Knoch, J. and Muller, M. R. 'Electrostatic doping-controlling the properties of carbon-based FETs with gates', *IEEE Transactions on Nanotechnology*, *13*, pp. 1044–1052, Nov. 2014.

[80] Knoch, J. and Appenzeller, J. 'A novel concept for field-effect transistors: The tunneling carbon nanotube FET', in *Proceedings of the 63rd DRC*, pp. 153–156, IEEE, June 2005.

[81] Appenzeller, J., Lin, Y.-M., Knoch, J. and Avouris, P. 'Band-to-band tunneling in carbon nanotube field-effect transistors', *Physical Review Letters*, *93*, pp. 196805-1-196805-4, Nov. 2004.

[82] Es-Sakhi, A. D. and Chowdhury, M. H. 'Multichannel tunneling carbon nanotube field effect transistor (MT-CNTFET)', in *IEEE International SoC Conference (SOCC)*, pp. 156–159, IEEE, Sept. 2014.

[83] Koswatta, S. O., Lundstrom, M. S. and Nikonov, D. 'Performance comparison between p-i-n tunneling transistors and conventional MOSFETs', *IEEE Transactions on Electron Devices*, *56*, pp. 456–465, Mar. 2009.

[84] Koswatta, S. O., Nikonov, D. and Lundstrom, M. S. 'Computational study of carbon nanotube p-i-n tunnel FETs', in *IEDM Technical Digest*, pp. 518–521, IEEE, 2005.

[85] Sarker, M. S., Islam, M. M., Alam, M. N. K. and Islam, M. R. 'Gate dielectric strength dependent performance of CNT MOSFET and CNT TFET: A tight binding study', *Results in Physics*, *6*, pp. 879–883, 2016.

[86] Nakano, T., Ogawa, M. and Souma, S. 'Strain induced modulation of switching behavior in carbon nanotube tunneling field effect transistors', in *IEEE International Meeting for Future Devices*, pp. 1–2, IEEE, 2012.

5 Characterization of Silicon FinFETs under Nanoscale Dimensions

Rock-Hyun Baek and Jun-Sik Yoon

CONTENTS

5.1 INTRODUCTION: BACKGROUND OF CONVENTIONAL SILICON FinFETs

Silicon fin-shaped field-effect transistors (FinFETs) were first introduced in the 22-nm node by enhancing gate-to-channel controllability over planar MOSFETs [1]. As the self-aligned double patterning scheme is highly sophisticated, fin shape changes from tapered to rectangular in 14-nm node [2], and a much taller and thinner fin has been formed in 10-nm node [3]. Also, thorough design-technology co-optimization (DTCO), both front-end-of-line (FEOL) and middle-of-line (MOL) schemes are improved concurrently to improve the devices in the perspectives of performance (speed), power, and area (PPA). For example, removal of the dummy gate between active devices (single diffusion break) increases the device density greatly [4, 5]. Self-aligned gate and drain contacts reduce the M0 size by using materials with different etching selectivity [3, 6]. DTCO, with help from extreme ultraviolet lithography (EUV) for ultrafine patterning, makes conventional device scaling feasible down to 5-nm node [7, 8].

In this chapter, we address conventional bulk FinFETs from 10-nm, 7-nm to sub-5-nm nodes. DC/AC performance among three different nodes was compared in

DOI: 10.1201/9781003200987-5

terms of channel stresses, physical parameters, and capacitances. Finally, a novel process scheme possibly compatible under conventional process is introduced to scale down bulk FinFETs further in the 2-nm node.

5.2 SIMULATION METHODOLOGY

All the bulk FinFETs in this chapter were simulated using Sentaurus TCAD [9]. Drift-diffusion transport equations were calculated self-consistently with Poisson and electron/hole continuity equations. The Slotboom bandgap narrowing model was used for all the semiconductor regions [10]. Carrier mobility was calculated using the Lombardi model [11] for remote phonon and remote Coulomb scatterings; the inversion and accumulation layer model [12] for impurity, phonon, and surface roughness scatterings; and the thin-layer model [13] for thin fin channel. Quasi-ballistic effect was considered using low-field ballistic mobility model [14]. Carrier generation-recombination models were Shockley-Read-Hall [15], Auger [16], and Hurkx band-to-band tunneling [17]. Deformation potential theory with two-band k·p model for electrons and six band k·p model for holes was also used to consider stress-induced effects of band structure, effective mass, and effective density-of-states [18].

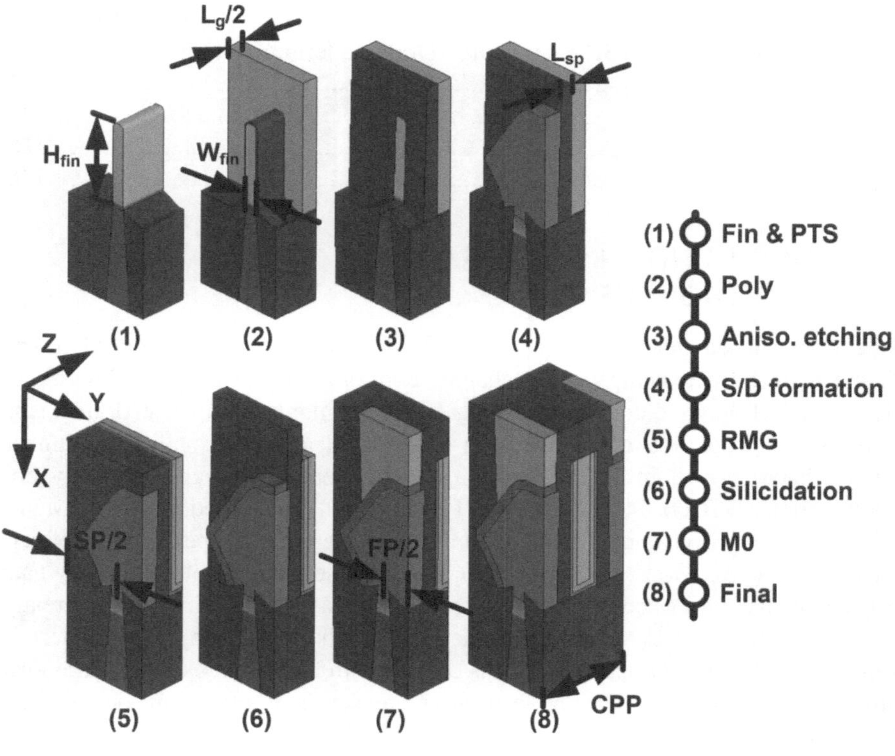

FIGURE 5.1 Simulated Process Flows of Bulk FinFETs

TABLE 5.1
Geometrical Parameters of Bulk FinFETs from 10-nm to Sub-5-nm Nodes

Parameters	Symbols	10 nm	7 nm	Sub 5 nm
Fin pitch	FP	34 nm	26 nm	21 nm
Contacted poly pitch	CPP	54 nm	48 nm	42 nm
N/P separation	SP	58 nm	58 nm	56 nm
Gate length	L_g	18 nm	15 nm	12 nm
Fin height	H_{fin}	46 nm		
Fin width	W_{fin}	7 nm	5, 6 nm	4, 5 nm
Spacer length	L_{sp}	7 nm	6 nm	5 nm
S/D doping	N_{sd}	P: $5{\cdot}10^{20}$ cm^{-3}, N: $2{\cdot}10^{20}$ cm^{-3}		
Punch-through-stopper doping	N_{stop}	$2, 5 \cdot 10^{18}$ cm^{-3}		

Figure 5.1 shows the simulated process flow of bulk FinFETs. This process flow is the same as the actual fabrication of bulk FinFETs [1–3]. After defining fin through patterning, punch-through-stopper (PTS) and shallow trench isolation (STI) regions were formed. Gate region is defined by patterning poly-Si region. After anisotropic etching dielectric and fin regions, source/drain (S/D) regions were formed using selective epitaxial growth (SEG) with different growth rates in different crystal directions of <100>, <110>, and <111> [19, 20]. After depositing dielectric regions, poly-Si gate is removed and replacement metal gate (RMG) region is formed consecutively in series of depositing 0.7-nm-thick interfacial layer (IL), 1.7-nm-thick HfO_2 (equivalent oxide thickness is 1 nm), and TiN. Dielectric constants of HfO_2 and spacer regions were 22 and 5, respectively. Contact resistivity is fixed at 10^{-9} Ω·cm^2 [21]. After silicidation using wrap-around contact (WAC) scheme and M0 formation, final structure is formed as in step 8. WAC reduces the parasitic resistances (R_{sd}) effectively by increasing the contact area [22]. Geometrical parameters of bulk FinFETs are defined in Figure 5.1, and their values in three different technology nodes are specified in Table 5.1. N/P separation (SP) indicates the distance between n-type and p-type FETs. SP, different from fin pitch (FP), is sufficiently long enough to prevent epi merging between n- and p-type FETs.

Beforehand, n-/p-type bulk FinFETs are calibrated to Intel's 10-nm node FinFETs [3] as shown in Figure 5.2. Calibrated device structures are also shown as the inset of Figure 5.2. First, S/D doping concentrations (N_{sd}) and profiles are controlled by changing annealing temperature and time to fit the subthreshold characteristics such as subthreshold swing (SS) and drain-induced barrier lowering (DIBL). N_{sd} of n-type and p-type devices are $2{\cdot}10^{20}$ cm^{-3} and $5{\cdot}10^{20}$ cm^{-3}, respectively. Then, ballistic coefficients and surface roughness scattering parameters were calibrated to fit the drain currents (I_{ds}) in the linear region. Finally, saturation velocity was calibrated to fit the I_{ds} in the saturation region.

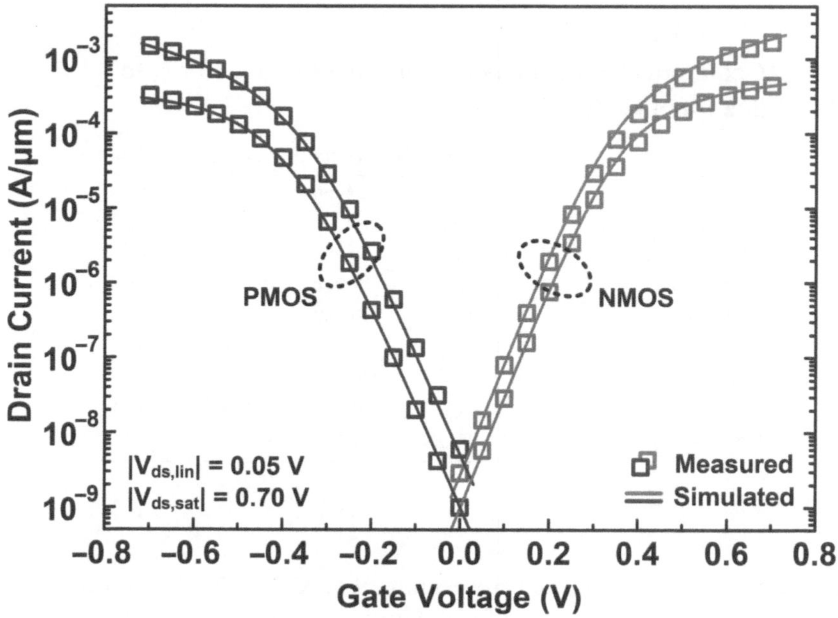

FIGURE 5.2 Transfer Characteristics of 10-nm-Node Bulk FinFETs

5.3 RESULTS AND DISCUSSION

5.3.1 Conventional Scaling of Bulk FinFETs Down to Sub-5-nm Nodes

Figure 5.3 shows the doping profiles of n-type bulk FinFETs in three different nodes. Since it is assumed that annealing temperature and time are the same irrespective of technology nodes, more S/D dopants penetrate into the active channel regions as the device is scaled down. PTS dopants also penetrate into the channels under the S/D annealing, but a few PTS dopants reside at the active regions in sub-5-nm node.

Although the scaling of gate length (L_g) and spacer length (L_{sp}) increases the S/D dopant penetration into the channels, stress components increase as the devices are scaled down as shown in Table 5.2 and Table 5.3. X, Y, and Z directions are specified in Figure 5.1. Among three stress components, S_{ZZ} (stress in the channel direction) induced by the S/D epi dominantly affects the DC performance [23]. As the device is scaled down, S_{ZZ} increases because the S/D epi size is not scaled much compared to gate length (L_g) and spacer length (L_{sp}). Also, smaller fin width (W_{fin}) increases all the stress components because the metal gate of bulk FinFETs holds the fin channels more tightly [24].

But for low power (LP) application, gate-to-channel controllability is very important to improve DC/AC performance [19, 20, 24, 25]. Especially, OFF currents (I_{off}) vary significantly as the over-etching depth (T_{over}) increases from 5 to 20 nm in sub-5-nm node as shown in Figure 5.4 (a). T_{over} change is one of variability sources under anisotropic etching prior to S/D formation [26, 27], so it should be considered

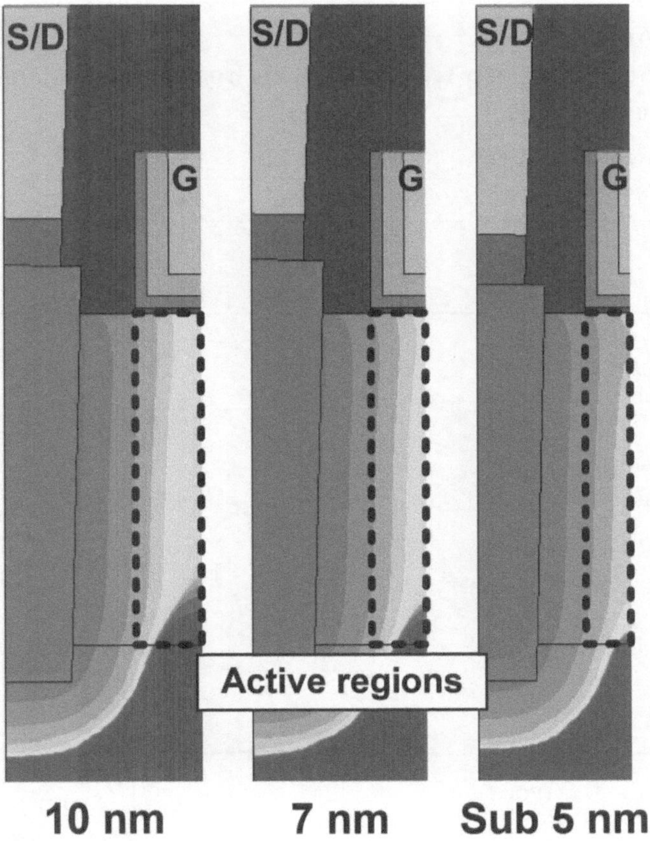

FIGURE 5.3 Doping Profiles of Bulk FinFETs from 10- to Sub-5-nm Nodes

TABLE 5.2
Stress Values of n-Type Bulk FinFETs from 10- to Sub-5-nm Nodes (in GPa)

Stress parameters	10 nm	7 nm		Sub 5 nm	
		W5	W6	W4	W5
S_{XX}	−0.167	−0.206	−0.205	−0.226	−0.225
S_{YY}	−0.292	−0.290	−0.275	−0.272	−0.253
S_{ZZ}	0.553	0.597	0.590	0.648	0.632

carefully. Increasing PTS doping concentration (N_{stop}) can have more T_{over} margin to maintain low I_{off}. Figure 5.4 (b) shows the effective currents (I_{eff}) of the bulk FinFETs having different N_{stop} at the fixed I_{off} for LP application. I_{eff} is calculated using [28]. As the N_{stop} increases from 2 to $5{\cdot}10^{18}$ cm^{-3}, decrease of sub-fin leakage current improves

TABLE 5.3
Stress Values of p-Type Bulk FinFETs from 10- to Sub-5-nm Nodes (in GPa)

Stress parameters	10 nm	7 nm		Sub 5 nm	
		W5	W6	W4	W5
S_{XX}	0.749	0.906	0.876	0.984	0.957
S_{YY}	0.295	0.296	0.278	0.277	0.255
S_{ZZ}	−1.340	−1.549	−1.490	−1.815	−1.727

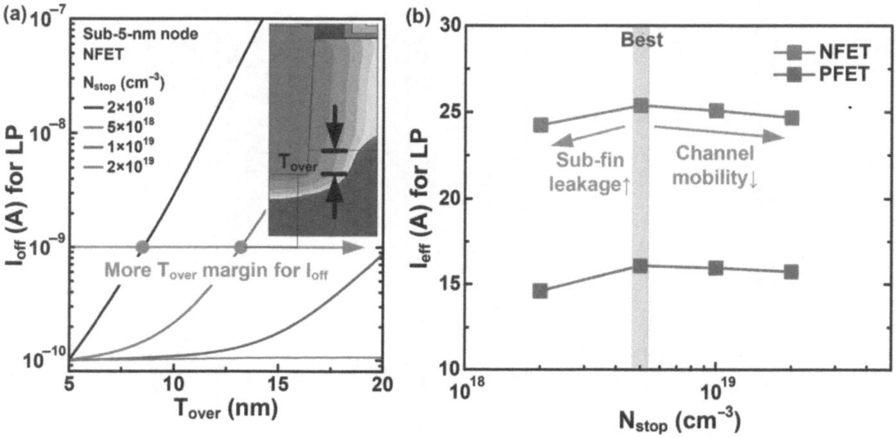

FIGURE 5.4 (a) OFF Currents (I_{off}) of Sub-5-nm Node FinFETs Having Different Over-etching Depth (T_{over}) and N_{stop} and (b) Effective Currents (I_{eff}) as a Function of N_{stop}

the I_{eff}. But as the N_{stop} increases over $5{\cdot}10^{18}$ cm^{-3}, more PTS dopants penetrate into the channels, decrease the carrier mobility, and thus I_{eff}. Thus, there is an optimal value of N_{stop} at $5{\cdot}10^{18}$ cm^{-3} to maximize the I_{eff} in sub-5-nm node.

Another important point to be considered is the variation of fin height (H_{fin}). Figure 5.5 shows the ON currents (I_{on}), gate capacitances (C_{gg}), and RC delay of sub-5-nm node bulk FinFETs. Although longer H_{fin} increases the effective width ($W_{eff} = 2H_{fin} + W_{fin}$) and thus current drivability under the same active area, larger C_{gg} compensates the DC performance improvement and thus almost same RC delay is obtained irrespective of H_{fin}. This phenomenon can be different depending on how abrupt S/D doping profile is. But under the calibration to Intel's 10-nm node FinFETs, RC delay is not improved in spite of longer H_{fin} (or larger fin aspect ratio). Thus, H_{fin} is fixed at 46 nm for all different technology nodes in this work.

Figure 5.6 (a) shows the DC performance of bulk FinFETs in three different nodes. I_{off} is fixed at 0.1 nA, according to the IEEE International Roadmap for Devices and Systems (IRDS) [29]. Operation voltage (V_{dd}) is fixed at 0.7 V. As the device is scaled

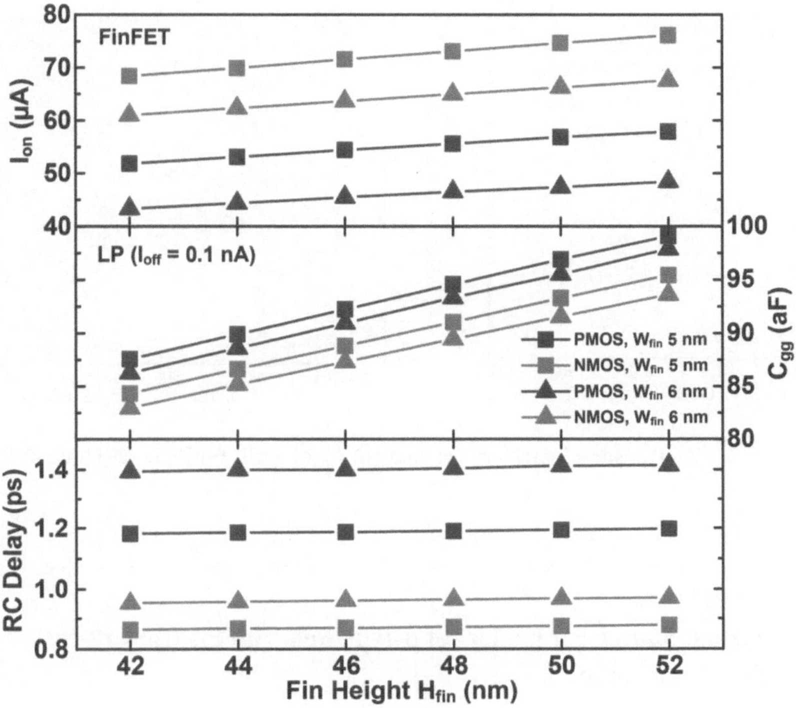

FIGURE 5.5 DC and AC Performance of Sub-5-nm Node Bulk FinFETs at Different W_{fin} and H_{fin}

down, SS increases in spite of the W_{fin} scaling due to larger penetration of S/D dopants into the channels (Figure 5.3). To enable high gate-to-channel controllability, much thinner W_{fin} is necessary in 7-nm and sub-5-nm nodes. Thinner W_{fin} can increase the I_{eff} at the same nodes as shown in Figure 5.6 (b). Especially, sub-5-nm node p-type FinFETs with the W_{fin} of 4 nm can outperform those in the 10-nm and 7-nm nodes by decreasing SS and DIBL. On the other hand, n-type FinFETs have smaller I_{eff} as the devices are scaled down in spite of smaller SS and DIBL.

To analyze the DC performance thoroughly, physical parameters are extracted using virtual-source (VS) model [30] as shown in Table 5.4 and Table 5.5. VS model can describe the nonplanar FETs like FinFETs and gate-all-around (GAA) FETs clearly [20, 31–33]. Inversion capacitance (C_{inv}) is extracted using the slope of inversion charge vs. gate voltage (V_{gs}) curve in the strong inversion region [34]. Here, α and β are the fitting parameters. Threshold voltages in the linear region ($V_{th,lin}$) are extracted using constant current method at $10^{-7}\cdot W_{eff}/L_g$ [35]. Larger C_{inv} with device scaling arises from S/D dopant penetration (Figure 5.3). As the devices are scaled down, both saturation velocity (v_{x0}) and carrier mobility (μ) decrease whereas parasitic resistivity ($W_{eff}*R_{sd}$) increases, thus degrading DC performance. But p-type FinFETs can compensate the DC performance degradation by enhancing the gate electrostatics through much small SS and DIBL compared to those in 10-nm node.

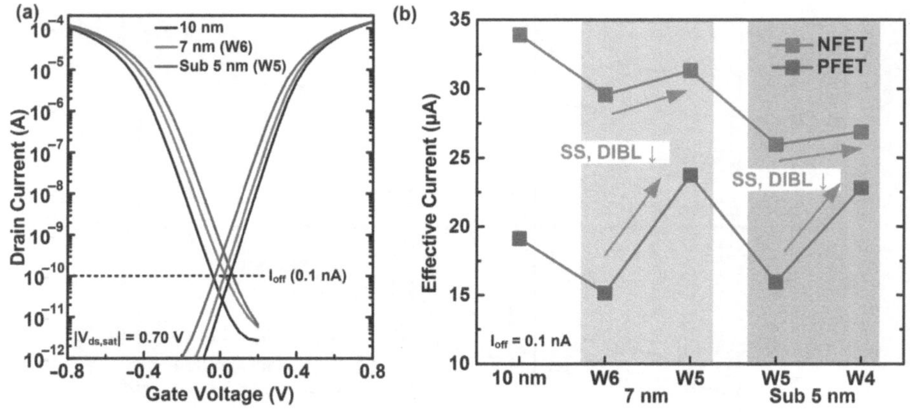

FIGURE 5.6 (a) Transfer Characteristics and (b) I_{eff} of Bulk FinFETs at Three Different Nodes

TABLE 5.4
Virtual-Source Parameters of p-Type Bulk FinFETs from 10- to Sub-5-nm Nodes

Virtual-source parameters	10 nm	7 nm		Sub 5 nm	
		W6	W5	W5	W4
C_{inv} (μF/cm²)	3.09	3.37	3.34	3.72	3.72
DIBL (mV/V)	73.5	84.3	50.8	83.0	51.5
SS (mV/dec)	74.8	76.6	68.8	75.1	68.9
W_{eff}*R_{sd} (Ω.μm)	223.3	248.5	267.7	248.4	259.8
v_{x0} (10^7 cm/s)	0.768	0.668	0.640	0.526	0.518
μ (cm²/(V.s))	87.2	60.0	60.2	33.8	37.5
$V_{th,lin}$ (V)	0.552	0.511	0.556	0.466	0.540
A	3.2	3.2	2.9	3.5	2.8
B	2.1	2.3	2.5	2.4	2.5

N-type FinFETs also decrease SS and DIBL by thinning W_{fin} in the 7-nm and sub-5-nm nodes, but those values are similar to 10-nm node while v_{x0} and μ decrease; thus the I_{eff} values are degraded as the devices are scaled down.

Figure 5.7 shows the C_{gg} and parasitic capacitances (C_{para}) of bulk FinFETs. C_{para} components are described in the schematic diagram. C_{para} consists of overlap capacitances (C_{ov}) within the active channel regions, outer-fringing capacitances (C_{of}) between S/D extension, S/D epi, and metal gate, and contact capacitances (C_{co}) between metal gate and S/D M0. The decomposition of C_{para} components has been done, showing that C_{of} and C_{co} are the dominant ones affecting the C_{para} [24]. P-type FinFETs have larger portion of C_{para} out of C_{gg} compared to n-type FinFETs because

TABLE 5.5
Virtual-Source Parameters of n-Type Bulk FinFETs from 10- to Sub-5-nm Nodes

Virtual-source parameters	10 nm	7 nm		Sub-5 nm	
		W6	W5	W5	W4
C_{inv} (μF/cm²)	3.03	3.30	3.31	3.66	3.67
DIBL (mV/V)	42.6	51.3	39.1	60.3	43.0
SS (mV/dec)	66.8	68.5	65.8	70.1	66.6
$W_{eff}*R_{sd}$ (Ω.μm)	121.3	147.0	164.2	179.7	191.2
v_{x0} (10^7 cm/s)	0.759	0.703	0.643	0.607	0.525
μ (cm²/(V.s))	109.1	80.7	71.9	51.9	42.5
$V_{th,lin}$ (V)	0.584	0.570	0.591	0.546	0.600
A	2.5	2.5	2.5	2.8	2.9
B	2.6	2.4	2.7	2.3	2.6

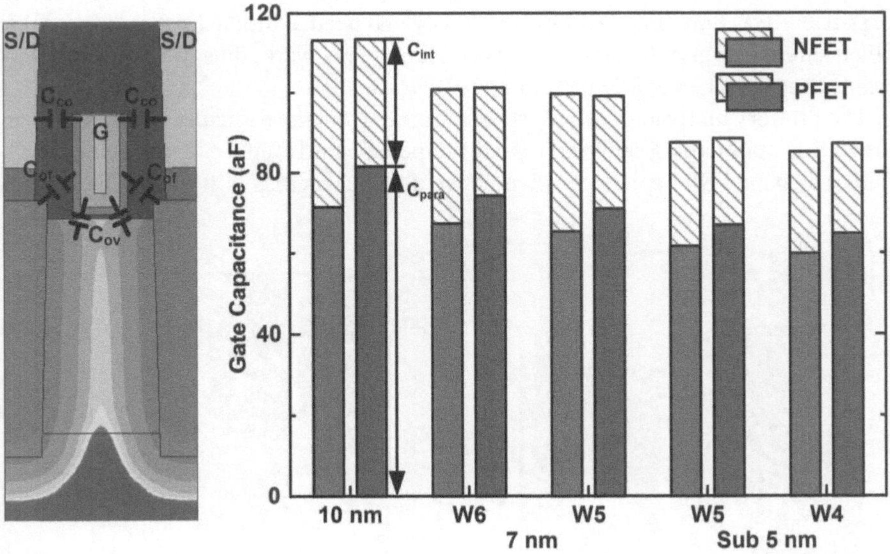

FIGURE 5.7 Schematic Diagram Describing Parasitic Capacitance Components (left) and Gate and Parasitic Capacitances (C_{para}) of Bulk FinFETs at Three Different Nodes (right)

p-type FinFETs have larger N_{sd} and thus larger S/D dopant penetration into the S/D extension and active channel regions. As the devices are scaled down, both L_g and L_{sp} decrease correspondingly and reduce intrinsic capacitances (C_{int}) and C_{para}. Shorter W_{fin} and thinner FP also reduce the C_{gg} due to smaller W_{eff} and shorter S/D M0 width, respectively.

5.3.2 New Process Scheme: Source/Drain Patterning to Enable Further Scaling

Conventional device scaling is feasible down to sub-5-nm node by shrinking CPP and FP along with L_g, L_{sp}, L_{sd}, and W_{fin} to maintain the gate-to-channel controllability. However, L_g scaling without optimizing S/D doping profile is limited by the short-channel effects (SCEs). Thinner W_{fin} below 4 nm is needed to decrease SS and DIBL, but too-high aspect ratio of fin channel can induce the fin breakdown [36]. Even sub-5-nm-node FinFETs have high aspect ratio of 11.5 for the W_{fin} and H_{fin} of 4 and 46 nm, respectively. GAA nanosheet FET is one of the promising candidates by wrapping around the channels for better controllability [22], but there is still a bottleneck to scale down CPP. Thus, a novel process scheme to scale down the device size without decreasing L_g is essential.

Meanwhile, DTCO enables further device scaling by changing FEOL as well as MOL concurrently. Buried power rail (BPR) has the ground and V_{dd} power delivery networks below the devices, providing larger metal-line margins to shrink the cell height (CH) [37]. Along with BPR, complementary FETs (CFETs) stack two different types of devices vertically to reduce the CH greatly [38]. Fork-sheet FETs pattern wide nanosheet channels into two for n- and p-type devices each [39]. But CFETs and fork-sheet FETs have high process complexity and need additional patterning masks, which surely increase the production cost. Here in this work, thus, we propose a new simple process scheme: S/D patterning (SDP) [20, 40].

The primary motivation of SDP scheme comes from the difficulty in scaling down due to S/D epi merging between p- and n-type devices [Figure 5.8 (a)]. Under SEG, diamond-shaped S/D epi larger than active channel regions is formed. Due to this

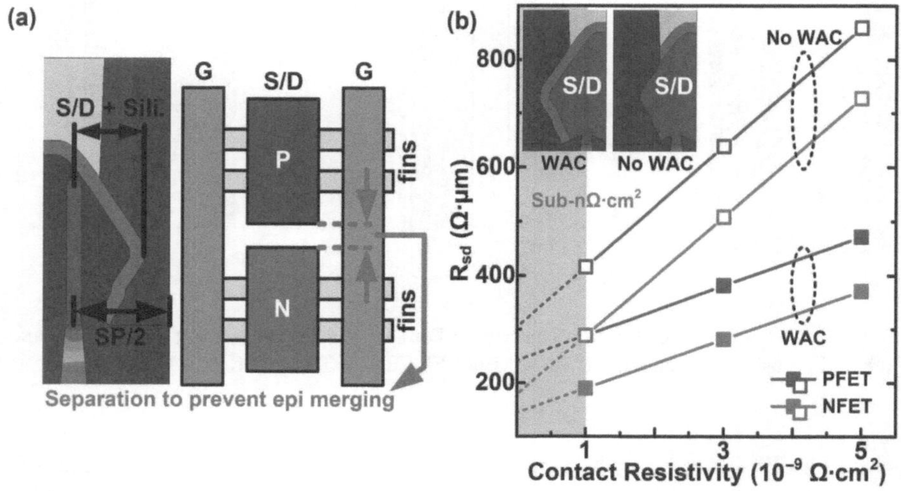

FIGURE 5.8 (a) Layout Design to Prevent Source/Drain epi Merging between p/n-type FETs and (b) Parasitic Resistances (R_{sd}) of Bulk FinFETs with and without Wrap-Around Contact (WAC)

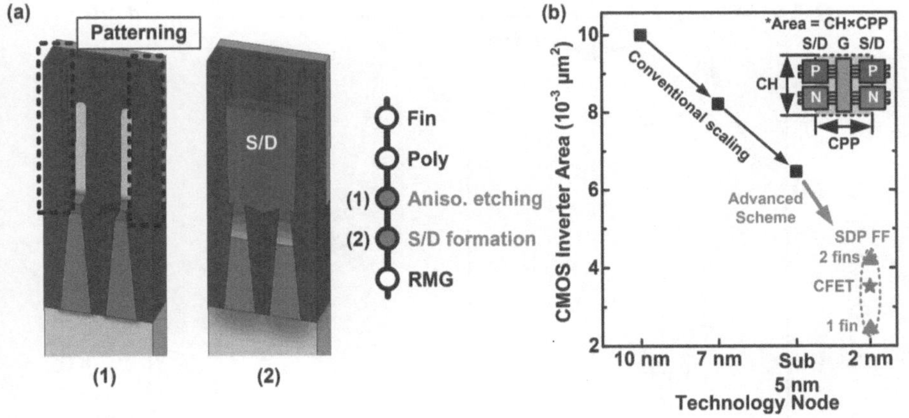

FIGURE 5.9 (a) Source/Drain Patterning (SDP) Scheme and (b) Device Scaling Projections toward 2 nm Node

circumstance, SP does not scale down from 10-nm node as shown in Table 5.1. Even with WAC scheme, additional silicide thickness is required to prevent epi merging. It is clarified that WAC effectively decreases the R_{sd} for all the contact resistivity values [Figure 5.8 (b)]. So, WAC scheme cannot be ruled out for the CH scaling.

Figure 5.9 (a) shows the additional process scheme for SDP. Under the anisotropic etching, spacer regions are patterned to prevent the lateral overgrowth of S/D epi. S/D epi is then formed by rectangular shape, so the S/D epi does not extend beyond the active channel regions. Through the SDP scheme, CMOS inverter area is scaled down successfully without changing the geometry of fin channels [Figure 5.9 (b)]. CMOS inverter area is calculated as the multiplication of CH and CPP. As the 1-fin channel is adopted instead of 2-fin, SDP FinFETs extend smaller CMOS inverter area than CFETs.

Figure 5.10 summarizes the RC delay of conventional and SDP FinFETs. RC delay is calculated as $C_{gg}V_{dd}/(2I_{eff})$. 7-nm-node FinFETs with the W_{fin} of 6 nm and sub-5-nm-node FinFETs with the W_{fin} of 5 nm have much smaller I_{eff} due to the degraded SCEs but similar C_{gg}, thus RC delay values increase. But thinner W_{fin} to 5 nm for 7-nm node and 4 nm for sub-5-nm node can achieve smaller RC delay, possible for conventional device scaling. SDP scheme shrinks the S/D epi, which decreases the C_{of}, C_{para}, and thus C_{gg} without losing S/D-induced channel stress and DC performance [20]. As the SDP scheme is adopted under the same device size as sub-5-nm node, both p- and n-type FinFETs are improved down to near 1 ps of RC delay along with CH scaling, promising for 2 nm node and beyond.

5.4 CONCLUSION

Scalability of conventional bulk FinFETs has been investigated using fully calibrated 3-D TCAD. Both p- and n-type bulk FinFETs possibly improve RC delay along with the device scaling by maintaining good gate-to-channel controllability through

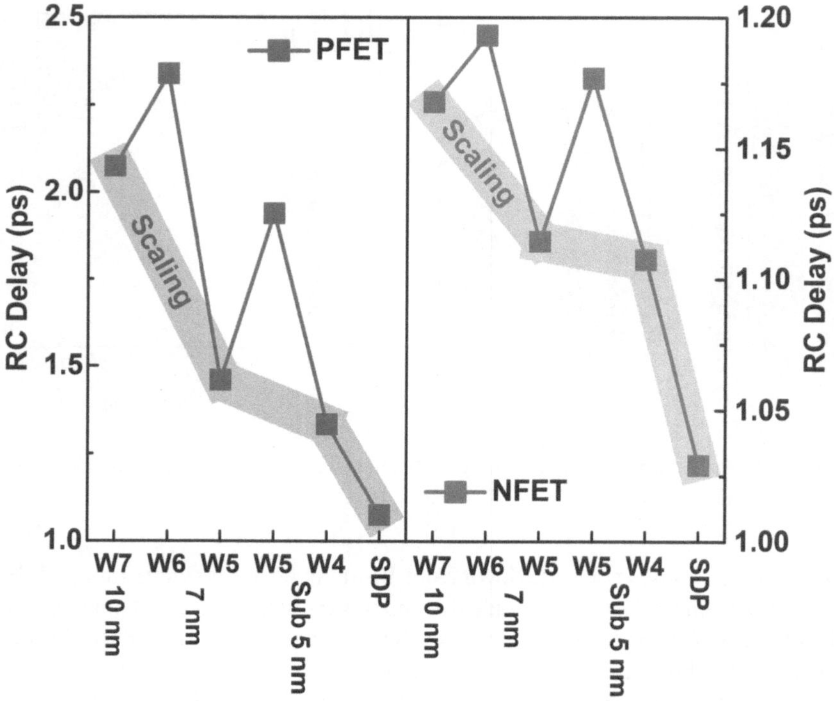

FIGURE 5.10 RC Delay of Conventional Bulk and SDP FinFETs

thinner W_{fin} and larger N_{stop} even though their performance parameters such as v_{x0} and μ decrease. For further device scaling in the CH direction, SDP scheme is introduced to shrink S/D epi. This scheme decreases the C_{gg} but keeps the I_{eff} almost the same, thus improving RC delay under feasible and CMOS compatible process.

ACKNOWLEDGMENT

This work was supported in part by the Brain Korea (BK) 21 FOUR program, in part by the Ministry of Trade, Industry & Energy under Grant 10080617, in part by the Korea Semiconductor Research Consortium Support Program for the Development of the Future Semiconductor Device, in part by the National Research Foundation of Korea Grant funded by the Ministry of Science, ICT under Grant NRF-2020M3F3A2A02082436 and NRF-2020R1A4A4079777, POSTECH-Samsung Electronics Industry–Academia Cooperative Research Center, and in part by the IC Design Education Center.

REFERENCES

[1] Auth, C. et al., "A 22nm high performance and low-power CMOS technology featuring fully-depleted tri-gate transistors, self-aligned contacts and high density MIM capacitors," in *VLSI Technology*, 2012, pp. 131–132.

[2] Natarajan, S. et al., "A 14nm logic technology featuring 2nd generation FinFET transistors, air-gapped interconnects, self-aligned double patterning and a 0.0588μm^2 SRAM cell size," in *IEEE IEDM*, 2014, pp. 71–73.
[3] Auth, C. et al., "A 10nm high performance and low-power CMOS technology featuring 3rd generation FinFET transistors, self-aligned quad patterning, contact over active gate and cobalt local interconnects," in *IEEE IEDM*, 2017, pp. 673–676.
[4] Wang, X. et al., "Design-technology co-optimization of standard cell libraries on Intel 10nm process," in *IEEE IEDM*, 2018, pp. 636–639.
[5] Kye, K. et al., "Smart scaling technology for advanced FinFET node," in *VLSI Technology*, 2018, pp. 149–150.
[6] Xie, R. et al., "Self-aligned gate contact (SAGC) for CMOS technology scaling beyond 7nm," in *VLSI Technology*, 2019, pp. 148–149.
[7] Wu, S.-Y. "Key technology enablers of innovations in the AI and 5G era," in *IEEE IEDM*, 2019, pp. 863–866.
[8] Yeap, G. et al., "5nm CMOS production technology platform featuring full-fledged EUV, and high mobility channel FinFETs with densest 0.021μm^2 SRAM cells for mobile SoC and high performance computing applications," in *IEEE IEDM*, 2019, pp. 879–882.
[9] Version O-2018.06, Synopsys, Mountain View, CA, USA, 2018.
[10] Klassen, D. B. M. et al., "Unified apparent bandgap narrowing in n- and p-type silicon," *Solid-State Electronics*, vol. 35, no. 2, pp. 125–129, Feb. 1992.
[11] Lombardi, C. et al., "A physically based mobility model for numerical simulation of nonplanar devices," *IEEE Transactions on Computer-Aided Design of Integrated Circuits and Systems*, vol. 7, no. 11, pp. 1164–1171, Nov. 1988.
[12] Mujtaba, S. A. "Advanced mobility models for design and simulation of deep submicrometer MOSFETs," PhD. Thesis Stanford University, CA, USA (1995).
[13] Reggiani, S. et al., "Low-field electron mobility model for ultrathin-body SOI and double-gate MOSFETs with extremely small silicon thicknesses," *IEEE Transactions on Electron Devices*, vol. 54, no. 9, pp. 2204–2212, Sept. 2007.
[14] Shur, M. S., "Low ballistic mobility in submicron HEMTs," *IEEE Electron Device Letters*, vol. 23, no. 9, pp. 511–513, Sept. 2002.
[15] Fossum, J. G. "Computer-aided numerical analysis of silicon solar cells," *Solid-State Electronics*, vol. 19, no. 4, pp. 269–277, Apr. 1976.
[16] Huldt, L. et al., "The temperature dependence of band-to-band Auger recombination in silicon," *Applied Physics Letters*, vol. 35, no. 10, pp. 776–777, Nov. 1979.
[17] Hurkx, G. A. M. et al., "A new recombination model for device simulation including tunnelling," *IEEE Transactions on Electron Devices*, vol. 39, no. 2, pp. 331–338, Feb. 1992.
[18] Bardeen, J. et al., "Deformation potentials and mobilities in non-polar crystals," *Physics Review*, vol. 80, no. 1, pp. 72–80, Oct. 1950.
[19] Yoon, J.-S. et al., "Bottom oxide bulk FinFETs without punch-through-stopper for extending toward 5-nm node," *IEEE Access*, vol. 7, pp. 75762–75767, June 2019.
[20] Yoon, J.-S. et al., "Source/Drain patterning FinFETs as solution for physical area scaling toward 5-nm node," *IEEE Access*, vol. 7, pp. 172290–172295, Nov. 2019.
[21] Wu, H. et al., "Parasitic resistance reduction strategies for advanced CMOS FinFETs beyond 7nm," in *IEEE IEDM*, 2018, pp. 35.4.1–35.4.4.
[22] Loubet, N. et al., "Stacked nanosheet gate-all-around transistor to enable scaling beyond FinFET," in *VLSI Technology*, 2017, pp. 230–231.
[23] Nainani, A. et al., "Is strain engineering scalable in FinFET era? Teaching the old dog some new tricks," in *IEEE IEDM*, 2012, pp. 427–430.
[24] Yoon, J.-S. et al., "Systematic DC/AC performance benchmarking of sub-7-nm node FinFETs and nanosheet FETs," *IEEE Journal of the Electron Devices Society*, vol. 6, pp. 942–947, Aug. 2018.

[25] Yoon, J.-S. et al., "Optimization of nanosheet number and width of multi-stacked nanosheet FETs for sub-7-nm node system on chip applications," *Japanese Journal of Applied Physics*, vol. 58, no. SBBA12, pp. 1–5, Mar. 2019.

[26] Jeong, J. et al., "Comprehensive analysis of source and drain recess depth variations on silicon nanosheet FETs for sub 5-nm node SoC application," *IEEE Access*, vol. 8, pp. 35873–35881, Feb. 2020.

[27] Yoon, J.-S. et al., "Punch-through-stopper free nanosheet FETs with crescent inner-spacer and isolated source/drain," *IEEE Access*, vol. 7, pp. 38593–38596, Apr. 2019.

[28] Na, M. H. et al., "The effective drive current in CMOS inverters," in *IEEE IEDM*, 2002, pp. 121–124.

[29] *IEEE International Roadmap for Devices and Systems*, 2020.

[30] Khakifirooz, A. et al., "A simple semiempirical short-channel MOSFET current-voltage model continuous across all regions of operation and employing only physical parameters," *IEEE Transactions on Electron Devices*, vol. 56, no. 8, pp. 1674–1680, Aug. 209.

[31] Majumdar, A. et al., "Room-temperature carrier transport in high-performance short-channel silicon nanowire MOSFETs," in *IEEE IEDM*, 2012, pp. 179–182.

[32] Lin, M.-H. et al., "Tackling fundamental challenges of carrier transport and device variability in advanced Si nFinFETs for 7nm node and beyond," in *IEEE IEDM*, 2018, pp. 644–647.

[33] Lee, S. et al., "Observation of mobility and velocity behaviors in ultra-scaled L_G = 15 nm silicon nanowire field-effect transistors with different channel diameters," *Solid-State Electronics*, vol. 164, no. 107740, pp. 1–5, Feb. 2020.

[34] Yoon, J.-S. et al., "Extraction of source/drain resistivity parameters optimized for double-gate FinFETs," *Japanese Journal of Applied Physics*, vol. 54, no. 04DC06, pp. 1–4, Jan. 2015.

[35] Yoon, J.-S. et al., "Statistical variability study of random dopant fluctuation on gate-all-around inversion-mode silicon nanowire field-effect transistors," *Applied Physics Letters*, vol. 106, no. 103507, pp. 1–5, Mar. 2015.

[36] Kim, D.-W. "CMOS transistor architecture and material options for beyond 5nm node," VLSI Short Course, 2018, pp. 1–57.

[37] Prasad, D. et al., "Buried power rails and back-side power grids: Arm CPU power delivery network design beyond 5nm," in *IEEE IEDM*, 2019, pp. 446–449.

[38] Ryckaert, J. et al., "The complementary FET (CFET) for CMOS scaling beyond N3," in *VLSI Technology*, 2018, pp. 141–142.

[39] Weckx, P. et al., "Novel forksheet device architecture as ultimate logic scaling device towards 2nm," in *IEEE IEDM*, 2019, pp. 871–874.

[40] Lee, J. et al., "TCAD-based flexible fin pitch design for 3-nm node 6T-SRAM using practical source/drain patterning scheme," *IEEE Transactions on Electron Devices*, 2021, Accepted.

6 Germanium or SiGe FinFETs for Enhanced Performance in Low Power Applications

Nilesh Kumar Jaiswal and V. N. Ramakrishnan

CONTENTS

6.1 INTRODUCTION

Over the past few years, the Internet of Things (IoT) has become an increasingly popular subject of discussion. The IoT is based on the idea of connecting every computer to the Internet. This includes everything from smartphones to kitchen appliances and almost anything else we can think of. According to the report published, the estimated 50 billion connected devices will exist by 2030 [1]. Figure 6.1 shows a progress of personal devices such as smartphones, tablets, personal computers, smart watches, connected TVs, and so on. However, because of the world's population, this growth is limited. All of these connected devices in areas like home appliances, medical clinics, and transportation are driving the real development. Essentially, there will be limitless gadgets. All of these gadgets should use less energy and have a longer battery life, but they should not lose efficiency.

Since the breakthrough of semiconductors and throughout the history of integrated circuit and device design, Moore's law has been driving the semiconductor industry

DOI: 10.1201/9781003200987-6

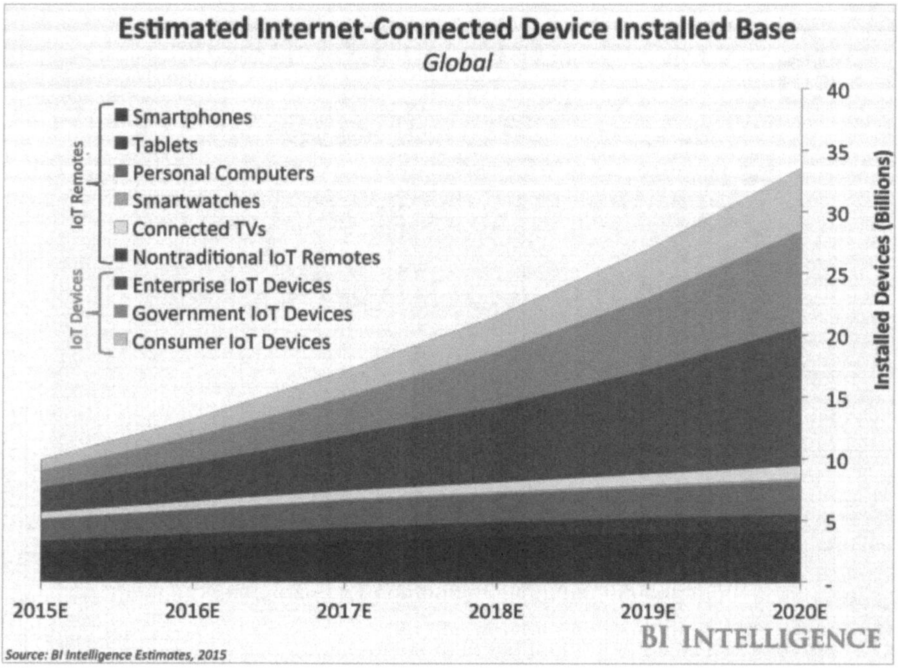

FIGURE 6.1 Estimated IoT-Based Device Connections

where the number of transistors on a piece of Si will double every 18–24 months at a constant cost [2], as shown in Figure 6.2. With every new technology node, tremendous advantages can be expected: 50% area reduction, 29% cost reduction, 20% power reduction, and 25% performance increase. As Moore's law is an economics law in nature, the transistor scaling can of course only be maintained if these benefits outweigh the manufacturing efforts that come along with the increased complexity of scaling. Transistors have been gradually improving from the first Ge transistor to planar Si MOSFET, and afterward to stressed SiGe source/drain (S/D). Planar MOSFETs have increased in efficiency and transistor density as they have been scaled up. However, more nanometer-scale planar transistor scaling is extremely difficult to achieve due to the severe increase in the leakage current (I_{off}). In fact, as the channel length of planar Si MOSFETs is decreased, the drain bias begins to affect the electrostatics in the channel, and the gate loses control of the channel, resulting in increased leakage current between the drain and source. Thinner and higher-gate oxides can mitigate this problem by raising gate-channel capacitance; however, gate oxide thickness is limited by increased gate leakage and the gate-induced drain leakage impact [3–5]. However, future CMOS technologies in the nanoscale regime will face numerous technical challenges, the most significant of which are SCEs, which appear to degrade analog figures of merit [6–7]. The FinFET is a transistor design, which attempts to overcome the addressing constraints like low gate leakage, low power consumption, high speed, and small area when compared to planar Si MOSFET at a lower cost.

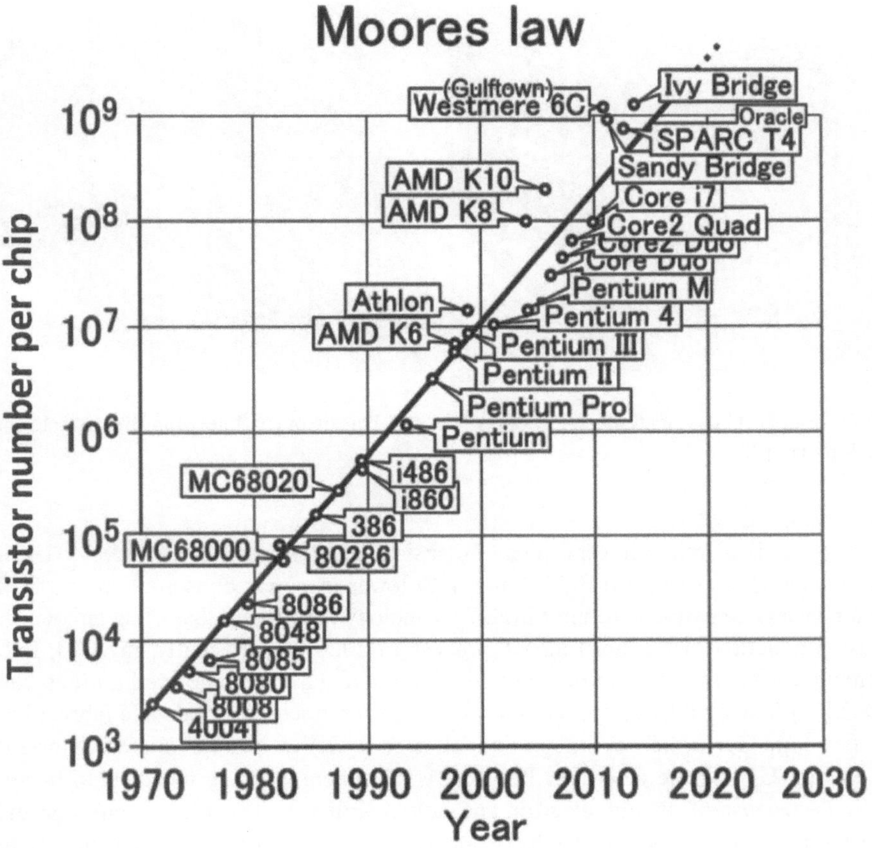

FIGURE 6.2 Moore's Law for Semiconductors [8]

Compared to a traditional planar Si MOSFET, the FinFET is a 3-D transistor with three gates that has been found to be the most promising alternative due to its improved performance and ease of fabrication, which is compatible with and easily incorporated into standard CMOS fabrication method [9–10]. The very first (25 nm node) Si channel-based FinFET transistor operating on a mere 0.7 volts was demonstrated in December 2002 by Taiwan Semiconductor Manufacturing Co. (TSMC). In 2001, Intel produced first commercial FinFET technology (22-nm node) where the gate surrounds the channel on three sides [11] as shown in Figure 6.3. Intel used strained Si channels in its second FinFET generation (14-nm node) [12]. Samsung announced the first-time production of 14-nm and 10-nm FinFET in 2015 and October 2016, respectively, for mobile applications. A 14-nm shows 40% reduction in power consumption and 27% improvement in efficiency along with 30% lower area consumption. After that, TSMC, Samsung, and Global Foundries demonstrated 7-nm FinFET using extreme ultraviolet (EUV) lithography instead of multiple-patterning lithography in 2017 [12]. This developed in a 25% reduction in the number of mask

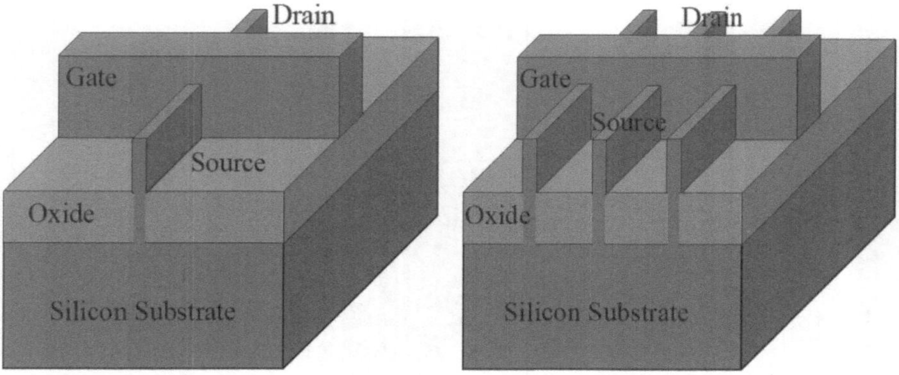

FIGURE 6.3 Cross-Sectional View of (a) Tri-Gate Transistor (by Intel) and (b) Multiple-Fin Technology [9]

steps needed, as well as lower critical dimension variability and higher fidelity. Using these techniques, 7-nm FinFET shows 45% lower in power consumption and 20% faster speed compared to 10-nm FinFET technology [13]. Again, Samsung announced mass production of a 5-nm FinFET process in 2020 [14]. Recently, in 2020, Intel announced that the plans to fabricate the 7-nm node have been delayed until at least 2022 [15]. Thus, FinFETs are the basis for modern semiconductor device fabrication.

For high performance, integration of high-mobility channel material instead of Si FinFET is quite desirable. Pure Ge is an appealing material for both n- and p-FETs because of its high electron and hole mobility, and several research groups have shown that Ge substrates have superior transport properties. Because of its highest registered hole mobility and CMOS compatibility, Ge FinFET is an exciting candidate for continuously increasing the I_{on}. However, as the melting temperature of Ge is much lower than Si, most Ge devices should be made at significantly lower temperature than Si. This may result in some penalty on performance and reliability of Si devices that require higher processing temperatures, especially for mobile and SoC applications. Ge has been pointed out as a best option for the pFinFET due to its high hole mobility and capability of providing higher velocity than uniaxial-strained Si under a similar strain level. Figure 6.4 shows I_{on} and I_{off} achieved for Ge p-type FinFET for different fin widths [16]. A 4-nm of fin width is required to meet 100 nA/μm I_{off} specifications. On the other hand, Si FinFET needs 5 nm of fin width based on electrostatic control requirements. Figure 6.4 also shows enhanced I_{on} achieved by Ge FinFET over Si FinFET, which highlights the advantage of Ge as a high-mobility channel material. However, the reduced density-of-state (DOS) of these high-mobility materials than Si would cause the smaller amount of inversion carriers at the same level of gate voltage overdrive and gate oxide capacitance. The high mobility-based channel not only provides higher carrier thermal velocity from material benefit but also provides a unique advantage of realizing global uniaxial strain engineering [17]. The strain configuration of epitaxial high mobility channel pseudomorphically grown in Si substrate can be converted from biaxial to uniaxial,

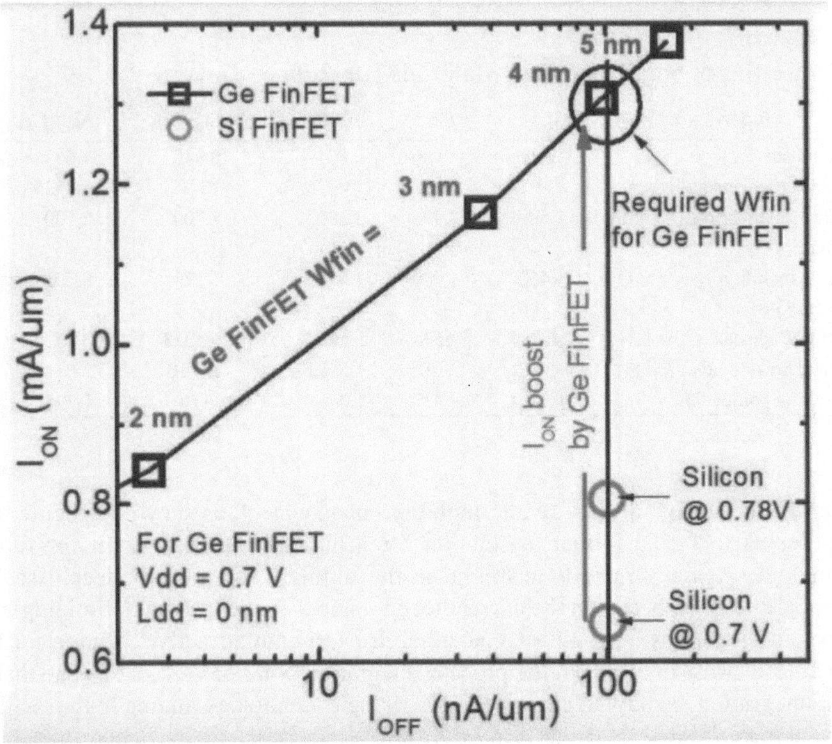

FIGURE 6.4 I_{on} and I_{off} Analysis of FinFET Devices [16]

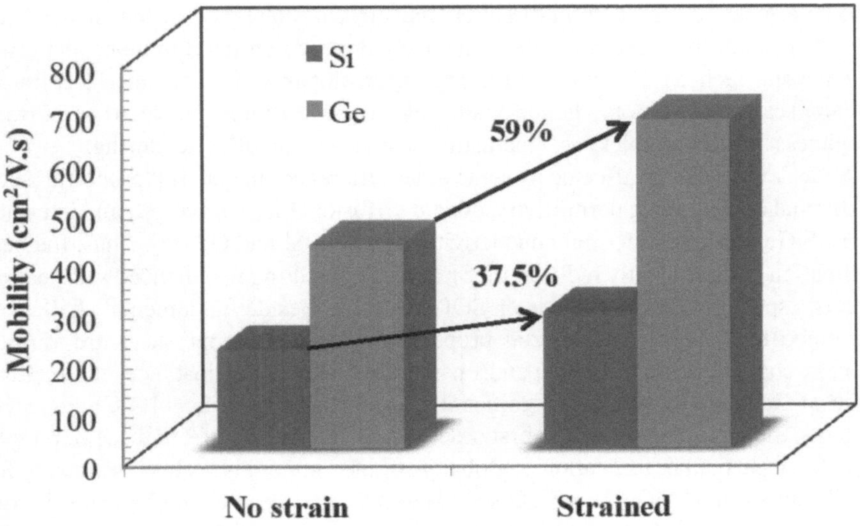

FIGURE 6.5 Strain-Free and Strained Devices

TABLE 6.1
Properties of Semiconductor Materials [19–20]

Properties	Si	Ge	$Si_{0.25}Ge_{0.75}$	$Si_{0.5}Ge_{0.5}$	$Si_{0.75}Ge_{0.25}$
Bandgap (eV)	1.12	0.66	0.804	0.945	1.05
Dielectric constant	11.7	16.2	14.975	13.95	12.925
Electron mobility (cm^2/V-s)	1,500	3,900	3,300	7,700	2,100
Hole mobility (cm^2/V-s)	450	1,900	1,537.5	1,175	812.5
Lattice constant (A°)	5.4310	5.6575	5.5960	5.5373	5.4310
Breakdown field (V/cm)	~3×10^5	~10^5	1.5×10^5	2×10^5	2.5×10^5
Melting point (°C)	1,415	937	1,056.5	1176	1,295.5

which further helps in velocity and mobility enhancement, as shown in Figure 6.5. This seems to be an attractive solution for achieving uniaxial strain for future technology. Again, strained-Ge might be the ultimate solution for deeply scaled nanoscale devices, based on the current technological challenges on optimizing gate stacks and junctions, etc., on a Ge channel. It can be an attractive solution for the near future. Most of all, from the process integration point of view, Ge is one that is most integration friendly to current Si CMOS logic technology among high-mobility channel materials.

Another option is to replace the Si with a high-mobility material such as SiGe. Mobility is inversely proportional to the scattering rate and the conductivity of effective mass. The effective mass of SiGe is a strong function of Ge fraction, strain state (compression or tension), and strain type (uniaxial, biaxial, or combined). It can also be affected by the current direction and surface orientation of the channel. On the other hand, the scattering rate collectively depends on the dominant scattering mechanism such as Coulomb scattering (from dopants or gate interface traps), phonon scattering, remote phonon scattering (in case of high-κ dielectrics), surface roughness scattering, thickness fluctuation scattering, and alloy scattering.

SiGe devices can be affected by some other differences in material properties such as thermal conductivity, permittivity, dopant diffusion, thermal budget, and interface traps. SiGe has lower thermal conductivity than both Si and Ge [18]. Thus, the heat dissipation is significantly reduced for SiGe FETs, resulting in enhanced self-heating effects, especially for those made on SOI substrates. Beside fundamental challenges attributed to the physical material properties of SiGe systems, there are various process challenges that also depend on the Ge content. For instance, the melting point of Ge is 938°C, which is significantly lower than that of Si (1414°C). For SiGe systems, the melting point at the first order can be interpolated from those of pure Si and Ge. In particular, utilization of globally strained materials such as strained Si for nFET and strained SiGe for pFETs sounds to be a near-term technological solution, while replacing the channel with other candidates such as III-V (SiGe) is among the long-term solutions for advanced high-performance technologies.

6.2 Ge P-CHANNEL FinFETs

Germanium is a commonly known semiconductor material. Indeed, the bandgap of Ge is about half of the Si bandgap: 0.66 eV and allows for low-voltage operation [18]. This is clearly reflected in the reduction of threshold voltage. The electron and hole mobility of germanium are greater (2.6 times and 4 times, respectively) than that of silicon, which is the most commonly used semiconductor material in the formation of integrated circuits, as shown in Table 6.1. Hence, germanium is an excellent material for forming integrated circuits. An additional advantageous feature of germanium is that its hole and electron motilities have greater stress sensitivity than that of silicon. For example, Figure 6.6 (a) illustrates the hole mobility of germanium and silicon as a function of uni-axial compressive stresses. It is noted that with the increase in the compressive stress, the hole mobility of germanium increases at a faster rate than silicon, indicating that germanium-based PMOS devices have a greater potential to have high-drive currents than silicon-based PMOS devices. Similarly, Figure 6.6 (b) illustrates the electron mobility of germanium and silicon as functions of uni-axial tensile stresses. It is noted that with the increase in the tensile stress, the electron mobility of germanium increases at a faster rate than that of silicon, indicating that germanium-based NMOS devices have a greater potential to have high-drive currents than silicon-based NMOS devices.

As reported in [21], in contrast to Si [22], where going from a (001) surface orientation to (110) leads to a degraded electron mobility, a six times gain can be seen in Ge when going from (001) to (110). This is especially important in FinFET devices, where (110) is the dominant surface orientation. Yu et al. [23] reported Ge p-channel-based FET presents even higher effective hole mobility than the (100) to (110). Thus, combining the advantages of the FinFET structure and the high mobility of Ge may be an approach to future high-performance FETs. After its introduction by Intel at 22 nm technology node [24], the FinFET device architecture is expected to be introduced by most. Companies at the 14–16 nm technology node and thus integration of high-mobility channels will need to be compatible with the 3-D aspects and higher topography of FinFETs.

High-performance Ge devices have been fabricated on Ge bulk, as well as germanium-on-insulator (GeOI) substrates [25–34]. These substrates can be fabricated either by bonding of a Ge donor wafer to an oxidized Si handle wafer or by the Ge condensation method, where a SiGe layer grown on a standard silicon-on-insulator (SOI) wafer is oxidized, thereby turning the SiGe layer into a Ge-rich layer [35]. Ge possesses lower bandgap, which induces higher band-to-band tunneling and junction leakage. To reduce power consumption, GeOI substrate is an option to suppress OFF-state leakage current. Liu et al. reported [36] first time Ω-type gate Ge p-channel FinFET using Si_2H_6 for passivation and NiGe used for source/drain metallization. The GeOI substrate used for fabrication uses sub-400°C process modules. The device performance shows that high I_{on} of ~494 μA/cm, I_{on}/I_{off} ratio of $3x10^4$, peak saturation transconductance of 540 μS/μm. An alternative high-performance Ge FinFETs device has been fabricated on Ge bulk, as well as silicon-on-insulator (SOI) substrates. Chung et al. demonstrated excellent short-channel control with 40-nm Ge fins. For the first time, both n- and p-Ge FinFETs are shown with gate-last process [37].

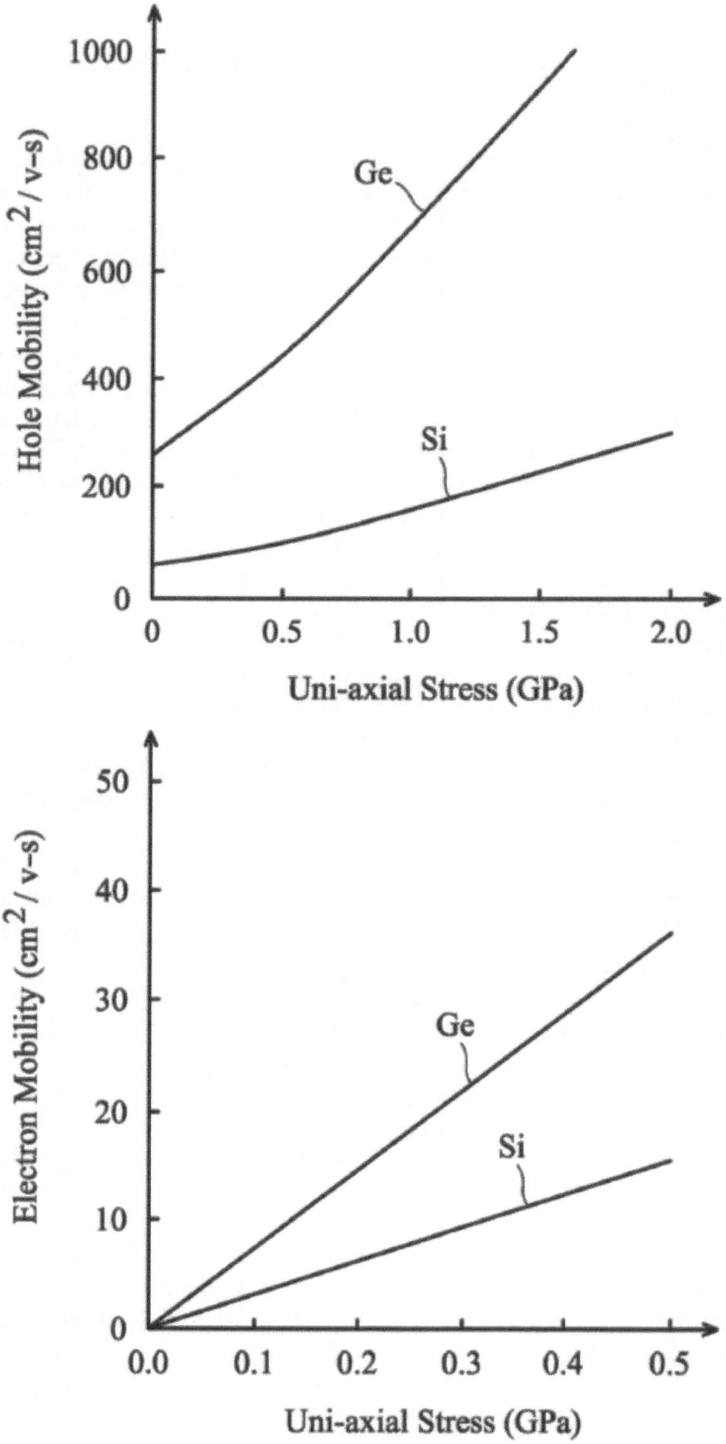

FIGURE 6.6 (a) Hole and (b) Electron Mobilities of Ge and Si as Function of Stress

Another high-performance Ge FinFET device has been fabricated on an Si bulk. Due to the excellent drive current and integration friendliness with the Si substrate, Ge pFETs are viable for future high-speed digital-logic applications [38–41]. Van Dal et al. [42–43] demonstrated Ge p-channel FinFET devices on an Si bulk using the aspect-ratio-trapping (ART) technique that used the process flow. Excellent subthreshold characteristics (long-channel subthreshold swing SS = 76 mV/dec at 0.5V), good SCE control, and high transconductance (1.2 mS/μm at 1V, 1.05 mS/μm at 0.5V) are achieved. The Ge FinFET presented in this work exhibits the highest g_m/SS at V_{dd}=1V reported for nonplanar unstrained Ge pFETs to date. Duriez et al. [44] demonstrated further improvement by implementing an ultra low capacitance equivalent thickness (CET = 8Å), low Dit gate stack using a replacement poly gate (RPG) process, scaling down the fin width to industry-relevant dimensions and adopting the <110> channel direction. The replacement gate high-k/metal gate p-channel Ge FinFETs integrated onto 300-mm Si wafers for which the best device shows record peak $g_{m,ext}$ = 2.7 mS/μm and I_{on} = 497 μA/μm at I_{off} = 100 nA/μm, all at V_{ds} = −0.5V. Chen et al. [45] fabricated body-tied Ge p-FinFETs directly on the Si substrate with a high-k/metal gate stack using the top-down method. The body-tied FinFET used a fin width of 40 nm and a mask channel length of 120 nm. A driving current of 22 μA/μm at V_g = −2V and I_{off} of 3 nA/μm at V_g = 2V, SS 228 mV/dec and drain-induced barrier lowering of 288 mV/V are demonstrated.

6.3 Ge N-Channel FinFETs

Ge n-type MOSFET has been used and faced several technical challenges. It is hard to implement a shallow S/D junction with low resistivity, because Ge has a low donor solubility and a high annealing diffusion rate, as well as a high density of gap-states, causing heavy Fermi-level pinning to the valence band edge. However, using techniques such as laser annealing and Sb co-doping with phosphorus, several studies have centered on introducing heavily doped n-type Ge with activation rates of donors greater than 10^{20} cm^{-3} [46–47]. Also, n^+ germanide formation has been documented as a method of lowering the resistivity in S/D of n-type Ge MOSFETs [48–49]. It produces, however, outcome in a high SBH of more than 0.3 eV, which results in low contact resistivity. Furthermore, since the S/D area in n-type Ge MOSFETs is aggressively scaled down, the consumption of Ge in the S/D area by germanide process can severely increase the contact resistance [50]. Another structure using MIS has been suggested as a promising contact scheme to lower the contact resistivity in n-type Ge MOSFET [50–52]. Using the same techniques, J-K. Kim et al. investigated Ge n-channel FinFET, and the study has shown its remarkable progress in suppressing the sensitivity of contact resistivity [53].

Many research works have been reported at stressed S/D SiC transistors using n-type channel [54–56]. Growing the carbon (C) mole fraction in the S/D epi-layer is one way to get high strain levels in the Si channel area for the drive current while also improving electron mobility in n-type transistors. Othman et al. provides insight into the use of stressor effects, namely, channel, stress-relaxed buffer (SRB), and S/D stressors by incorporating Ge in the layer, and their impact on the electrical characteristics of the 7-nm stress-engineered FinFET. While C is normally used as

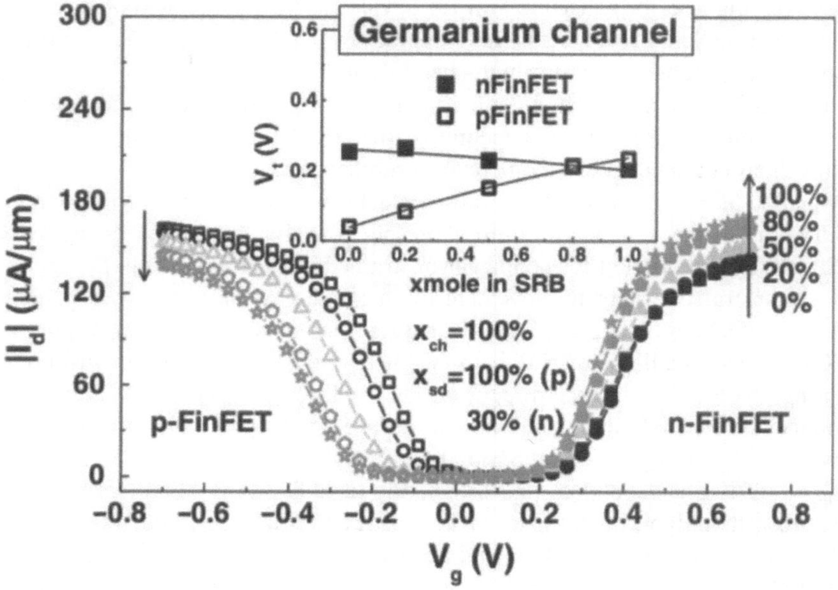

FIGURE 6.7 Transfer Characteristics for Ge-Based FinFET [57]

the dopant in the S/D region for n-FinFET, they used Ge doped n-type FinFET [57]. Figure 6.7 and its insets portray the transfer characteristics for Ge-based n-FinFET in S/D epitaxial are varied from 0% to 100%. The drain current is shown to increase back to ~50 μA/μm and the current is reduced back to ~10 μA/μm. The drain current fluctuation is mainly due to the highly doped S/D epitaxial region, which will contribute to the depraved performance to mobility degradation followed by the scattering effects [58–60].

Ge-based FinFET devices have emerged as non-disruptive performance boosters for future technology nodes due to the higher intrinsic mobility of Ge compared to Si. Though Ge p-type FinFET devices have received a lot of attention, the production of high-performance, reliable Ge n-type FinFET devices has lagged. Pre-gate stack process optimization has significantly improved efficiency and performance in Ge n-type FinFETs, with a 100% increase in PBTI and an improvement in the g_m with SS benchmark [61]. This happened due to an improved pre-cleaning and an optimized dummy gate oxide deposition and removal process, as part of a replacement metal-gate (RMG) process flow.

6.4 SELF-HEATING IN Ge FinFETs

Self-heating is the accumulation of heat that the device acquires in its channel when it is under specific operating conditions. Due to the increasingly small dimensions and the use of new materials with low thermal conductivity, the performance of advanced transistors is increasingly affected by self-heating. Transistors belonging to SOI technology are thermally isolated from the substrate through the buried oxide.

As a result, the dissipation of excess heat generated inside the device is less efficient, which results in a substantial increase in the temperature of the device, as highlighted by McDaid et al. [62] and Jomaah et al. [63]. Self-heating effect has become an important factor in device manufacture [64–65] and circuit design [66–67]. It can lead to serious device performance degradation and reliability problems [68–69].

Nowadays, self-heating is an important issue for smartphones, virtual reality handsets, and data centers, when the dimension of the device is scaled down. The formation of charged traps inside the gate oxide layer due to the high electric field causes many issues in the IC, including system performance degradation. As a result, when the variations exceed a certain limit, these issues weaken the reliability of the scaled devices and can ultimately lead to an IC failure. As we moved into the FinFET era, it is one of the most promising device structures for future scaling-down demands. In the FinFET, vertical fins are wrapped in an oxide layer and interconnecting between fins and bulk Si substrate with very narrow channel lengths depending on the design. Due to its complex geometry and lack of a heat release path, self-heating is a significant issue in FinFETs. Self-heating FinFETs cause a slew of issues, including a rise in the temperature of the metal interconnects, which increases the electro-migration effect [70]. Many different research groups have recently investigated this effect using theoretical calculations [71], experimental setups for observing the self-heating effect [72–75], and a brief discussion of the effect on the Si unit [76]. On the opposite, Ge is a promising candidate due to its high electron and hole mobility compared to an Si-based device, as we can see in Equation 6.1. However, Ge bulk thermal conductivity is 53 W/(m.K), three times lower than Si (148 W/(m.K)) [77], which means a weaker ability to dissipate heat out of hotspots. A more serious self-heating effect can be expected in Ge FinFETs than in Si FinFETs. Bury et al. demonstrated implementing of self-heating in p-based Ge FinFET devices using SiGe buffer. The simulations and measurements show a 115% increase in self-heating compared to Si FinFET [78].

$$R_{th} = \frac{\Delta T}{\Delta Q} \tag{6.1}$$

where T denotes the peak temperature and Q the input power of the device.

Yin et al. [79] evaluated self-heating effect in 14-nm Ge p-channel FinFETs on Si substrate using TCAD simulation. It shows on-state degradation by 9.7% when Fin pitch is increased from 47 nm to 14 nm. Because of the carriers releasing more energy to the lattice when distance is longer, it leads to an increase in lattice temperature and a larger current degradation considering the same input power. Moreover, FinFETs with a taller fin height also have a higher lattice temperature. While studying the temperature in the multi-fin FinFETs, the device will first increase temperature then saturate with the increasing fin numbers. At the end, thermal resistances in Ge p-channel single-fin FinFETs have exponential relationships with the S/D extension length (0.52 index) and the fin height (0.52 index).

Due to ever-increasing technology frequency targets, required drive current per footprint is increasing node to node. This results in increasing input power and, combined with thermal resistance increase, results in higher self-heating. The thermal resistance is defined as Equation 6.1. Jang et al. [76] fabricated FinFET to study the self-heating effect from 14 to 7-nm node and validated by simulations. Using Si-FinFET, thermal resistance increases over the decrease in device dimensions. Heat confinement is expected to increase by 20% from 14 to 10-nm node in Si-FinFET. For the 7-nm node, thermal resistance is 57% higher than Si-channel when a strained Ge-channel on SiGe buffer layer is introduced. This is mainly due to the drastically reduced thermal conductivity of the SiGe alloy used to strain the Ge-channel on top of Ge's lower thermal conductivity than Si.

Thermal conductivity is the next important parameter of self-heating effect. It is dependent on different parameters such as operated temperature, material thickness, and impurity concentration. Figure 6.8 (a)–(c) shows the values of thermal conductivity based on mentioned parameters with respect to Si and Ge devices [80–81]. Liao et al. reported [81] the comparison between different technology nodes, Fin pitches, and Fin heights with respect to temperature in Si and Ge chip, as shown in Figure 6.9 (a)–(c). As a result, the chip temperature is raised when device dimension will be scaled down continuously. The maximum temperature in the Ge FinFET is found to be ~50°C higher than Si FinFET due to the poor material property of thermal conductivity in the Ge material. The high operated temperature in the chip will lead to the device performance variation [82], the increase of the I_{off}, [83], and the worse device reliability [82]. This phenomenon is called self-heating effect and should be avoided in the real product in the industry. Although the increase of the Fin heights in the FinFET device can increase the heat dissipation surface to further reduce the thermal resistance, the maximum temperature in the chip is still found to be increased as shown in Figure 6.9 (c), due to the simultaneous increase of the ON-state current (I_{on}) per Fin, resulting in the increased power.

Liao et al. again simulated both Si and Ge FinFETs with different applied voltage with respect to maximum temperature as shown in Figure 6.10 [81]. It shows that the maximum temperature in the pure Ge FinFET can be kept below as ~170°C, which is the temperature in the current pure Si FinFET, when the operated voltage in the pure Ge FinFET can be kept below to 0.8 V. Therefore, we conclude that the simulated results shown indicate that the high mobility material (Ge) will be difficult to use in the next-generation scaled technology node devices, unless these devices can be operated at the low voltage bias (less than 0.8 V) from the SHE point of view.

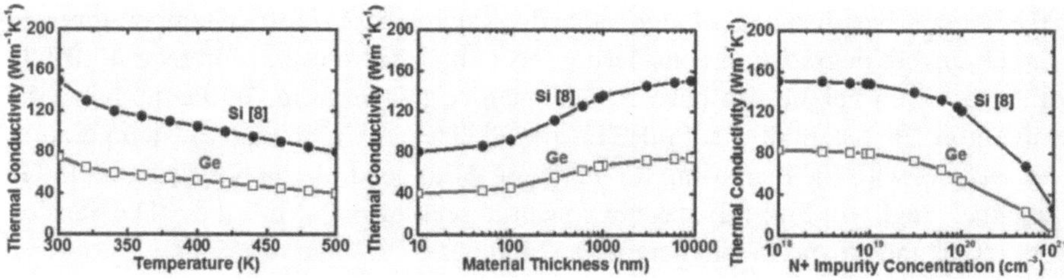

FIGURE 6.8 The Function of Thermal Conductivity with (a) Temperature, (b) Material Thickness, and (c) Concentration in both Si and Ge Materials

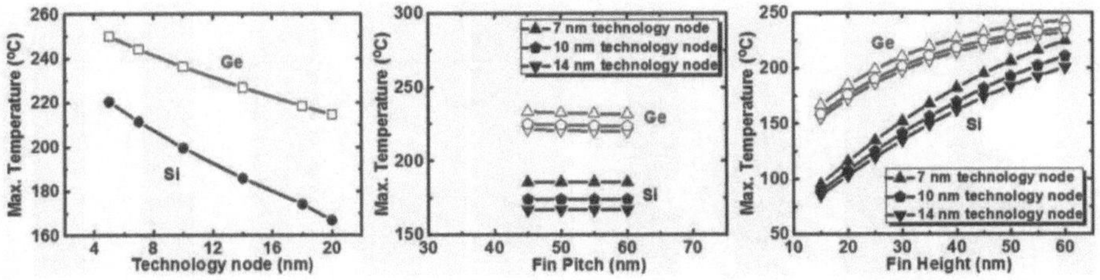

FIGURE 6.9 The Simulated Temperature Functionality with (a) Technology Nodes, (b) Fin Pitches, and (c) Fin Heights in Si and Ge FinFET Devices

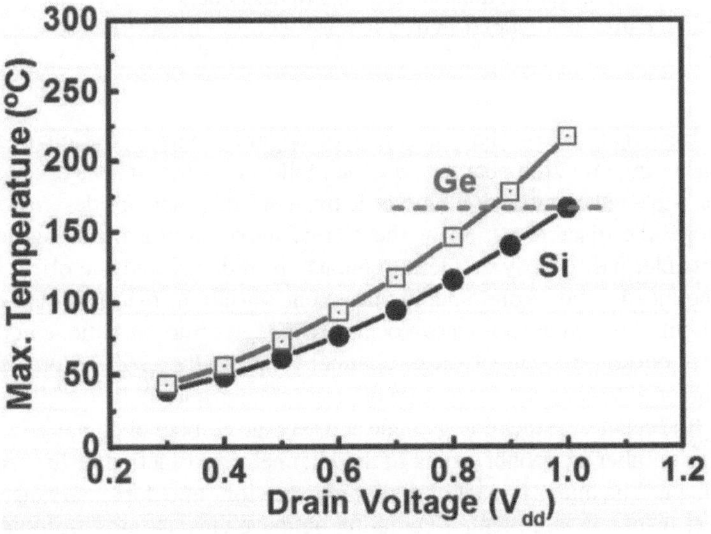

FIGURE 6.10 The Dependency of the Maximum Device Temperature (Tmax) with Different Operated Voltage Bias in both Si and Ge FinFETs

6.5 RANDOM DOPANT FLUCTUATION IN GE FINFETS

When semiconductor devices are fabricated in the nano-meter regime, they are prone to variability in physical parameters due to technological limitation. FinFETs offer better performance in comparison to the conventional CMOS transistors. However, FinFETs are superior in comparison to MOSFETs due to their capability of current flow through the undoped channel. FinFETs' lightly doped channel makes the channel less susceptible to the random fluctuation of dopants (RDF), as shown in Figure 6.11. But FinFET is affected by the variations and is vulnerable to other variability issues. Usually, the channel of FinFET is undoped, but in the case of low-power applications, the Fin channel is still doped to control the threshold voltage, hence making RDF an important aspect in FinFET modeling.

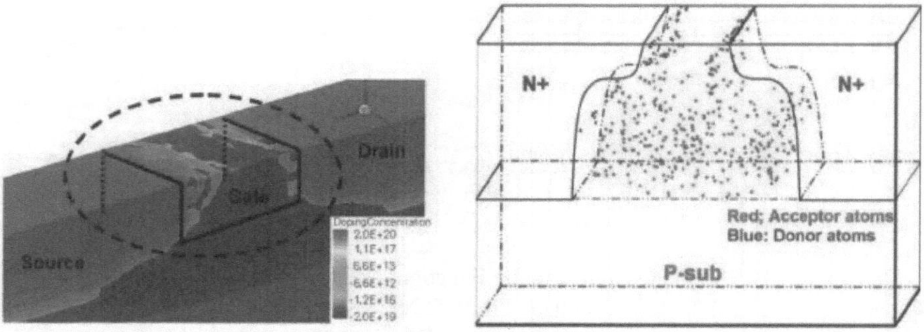

FIGURE 6.11 (a) Example of Atomistic Doping Profiles in Bulk MOSFET and (b) Isometric View of the RDF in Bulk MOSFET [84]

RDFs are generated by the placement of dopant atoms in the channel, which occurs during implantation phases and which follows statistical laws of nature, making it impossible to achieve a doping profile that perfectly matches design conditions. Furthermore, the discrete nature of the charge makes uniform doping concentration impossible, particularly as measurements are reduced and the effect becomes more pronounced. Aside from random placement, variations in the amount of dopant atoms present in the channel area can occur. Although minor variations in this number are unimportant in relatively large channel volumes, they will become critical in deca-nanometer devices with moderate doping concentrations.

Figure 6.12 relates the technology node and average number of dopants. Decreasing the average number of dopant atoms in the channel as a function of the technology node is shown. The major impact of random fluctuations and dopant arrangements in the channel results in significant fluctuation on drive current and threshold voltage. Understanding and modeling RDF has been a popular subject of research in recent past, since RDF is considered one of the main sources of variability in FET devices.

Keyes studied the consequences of spontaneous variations in the number of impurity atoms as one of the issues for continued transistor scaling in the early 1970s [86]. Experiments confirmed his theory for a wide range of fabricated devices [87]. RDF-induced V_T variance in deep sub-micron was then studied and predicted using analytical models and statistical device simulations [88–89]. It is well known that RDF-induced V_T is inversely proportional to transistor channel width and length. Recently, in a 100,000-sample 3-D simulation study [90], the complete V_T distribution caused by RDF was constructed through the discrete convolution of a Poisson distribution with the mean (N) of the number of dopants in channel region, and a Gaussian distribution of V_T for a fixed N, as shown in Figure 6.13. As the channel doping and/or halo doping in conventional planar-bulk/PD-SOI MOSFETs is increased with scaling to suppress short-channel effects, RDF-induced variation will worsen. The use of a lightly doped (fully depleted) SOI MOSFET structure with a thin (~10 nm-thick) buried oxide (BOX) and a heavily doped substrate has been reported to be effective for suppressing this variation [91].

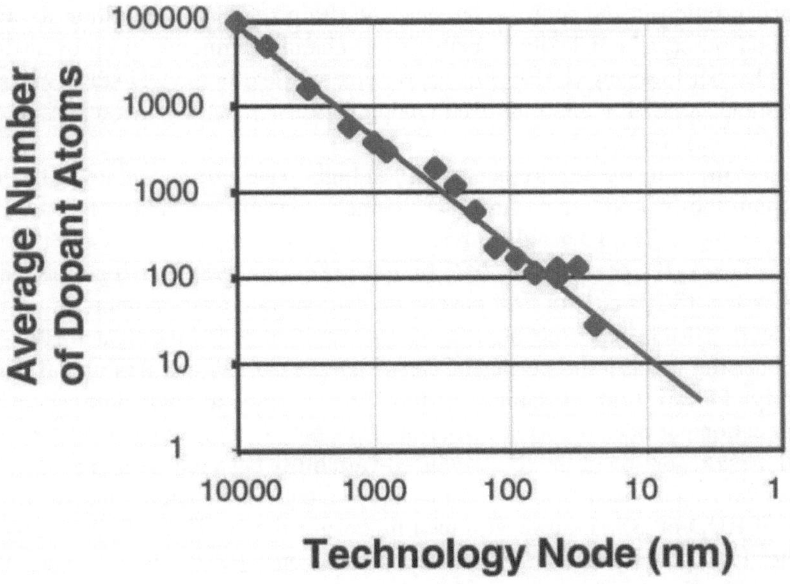

FIGURE 6.12 Number of Dopants vs Technology Node [85]

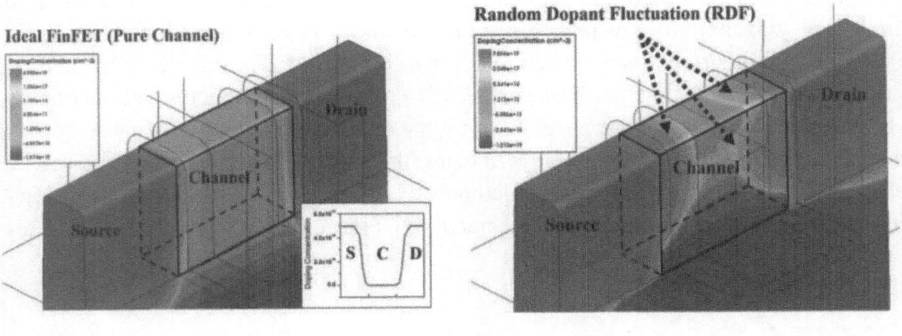

FIGURE 6.13 (a) Doping Profile and (b) the Atomistic Doping Profiles in a 7-nm n-Type Ge FinFETs

6.6 Ge Junctionless FinFETs

Some of the process challenges for further downscaling of device dimensions for conventional transistor poses a great challenge, which do not need any junctions, and this is the main advantage of junctionless (JL) transistor. Unlike the conventional MOSFET, JL transistor has heavy channel doping and is fully depleted in the OFF-state. JL transistors do not have pn junction in the channel path, as shown in Figure 6.14 [92]. For this purpose, a gate metal, which has a large work function difference to that of the channel, is needed. JL transistor has identical doping type and concentration in the source, channel, and drain regions. In addition to simple fabrication process, a JLT shows better short-channel immunity, improved drain-induced barrier lowering (DIBL) [93], lower electric field in the ON-state [94], steep subthreshold slope [95], and improved analog performance [96] in comparison with its conventional counterpart.

To meet the industry requirements, such as high-drive current and low gate delay, germanium shows promising future as a channel material, particularly for p-channel devices, owing to its higher bulk carrier mobility and compatibility with Si process line. Excellent I_{on}/I_{off} performance [97], high-drain current [98], and impressive short-channel effect (SCE) control [99] have been demonstrated for germanium-on-insulator (GeOI) p-MOSFET. Recently, JLTs on GeOI have been investigated [100–102] to combine the intrinsic properties of Ge with the excellent features of JLT. Better immunity of Ge JL FinFET against random-discrete-dopant-induced threshold voltage fluctuation has been reported earlier by University of Calcutta group [103]. They studied the comparison of the RDD induced variability between Ge and Si JL p-FinFETs for varying device parameters and supply voltage, as shown in Figure 6.15. The impact of RDD on SS is, however, much higher for the Ge JL FinFET mainly due to higher leakage currents in such devices arising out of the lower bandgap of Ge than Si. Technology scaling reduces SS, although it has almost no effects on VT for both type of devices. Also, the reduction in SS is much higher for Ge devices than Si devices making Ge as an attractive channel material for p-type JL devices than Si particularly for scaled device dimensions [104].

Usuda et al. proposed the fabrication of a poly-Ge JL-FinFET by flash-lamp annealing, substantially pushing this approach into 3-D-IC applications [106, 107]. Previously, another approach to fabricate poly-Ge JL-FinFETs was demonstrated by laser annealing, behaving the characteristics of a p-type semiconductor without doping [108]. As a result, a high concentration of n-type dopants is inevitable, because doping n-type dopants into p-type poly-Ge will cause the loss of electron concentration. The electrical characteristics of laser-enabled poly-Ge JL-FinFETs with Fin channels having different aspect ratios were investigated [109]. Figure 6.16 schematically illustrates

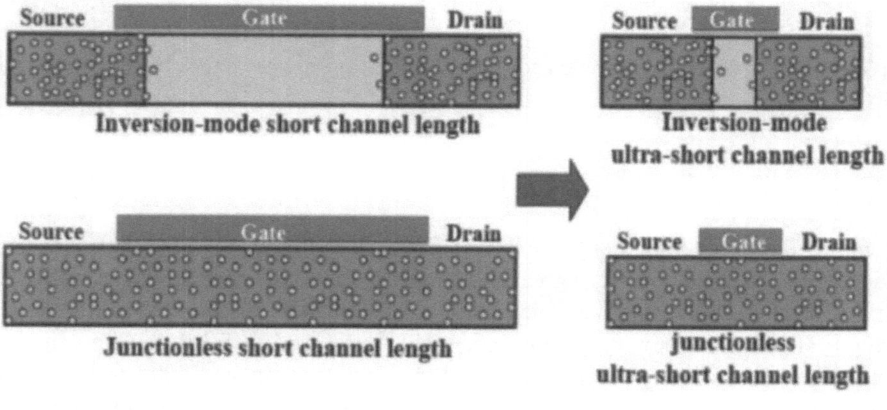

FIGURE 6.14 (a) With and (b) without Junction-Based Transistor [105]

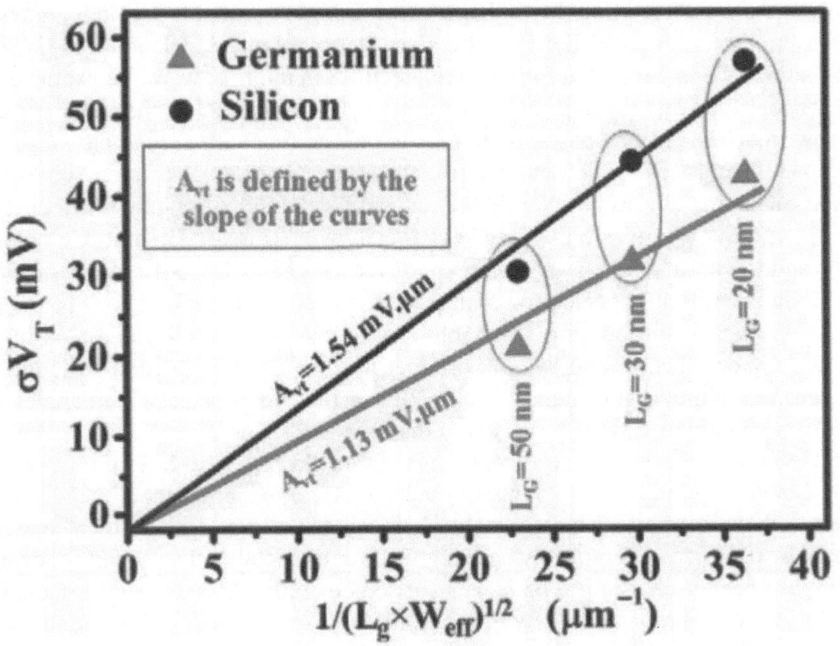

FIGURE 6.15 Comparison of Ge and Si JL FinFET [104]

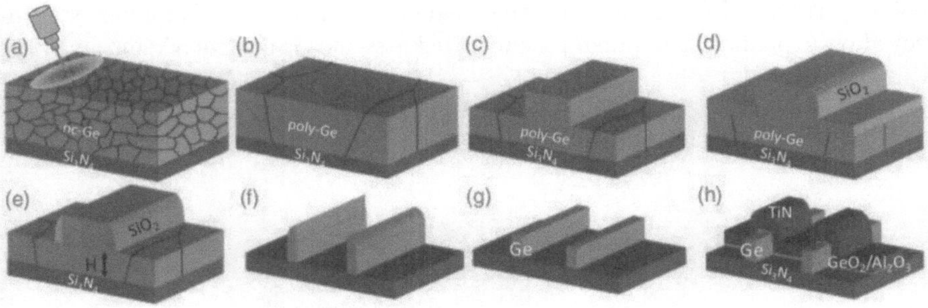

FIGURE 6.16 The Fabrication Processes of N-Type Poly-Ge JL-FinFET [110]

the fabrication processes of an n-type JL-FinFET with an in-situ-doped poly-Ge Fin channel, a gate stack of plasma-oxidized $GeO_2 = Al_2O_3$, and a TiN gate electrode.

6.7 SiGe FinFETs

SiGe is a promising replacement for the conventional Si PMOS channel material in future CMOS nodes. SiGe PMOS offers excellent NBTI reliability [111] and enables CMOS integration based on a single high-k dielectric/metal gate stack by using dual-channel technology [112]. In addition, compressively strained SiGe on Si can provide strain-induced mobility enhancement in highly scaled technology nodes in which proximity stressors are less effective due to volume restrictions [113]. SiGe channels can be integrated in FinFET architectures in multiple ways, for example, by Ge condensation and subsequent fin patterning [114] or by SiGe epitaxial growth in narrow trenches. An alternative method is to clad Si fins with a thin epitaxial SiGe layer. This approach, which closely resembles the channel fabrication scheme of planar SiGe quantum-well (QW) devices except for the Si topography [115], enables the incorporation in FinFETs of SiGe channels with (i) relatively high Ge content [116] and (ii) an in-situ grown Si capping layer for channel passivation. Moreover, the strain stability is expected to be high, since the epitaxial SiGe is well below the critical thickness. Furthermore, the mobile holes can be effectively confined in the $Si_{1-x}Ge_x$ region when Si and $Si_{1-x}Ge_x$ are met to form a heterojunction structure by the help of a large valence band offset (VBO) at their interface [117]; the holes are effectively confined into the place where they can run faster.

SiGe FinFET devices have been proposed as an alternative to Si FinFET devices. SiGe is able to provide p-FET threshold voltage of, for example, about 0.25 V, compared to Si counterparts, which give threshold voltage of greater than 0.5 V. In addition, SiGe pFET devices typically have higher channel mobility than Si pFET devices. Boron has been used as a p-type dopant for source/drain and extension formation in pFETs. However, boron diffusivity is lower in SiGe than in Si and, as a result, the SiGe devices are likely to be underlapped and have high access resistance. Yu et al. proposed SiGe shell channel p-type FinFETwith an emphasis on high-speed operation capability in the tera-hertz (THz) regime, as shown in Figure 6.17. They varied the SiGe channel thickness, Ge fraction in the $Si_{1-x}Ge_x$channel, and gate length. The simulations with multiple drift and diffusion models and quantum mechanical models for higher accuracy allow to predict the minimum channel thickness and Ge content of 2 nm and 40%,

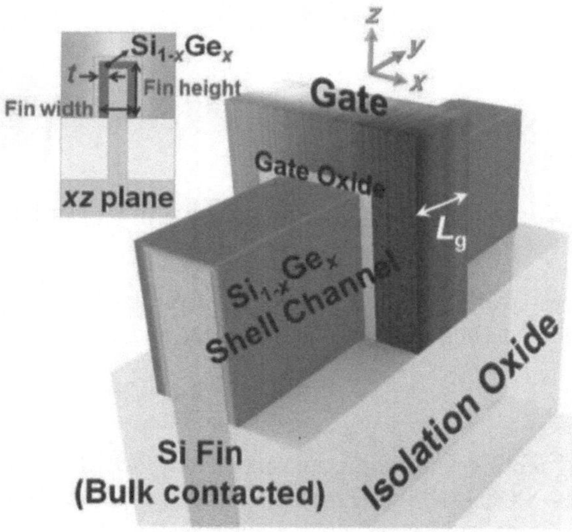

FIGURE 6.17 Cross-Sectional View of SiGe Shell Channel p-Type FinFET

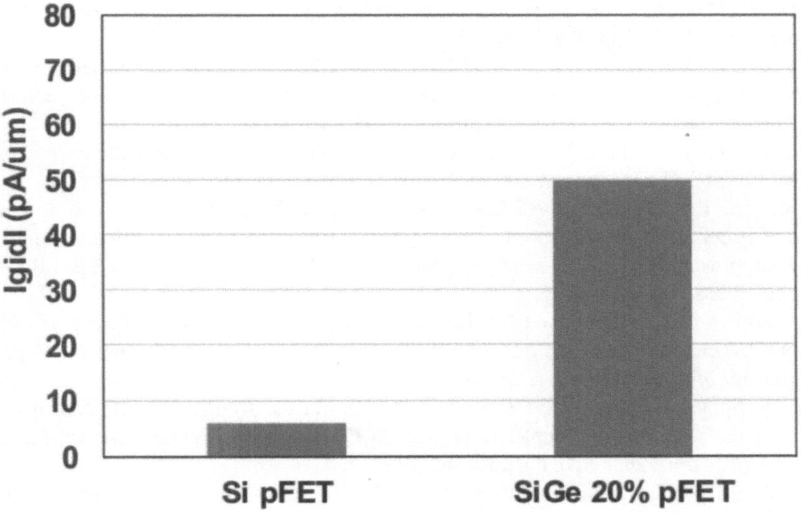

FIGURE 6.18 Comparison of the GIDL

respectively, for suppressing the hole leakage outflowing over the VBO. Also, scalability was checked down to 5 nm for achieving the upcoming logic technology nodes. Cutoff frequency (f_T) and maximum oscillation frequency (f_{max}) are obtained to be 240 GHz and 1.04 THz at a low drive voltage as –0.7 V, respectively [118].

Leakage control is one of the key factors for low power applications. Replacing conventional Si with SiGe as a channel material modulates interface trap density (Dit) and bandgap, which are crucial for the two major transistor leakage components, subthreshold leakage and gate-induced drain leakage (GIDL). Dit and GIDL in low Ge content Si/SiGe FinFET [119, 120] are examined. Dit is degraded with increased integration such as CMOS enablement (nFET implementation), multi-V_t, and Fin pitch/height scaling. Si cap passivation [121, 122] is identified as a major knob to improve Dit. A pFET Dit reduction of 80% is demonstrated, which improves both subthreshold leakage and channel hole mobility resulting in 20% pFET DC performance improvement. GIDL in SiGe pFET is higher than Si by a factor of 9, though GIDL is limited to 50pA/μm. It is demonstrated that SiGe GIDL can be reduced to a level similar to Si by carefully designing junction gradient and location along with physical gate length. G. Tsutsui et al. examined the leakage in Si/SiGe CMOS FinFET. Si cap passivation effectively improves SiGe pFET Dit, subthreshold slope, and mobility, which improves pFET DC performance by 20%. SiGe GIDL is higher than Si by a factor of 9, though GIDL is limited to 50pA/μm [123], as shown in Figure 6.18.

REFERENCES

[1] "How the Internet of Things changes everything" [online]. Available at: https://martechtoday.com/internet-things-changes-everything-202482.

[2] "International Technology Roadmap for Semiconductors", 2010. Available at: http://www.itrs.net/

[3] Hu, C., "Gate oxide scaling limits and projection", *IEEE International Electron Devices Meeting (IEDM),* pp. 319–322, 1996.
[4] Yeo, Y.C., T.J. King, and C. Hu, "MOSFET gate leakage modeling and selection guide for alternative gate dielectrics based on leakage considerations", *IEEE Transactions on Electron Devices,* vol. 50, no. 4, pp. 1027–1035, 2003.
[5] Chen, J., T.Y. Chan, I.C. Chen, et al., "Sub-breakdown drain leakage current in MOSFET", *IEEE Electron Device Letters,* vol. 8, no. 11, pp. 515–517, 1987.
[6] Silveira, F., D. Flandre, and P.G.A. Jespers, "A g_m/I_D based methodology for the design of CMOS analog circuits and its application to the synthesis of a silicon-on-insulator micropower OTA", *IEEE Journal of Solid-State Circuits*, vol. 31, no. 9, pp. 1314–1319, Sept. 1996.
[7] Kranti, A., T.M. Chung, and J.-P. Raskin, "Analysis of static and dynamic performance of short channel double gate SOI MOSFETs for improved cut-off frequency", *Japanese Journal of Applied Physics*, vol. 44, no. 4B, pp. 2340–2346, 2005.
[8] https://phys.org/news/2015-08-silicon-limits-power-electronics-revolution.html.
[9] Auth, C., "22-nm fully-depleted tri-gate CMOS transistors", *IEEE Custom Integrated Circuits Conference (CICC)*, pp. 1–6, 2012.
[10] Guillorn, M., J. Chang, A. Bryant, et al., "FinFET performance advantage at 22nm: An AC perspective", *Symposium on VLSI Technology*, pp. 12–13, 2008.
[11] Morikawa, Y., T. Murayama, Y.N.T. Sakuishi, et al., "Total cost effective scallop free Si etching for 2.5 D & 3D TSV fabrication technologies in 300mm wafer", *IEEE 63rd Electronic Components and Technology Conference (ECTC)*, pp. 605–607, 2013.
[12] Natarajan, S., M. Agostinelli, S. Akbar, et al., "A 14nm logic technology featuring 2nd-generation FinFET, air-gapped interconnects, self-aligned double patterning and a 0.0588 μm 2 SRAM cell size", in *2014 IEEE International Electron Devices Meeting (IEDM), IEEE*, pp. 3–7, 2014.
[13] "7nm Technology" [Online]. Available at: www.tsmc.com/english/dedicatedFoundry/technology/7nm.htm.
[14] Asif, S. "Samsung to start mass production of 5nm chips in Q2 2020" [Online]. Available at: www.sammobile.com/news/samsung-start-mass-production5nm-chips-q2-2020/.
[15] Fox, C., "Intel's next-generation 7nm chips delayed until 2022", *Technology reporter*, 24 July 2020. https://www.bbc.com/news/technology-53525710
[16] Mittall, S., S. Gupta, A. Nainani, M.C. Abraham, K. Schuegraf, S. Lodha, and U. Ganguly, "Epi Defined (ED) FinFET: An alternate device architecture for high mobility Ge channel integration in PMOSFET", *IEEE 5th International Nanoelectronics Conference (INEC)*, 2013.
[17] Irisawa, T., T. Numata, T. Tezuka, K. Usuda, S. Nakaharai, N. Hirashita, N. Sugiyama, E. Toyoda, and S. Takagi, "High performance multi-gate pMOSFET using uniaxially-strained SGOI channels", *IEEE International Electron Devices Meeting (IEDM)*, pp. 709–712, 2005.
[18] Claeys, Cor, and Eddy Simoen, *Germanium-based Technologies from Materials to Devices*, Ist Edition, Elsevier, London, 2007.
[19] Sze, S.M., *Physics of Semiconductor Devices*, John Wiley & Sons, New York, 1981.
[20] Wolf, S., and R. Tauber, *Silicon Processing for the VLSI Era*, Lattice Press, Sunset Beach, CA, 1986.
[21] Yang, Y.-J., W.S. Ho, C.-F. Huang, S.T. Chang, and C.W. Liu, "Electron mobility enhancement in strained-germanium n-channel metal-oxide-semiconductor field-effect transistors", *Applied Physics Letters*, vol. 91, p. 102103, 2007.
[22] Yang, M., E.P. Gusev, M. Ieong, O. Gluschenkov, D.C. Boyd, K.K. Chan, P.M. Kozlowski, et al., "Performance dependence of CMOS on silicon substrate orientation for ultrathin oxynitride and HfO_2 gate dielectrics", *IEEE Electron Device Letters*, vol. 24, no. 5, p. 3, 2003.

[23] Yu, D.S., A. Chin, C.C. Liao, C.F. Lee, C.F. Cheng, M.F. Li, W.J. Yoo, and S.P. McAlister, "Three-dimensional metal gate-high-k-GOI CMOSFETs on 1-poly-6-metal 0.18-µm Si devices", *IEEE Electron Device Letters*, vol. 26, no. 2, pp. 118–120, Feb. 2005.

[24] "Intel reinvents transistors using new 3-D structure, 4 May" [online]. http://news-room.intel.com/community/intel_newsroom/blog/2011/05/04/intel-reinvents-transistors-using-new-3-d-structure.

[25] Zimmerman, P., et al., "High performance Ge pMOS devices using a Si-compatible process flow", *2006 International Electron Devices Meeting*, p. 655, 2006.

[26] Mitard, J., et al., "Impact of EOT scaling down to 0.85nm on 70nm Ge-pFETs technology with STI", *2009 Symposium on VLSI Technology*, pp. 82–83, 2009.

[27] Pillarisetty, R., et al., "High mobility strained germanium quantum well field effect transistor as the p-channel device option for low power (Vcc = 0.5 V) III–V CMOS architecture", *2010 International Electron Devices Meeting*, pp. 6.7.1–6.7.4, 2010.

[28] Hutin, L., et al., "GeOI pMOSFETs scaled down to 30 nm gate length with record off-state current", *IEEE Electron Device Letter,* vol. 31, p. 234, 2010.

[29] Zhang, L., et al., "Parallel core-shell metal-dielectric-semiconductor germanium nanowires for high-current surround-gate field-effect transistors", *Nano Letters,* vol. 6, pp. 2785–2789, 2006.

[30] Xiang, J., et al., "Ge/Si nanowire heterostructures as high-performance field-effect transistors", *Nature*, vol. 441, pp. 489–493, 2006.

[31] Feng, J., et al., "P-Channel germanium FinFET based on rapid melt growth", *IEEE Electron Device Letters*, vol. 28, no. 7, pp. 637–639, July 2007.

[32] Hsu, S.-H., et al., "Nearly defect-free Ge gate-all-around FETs on Si substrates", *2011 International Electron Devices Meeting*, pp. 35.2.1–35.2.4, 2011.

[33] Feng, J., et al., "High-performance gate-all-around GeOI p-MOSFETs fabricated by rapid melt growth using plasma nitridation and ALD $\hbox{Al}_{2}\hbox{O}_{3}$ Gate dielectric and self-aligned NiGe contacts", *IEEE Electron Device Letters*, vol. 29, no. 7, pp. 805–807, July 2008.

[34] Peng, J.W., et al., "CMOS compatible Ge/Si core/shell nanowire gate-all-around pMOSFET integrated with HfO2/TaN gate stack", *2009 IEEE International Electron Devices Meeting (IEDM)*, pp. 1–4, 2009.

[35] Nakaharai, S., T. Tezuka, N. Sugiyama, Y. Moriyama, and S. Takagi, "Characterization of 7-nm-thick strained Ge-on-insulator layer fabricated by Ge-condensation technique", *Applied Physics Letters*, vol. 83, no. 17, pp. 3516–3518, 2003.

[36] Liu, Bin, Xiao Gong, G Han, Phyllis Shi Ya Lim, Yi Tong, Qian Zhou, Yue Yang, Nicolas Daval, Matthieu Pulido, Daniel Delprat, Bich-Yen Nguyen, and Yee-Chia Yeo, "High performance Ω-Gate Ge FinFET featuring low temperature Si_2H_6 passivation and implantless Schottky-Barrier NiGe metallic source/drain", *2012 IEEE Silicon Nanoelectronics Workshop (SNW)*, pp. 1–2, 2012.

[37] Chung, Cheng-Ting, Che-Wei Chen, Jyun-Chih Lin, Che-Chen Wu, Chao-Hsin Chien and Guang-Li Luo, "First experimental Ge CMOS FinFETs directly on SOI substrate", *2012 International Electron Devices Meeting*, pp. 16.4.1–16.4.4, 2012.

[38] Mitard, J., B. De Jaeger, F.E. Leys, G. Hellings, K. Martens, G. Eneman, D.P. Brunco, R. Loo, J.C. Lin, D. Shamiryan, T. Vandeweyer, G. Winderickx, E. Vrancken, C.H. Yu, K. De Meyer, M. Caymax, L. Pantisano, M. Meuris, and M. Heyns, "Record I_{ON}/I_{OFF} performance for 65 nm Ge pMOSFET and novel Si passivation scheme for improved EOT scalability", *IEDM Technical Digest*, 2008, pp. 1–4.

[39] Witters, L., S. Takeoka, S. Yamaguchi, A. Hikavyy, D. Shamiryan, M.J. Cho, T. Chiarella, L.-Å. Ragnarsson, R. Loo, C. Kerner, Y. Crabbe, J. Franco, J. Tseng, W.-E. Wang, E. Rohr, T. Schram, O. Richard, H. Bender, S. Biesemans, P. Absil, and T. Hoffmann, "8 Å Tinv gate-first dual channel technology achieving low-Vt high performance CMOS", in *VLSI Symposium on Technical Digest*, 2010, pp. 181–182.

[40] Hellings, G., J. Mitard, G. Eneman, B. De Jaeger, D. Brunco, D. Shamiryan, T. Vandeweyer, M. Meuris, M. Heyns, and K. De Meyer, "High performance 70-nm germanium pMOSFETs with boron LDD implants", *IEEE Electron Device Letters*, vol. 30, no. 1, pp. 88–90, Jan. 2009.
[41] Feng, J., R. Woo, S. Chen, Y. Liu, P. Griffin, and J. Plummer, "P-channel germanium FinFET based on rapid melt growth", *IEEE Electron Device Letters*, vol. 28, no. 7, pp. 637–639, July 2007.
[42] van Dal, M.J.H., G. Vellianitis, G. Doornbos, B. Duriez, T.M Shen, C.C. Wu, R. Oxland, K. Bhuwalka, M. Holland, T.L. Lee, C. Wann, C.H. Hsieh, B.H. Lee, K.M. Yin, Z.Q. Wu, M. Passlack, and C.H. Diaz, "Demonstration of scaled Ge p-channel FinFETs integrated on Si", *2012 International Electron Devices Meeting*, pp. 23.5.1–23.5.4, 2012.
[43] van Dal, M.J.H., G. Vellianitis, G. Doornbos, B. Duriez, T.M Shen, C.C. Wu, R. Oxland, K. Bhuwalka, M. Holland, T.L. Lee, C. Wann, C.H. Hsieh, B.H. Lee, K.M. Yin, Z.Q. Wu, M. Passlack, and C.H. Diaz, "Germanium p-Channel FinFET Fabricated by Aspect Ratio Trapping", *IEEE Transactions on Electron Devices*, vol. 61, no. 2, Feb. 2014.
[44] Duriez, B., G. Vellianitis, M.J.H. van Dal, G. Doornbos, R. Oxland, K.K. Bhuwalka, M. Holland, Y.S. Chang, C.H. Hsieh, K.M. Yin, Y.C. See, M. Passlack, and C.H. Diaz, "Scaled p-channel Ge FinFET with optimized gate stack and record performance integrated on 300mm Si wafers", *2013 IEEE International Electron Devices Meeting*, pp. 20.1.1–20.1.4, 2013.
[45] Chen, Che-Wei, Cheng-Ting Chung, Guang-Li Luo, and Chao-Hsin Chien, "Body-tied Germanium FinFETs directly on a silicon substrate", *IEEE Electron Device Letters*, vol. 33, no. 12, Dec. 2012.
[46] Kim, J., S.W. Bedell, and D.K. Sadana, "Improved germanium n/p junction diodes formed by coimplantation of antimony and phosphorus", *Applied Physics Letters*, vol. 98, no. 8, pp. 082112-1-082112-3, Feb. 2011.
[47] Thareja, G., et al., "High n-type antimony dopant activation in germanium using laser annealing for n^+/p junction diode", *IEEE Electron Device Letters,* vol. 32, no. 7, pp. 838–840, July 2011.
[48] Han, D., et al., "Studies of Ti- and Ni-Germanide Schottky contacts on n-Ge (100) substrates", *Microelectronics Engineering*, vol. 82, no. 2, pp. 93–98, Oct. 2005.
[49] Li, R., et al., "Metal-germanide Schottky source/drain transistor on germanium substrate for future CMOS technology", *Thin Solid Films*, vol. 504, nos. 1–2, pp. 28–31, May 2006.
[50] Paramahans, P., et al., "ZnO: An attractive option for n-type metal-interfacial layer-semiconductor (Si, Ge, SiC) contacts", in *Proceeding of the Symposium on VLSI Technology*, June 2012, pp. 83–84.
[51] Kim, J.-K., et al., "Analytical study of interfacial layer doping effect on contact resistivity in metal-interfacial layer-Ge structure", *IEEE Electron Device Letters*, vol. 35, no. 7, pp. 705–707, July 2014.
[52] Gupta, S., et al., "Contact resistivity reduction through interfacial layer doping in metal-interfacial layer-semiconductor contacts", *Applied Physics Letters*, vol. 113, no. 23, pp. 234505-1-234505-7, June 2013.
[53] Kim, Jeong-Kyu, et al., "The efficacy of metal-interfacial layer-semiconductor source/drain structure on sub-10-nm n-type Ge FinFET performances", *IEEE Electron Device Letters*, vol. 35, no. 12, Dec. 2014.
[54] Md Rezali, F.A., N.A.F. Othman, M. Mazhar, S.W.M. Hatta, and N. Soin, "Performance and device design based on geometry and process considerations for 14/16-nm Strained FinFETs", *IEEE Transactions on Electron Devices*, vol. 63, no. 3, pp. 974–981, Mar. 2016.
[55] Xiong, W., C.R. Cleavelin, P. Kohli, C. Huffman, T. Schulz, K. Schruefer, G. Gebara, K. Mathews, P. Patruno, Y.M. Le Vaillant, I. Cayrefourcq, M. Kennard, C. Mazure, K. Shin, and T.J.K. Liu, "Impact of strained-silicon-on-insulator (sSOI) substrate on FinFET mobility", *IEEE Electron Device Letters,* vol. 27, no. 7, pp. 612–614, July 2006.

[56] Lin, C.Y., S.T. Chang, J. Huang, W.C. Wang, and J.W. Fan, "Impact of source/drain $Si_{1-}yCy$ stressors on silicon-on-insulator N-type metal-oxide-semiconductor field-effect transistors", *Japanese Journal of Applied Physics,* vol. 46, p. 2107 (2007).

[57] Othman, N.A.F., S.F.A.M. Hatta, and N. Soin, "Impact of channel, stress-relaxed buffer, and S/D $Si_{1-x}Ge_x$ stressor on the performance of 7-nm FinFET CMOS design with the implementation of stress engineering", *Journal of Electronic Materials*, vol. 47, no. 4, 2018.

[58] Manoj, C.R., M. Nagpal, D. Varghese, and V.R. Rao, " Device design and optimization considerations for bulk FinFETs", *IEEE Transactions on Electron Devices*, vol. 55, p. 609, 2008.

[59] Rousseau, P.M., P.B. Griffin, S. Luning, and J.D. Plummer, "A model for mobility degradation in highly doped arsenic layers", *IEEE Transactions on Electron Devices*, vol. 43, no. 11, p. 2025–2027, Nov. 1996.

[60] Wang, J.S., W.P.N. Chen, C.H. Shih, C. Lien, P. Su, Y.M. Sheu, D.Y.S. Chao, and K.I. Goto, "Mobility modeling and its extraction technique for manufacturing strained-Si MOSFETs", *IEEE Electron Device Letters,* vol. 28, no. 11, pp. 1040–043, Nov. 2007.

[61] "Imec demonstrates optimized process flows for high-performance Ge-based devices" [online]. Available at: www.semiconductor-today.com.

[62] McDaid, L.J., et al., "Physical origin of negative differential resistance in SOI transistors", *Electronics Letters*, vol. 25, no. 13, pp. 827–828, June 1989.

[63] Jomaah, J., et al., "Impact of self-heating effects on the design of SOI devices versus temperature", *1995 IEEE International SOI Conference Proceedings.* Tucson: [sn]. 1995, pp. 114–115.

[64] Liao, M.H., et al., "Systematic investigation of self-heating effect on CMOS logic transistors from 20 to 5 nm technology nodes by experimental thermoelectric measurements and finite element modeling", *IEEE Transactions on Electron Devices*, vol. 64, no. 2, p. 646, 2017.

[65] Jin, M., et al., "Hot carrier reliability characterization in consideration of self-heating in FinFET technology", *IEEE IRPS,* 2A-2–1, 2016.

[66] Jiang, H., et al., "The impact of self-heating on HCI reliability in high-performance digital circuits", *IEEE Device Letters*, vol. 38, no. 4, p. 430, 2017.

[67] Jiang, H., et al., "Unified self heating effect model for advanced digital and analog technology and thermal-ware lifetime prediction methodology", *VLSI Technology*, T136, 2017.

[68] Si, M.W., et al., "Characterization and reliability of III-V gate-all-around MOSFETs", *IEEE IRPS*, 4A-1–1, 2015.

[69] Jiang, H., et al., "Investigation of self-heating effect on hot carrier degradation in multiple-fin SOI FinFETs", *IEEE EDL*, vol. 36, no. 12, 2015.

[70] Pae, S.W., et al., "Considering physical mechanisms and geometry dependencies in 14nm FinFET circuit aging and product validations", in *IEDM Technical Digest*, 2015, pp. 557–560.

[71] Rhyner, R., and M. Luisier, "Self-heating effects in ultra-scaled Si nanowire transistors", in *IEDM Technical Digest*, 2013, pp. 790–793.

[72] Haras, M., et al., "Fabrication of integrated micrometer platform for thermoelectric measurements", in *IEDM Technical Digest*, 2014, pp. 212–215.

[73] Makovejev, S., S.H. Olsen, V. Kilchytska, and J.-P. Raskin, "Time and frequency domain characterization of transistor self-heating", *IEEE Transactions on Electron Devices*, vol. 60, no. 6, pp. 1844–1851, June 2013.

[74] Shin, S.H., et al., "Direct observation of self-heating in III–V gate-all-around nanowire MOSFETs", in *IEDM Technical Digest*, 2014, pp. 510–513.

[75] Shin, S.H., et al., "Direct observation of self-heating in III–V gate-all-around nanowire MOSFETs", *IEEE Transactions on Electron Devices*, vol. 62, no. 11, pp. 3516–3523, Nov. 2015.

[76] Jang, D., et al., "Self-heating on bulk FinFET from 14nm down to 7nm node", in *IEDM Technical Digest,* 2015, pp. 289–292.

[77] Dames, C., and G. Chen, "Theoretical phonon thermal conductivity of Si/Ge superlattice nanowires", *Journal of Applied Physics*, vol. 95, pp. 682–693, 2004.

[78] Bury, E., B. Kaczer, J. Mitard, N. Collaert, N.S. Khatami, Z. Aksamija, D. Vasileska, K. Raleva, L. Witters, G. Hellings, D. Linten, G. Groeseneken, and A. Thean, "Characterization of self-heating in high-mobility Ge FinFET pMOS devices", 2015 Symposium on VLSI Technology Digest of Technical Papers.

[79] Yin, Longxiang, Lei Shen, Hai Jiang, Gang Du, and Xiaoyan Liu, "Impact of self-heating effects on nanoscale Ge p-channel FinFETs with Si substrate", *Science China Information Sciences*, vol. 61, pp. 062401:1–062401:9, June 2018.

[80] Haras, M., V. Lacatena, F. Morini, J.-F. Robillard, S. Monfray, T. Skotnicki, and E. Dubois, "Fabrication of integrated micrometer platform for thermoelectric measurements", in *IEDM Technical Digest*, 2014, pp. 212–215.

[81] Liao, M.-H., C.-P. Hsieh, and C.-C. Lee, "The systematic investigation of self-heating effect on CMOS logic transistors from 20 nm to 5 nm technology nodes by experimental thermo-electric measurements and finite element modeling", *IEEE Transactions on Electron Devices,* vol. 64, no. 2, pp. 646–648, 2017.

[82] Pae, S.W., H.C. Sagong, C. Liu, M.J. Jin, Y.H. Kim, S.J. Choo, J.J. Kim, H.J. Kim, S.Y. Yoon, H.W. Nam, H.W. Shim, S.M. Park, J.K. Park, S.C. Shin, and J.W. Park, "Considering physical mechanisms and geometry dependencies in 14 nm FinFET circuit aging and product validations", in *IEDM Technical Digest*, 2015, pp. 557–560.

[83] Koswatta, S.O., N. Mavilla, M. Bajaj, J. Johnson, S. Gundapaneni, C. Scott, G. Freeman, D. Poindexter, P.S. McLaughlin, S.W. Mittl, L. Sigal, J.D. Warnock, N. Zamdmer, S. Lee, R. Wachnik, C.-H. Lin, and E. Nowak, "Off-state self-heating, micro-hot-spots, and stress-induced device considerations in scaled technologies", in *IEDM Technical Digest,* 2015, pp. 539–542.

[84] Shin, Changhwan, Xin Sun, and Tsu-Jae King Liu, "Study of random-dopant-fluctuation (RDF) effects for the trigate bulk MOSFET", *IEEE Transactions on Electron Devices*, vol. 56, no. 7, pp. 1538–1542, 2009.

[85] Kuhn, Kelin, et al., "Managing process variation in Intel's 45nm CMOS technology", *Intel Technology Journal*, vol. 12, no. 2, 2008.

[86] Keyes, R.W., "Physical limits in digital electronics", *Proceedings of the IEEE*, vol. 63, no. 5, pp. 740–767, May 1975.

[87] Mizuno, T., J.-I. Okamura, and A. Toriumi, "Experimental study of threshold voltage fluctuation due to statistical variation of channel dopant number in MOSFETs", *IEEE Transactions on Electron Devices*, vol. 41, no. 11, pp. 2216–2221, Nov. 1994.

[88] Asenov, A., and S. Saini, "Suppression of random dopant-induced threshold voltage fluctuations in sub-0.1-μm MOSFET's with epitaxial and δ-doped channels", *IEEE Transaction Electron Devices,* vol. 46, no. 8, pp. 1718–1724, Aug. 1999.

[89] Cathignol, A., B. Cheng, D. Chanemougame, A.R. Brown, K. Rochereau, G. Ghibaudo, and A. Asenov, "Quantitative evaluation of statistical variability sources in a 45-nm technological node LP N-MOSFET", *IEEE Electron Device Letters,* vol. 29, no. 6, pp. 609–611, June 2008.

[90] Reid, D., C. Millar, G. Roy, S. Roy, and A. Asenov, "Analysis of threshold voltage distribution due to random dopants: A 100 000-sample 3-D simulation study", *IEEE Transactions on Electron Devices*, vol. 56, no. 10, pp. 2255–2263, Oct. 2009.

[91] Ohtou, T., N. Sugii, and T. Hiramoto, "Impact of parameter variations and random dopant fluctuations on short-channel fully depleted SOI MOSFETs with extremely thin BOX", *IEEE Electron Device Letters,* vol. 28, no. 8, pp. 740–742, Aug. 2007.

[92] Lilienfield, J.E., "Method and apparatus for controlling electric currents", US Patent, 1745175, 1925.

[93] Lee, C.-W., I. Ferain, A. Afzalian, R. Yan, N.D. Akhavan, P. Razavi, and J.-P. Colinge, "Performance estimation of junctionless multigate transistors", *Solid State Electronics*, vol. 54, pp. 97–103, 2010.

[94] Colinge, J.-P., C.-W. Lee, I. Ferain, N.D. Akhavan, R. Yan, P. Razavi, R. Yu, A.N. Nazarov, and R.T. Doria, "Reduced electric field in junctionless transistors", *Applied Physics Letters,* vol. 96, p. 073510, 2010.

[95] Lee, C.-W., A.N. Nazarov, I. Ferain, N.D. Akhavan, R. Yan, P. Razavi, R. Yu, R.T. Doria, and J.-P. Colinge, "Low subthreshold slope in junctionless multigate transistors", *Applied Physics Letters,* vol. 96, p. 102106, 2010.

[96] Doria, R.T., M.A. Pavanello, R.D. Trevisoli, M. de—Souza, C.-W. Lee, I. Ferain, N.D. Akhavan, R. Yan, P. Razavi, R. Yu, A. Kranti, and J.-P. Colinge, "Junctionless multiple-gate transistors for analog applications", *IEEE Transactions on Electron Devices,* vol. 58, pp. 2511–2519, 2011.

[97] Royer, C.L., L. Clavelier, C. Tabone, K. Romanjek, C. Deguet, L. Sanchez, J.M. Hartmann, M.C. Roure, H. Grampeix, S. Soliveres, G. Le Carval, R. Truche, A. Pouydebasque, M. Vinet, and S. Deleonibus, *Solid State Electronics,* vol. 52, pp. 1285–1290, 2008.

[98] Feng, J., R. Woo, S. Chen, Y. Liu, P.B. Griffin, and J.D. Plummer, "P-Channel germanium FinFET based on rapid melt growth", *IEEE Electron Device Letters,* vol. 28, no. 7, pp. 637–639, July 2007.

[99] van Dal, M.J.H., G. Vellianitis, B. Duriez, G. Doornbos, C.H. Hsieh, B.H. Lee, K.M. Yin, M. Passlack, and C.H. Diaz, "Germanium p-channel FinFET fabricated by aspect ratio trapping", *IEEE Transactions on Electron Devices*, vol. 61, no. 2, pp. 430–436, Feb. 2014.

[100] Zhao, D.D., T. Nishimura, C.H. Lee, K.K. Kita, A. Toriumi, "Junctionless Ge p-channel metal–oxide–semiconductor field-effect transistors fabricated on ultrathin Ge-on-insulator substrate", *Applied Physics Express,* vol. 4, pp. 031302-1-031302-3, 2011.

[101] Zhao, D.D., C.H. Lee, T. Nishimura, K. Nagashio, G.A. Cheng, and A. Toriumi, "Experimental and analytical characterization of dual- gated germanium junctionless p-channel metal–oxide–semiconductor field-effect transistors", *Japanese Journal of Applied Physics*, vol. 51, 04DA03–1–04DA03–7, 2012.

[102] Chen, C.-W., C.-T. Chung, J.-Y. Tzeng, P.-S. Chang, G.-L. Luo, and C.-H. Chien, "Body-tied germanium tri-gate junctionless PMOS- FET with in situ boron doped channel", *IEEE Electron Device Letters,* vol. 35, pp. 12–14, 2014.

[103] Nawaz, S.M., S.K. Dutta, and A. Mallik, "A comparison of random discrete dopant induced variability between Ge and Si junctionless p-FinFETs", *Applied Physics Express,* vol. 107, no. 3, p. 033506, 2015.

[104] Nawaz, Masum, S.K., Souvik Dutta, and Abhijit Mallika, "A comparison of random discrete dopant induced variability between Ge and Si junctionless p-FinFETs", *Applied Physics Letters*, vol. 107, p. 033506, 2015.

[105] Yu, R., "A study of silicon and germanium junctionless transistors", Ph.D Thesis, University College Cork, 2013.

[106] Kamata, Y., M. Koike, E. Kurosawa, M. Kurosawa, H. Ota, O. Nakatsuka, S. Zaima, and T. Tezuka, "Operation of inverter and ring oscillator of ultrathin-body poly-Ge CMOS", *Applied Physics Express,* vol. 7, p. 121302, 2014.

[107] Usuda, K., Y. Kamata, Y. Kamimuta, T. Mori, M. Koike, and T. Tezuka, "High-performance tri-gate poly-Ge junction-less p- and n-MOSFETs fabricated by flash lamp annealing process", *2014 IEEE International Electron Devices Meeting*, pp. 16.6.1–16.6.4, 2014. DOI: 10.1109/IEDM.2014.7047066.

[108] Huang, W.H., J.M. Shieh, C.H. Shen, T.E. Huang, H.H. Wang, C.C. Yang, T.Y. Hsieh, J.L. Hsieh, and W.K. Yeh, "Junction-less poly-Ge FinFET and charge-trap NVM fabricated by laser-enabled low thermal budget processes", *Applied Physics Letters,* vol. 108, p. 243502, 2016.

[109] Huang, W.H., J.M. Shieh, F.M. Pan, C.C. Yang, C.H. Shen, H.H. Wang, T.Y. Hsieh, S.Y. Wu, and M.C. Wu, "Charge-trap non-volatile memories fabricated by laser-enabled low-thermal budget processes", *Applied Physics Letters,* vol. 107, p. 183506, 2015.

[110] Huang, Wen-Hsien, et al., "Enabling n-type polycrystalline Ge junctionless FinFET of low thermal budget by in situ doping of channel and visible pulsed laser annealing", *Applied Physics Express*, vol. 10, p. 026502, 2017.

[111] Franco, J., et al., "Superior NBTI reliability of SiGe channel pMOSFETs: Replacement gate, FinFETs, and impact of Body Bias," *2011 International Electron Devices Meeting*, pp. 18.5.1–18.5.4, 2011.

[112] Witters, L., et al., "Dual-channel technology with cap-free single metal gate for high performance CMOS in gate-first and gate-last integration," *2011 International Electron Devices Meeting*, pp. 28.6.1–28.6.4, 2011.

[113] Eneman, G., et al., "Stress simulations for optimal mobility group IV p- and nMOS FinFETs for the 14 nm node and beyond," *2012 International Electron Devices Meeting*, pp. 6.5.1–6.5.4, 2012.

[114] Hashemi, P., et al., "High-performance Si1−xGex channel on insulator trigate PFETs featuring an implant-free process and aggressively-scaled fin and gate dimensions," *2013 Symposium on VLSI Technology*, pp. T18–T19, 2013.

[115] Hellings, G., et al., "Implant-Free SiGe Quantum Well pFET: A novel, highly scalable and low thermal budget device, featuring raised source/drain and high-mobility channel," *International Electron Devices Meeting,* pp. 10.4.1–10.4.4, 2010.

[116] Hikavyy, A., et al., "Growth of high Ge content SiGe on (110) oriented Si wafers", *Thin Solid Films*, vol. 520, no. 8, pp. 3179–3184, 2012.

[117] Claeys, C., and E. Simoen, *Germanium-Based Technologies,* Elsevier, Amsterdam, 2007.

[118] Yu, E., W-J. Lee, J. Jung, and S. Cho, "SiGe Heterojunction FinFET Towards Tera-Hertz Applications", *Journal of the Korean Physical Society*, vol. 72, no. 4, pp. 527–532, 2018.

[119] Guo, D., et al., "FINFET technology featuring high mobility SiGe channel for 10nm and beyond," *IEEE Symposium on VLSI Technology*, pp. 1–2, 2016.

[120] Tsutsui, G., et al., "Technology viable DC performance elements for Si/SiGe channel CMOS FinFTT," *2016 IEEE International Electron Devices Meeting (IEDM)*, pp. 17.4.1–17.4.4, 2016.

[121] Oh, J., et al., "Thermally robust phosphorous nitride interface passivation for InGaAs self-aligned gate-first n-MOSFET integrated with high-k dielectric," *2009 IEEE International Electron Devices Meeting (IEDM)*, pp. 1–4, 2009.

[122] Witters, L., et al., "Strained germanium gate-all-around PMOS device demonstration using selective wire release etch prior to replacement metal gate deposition," *2017 Symposium on VLSI Technology*, pp. T194–T195, 2017.

[123] Tsutsui, Gen et al., "Leakage aware Si/SiGe CMOS FinFET for low power applications", *2018 IEEE Symposium on VLSI Technology*, Honolulu, HI, USA, pp. 87–88, 2018.

7 Switching Performance Analysis of III-V FinFETs

Arighna Basak, Arpan Deyasi, Kalyan Biswas, and Angsuman Sarkar

CONTENTS

7.1 BACKGROUND AND DRIVING FORCES

During the last decades, field-effect transistor technology has changed at an amazing rate and improved its performance gradually. Mostly utilized transistor technology has been based on silicon transistors, but the technology is nearing its limits following scaling rules. Today, lots of efforts are made to expand metal oxide semiconductor field-effect transistors (MOSFETs) technology using III-V group semiconductor materials. These III-V semiconductor-based devices are likely to take the technology more due to enhanced material properties such as higher electron mobility [1].

To attain high currents per unit area through the channel, high electron concentration in the channel as well as high drift velocity for electron is required. These high electron concentration and higher mobility help to achieve better transconductance

DOI: 10.1201/9781003200987-7

and lower source and drain parasitic resistances. Therefore, integration of channel materials in CMOS technology having high mobility became a topic of intense research [2–3]. Most III-V group materials of the periodic table (compound materials like InGaAs, InAs, GaAs, and InSb) can present enormously high electron mobility. InGaAs-based channel materials have an added advantage that allows tuning their bandgap by changing the mole fraction of the composition. When III-V group materials are appropriate for n-channel MOSFET operations, Ge, a group IV material and having relatively higher hole mobility, is ideal for p-channel MOSFET. However, mobility in these channel materials is inversely proportional to bandgap (E_G). To achieve higher mobility lower E_G is desirable but that in turn increases the OFF-state leakage current due to tunneling effect from drain to bulk or drain to source. Smaller E_G also degrades scalability of the device and worsens SCEs. Therefore, proper study and careful analysis are very much important to integrate these high-mobility channel materials and improve performance for the emerging nanoscale devices.

To allow higher ON current and transconductance of the MOSFET device as compared to its Si-built counterpart, high-mobility channel materials are considered. InGaAs-based channel materials have added advantages in terms of tunable bandgap by varying the Indium mole fraction of the compound semiconductor [4]. Higher electron mobility and better performance are also possible using a suitable barrier layer [5]. Therefore, study and performance investigation of InGaAs/InP material-based MOSFETs varying barrier layer properties and channel composition are vital.

7.2 INTRODUCTION

Since birth, there has been a rapid expansion of the IC industry in the last 50 years, which is vividly introspected by a rapid reduction of separating distance between any two elements inside the chip (pitch), and which results in faster data transfer and processing rate. Functionalities are sequentially added to satisfy the ever-increasing customer requirement, and as a result of that, the size of the chip starts progressively increasing [6]. Simultaneously, usage of power is drastically reduced, thanks to the development ofthe fabrication industry and chronological progression of low-dimensional devices. From a customer pointofview, the price of logic gates or memory elements dropped severely. Miniaturized electronic components are available in the market, but with increased complexity, and the majority of them have the short-channel problem [7]. In the truest sense, it may be noted that technology is forced to be manifested due to the surge of economy.

In the last 20 years, several methodologies emergedto tackle the problem of short-channel effect, which has risen owing to the reduction of gate length. A key wayout is to incorporate more gate terminals so that gate control can somewhat neutralize the divergence behavior of channel electrons. This concept results in various novel architectures like double-gate MOSFET [8–9], tri-gate MOSFET [10–11], gate-all-around FET [12–13], CNTFET [14–15], etc. Another technique is obviously to introduce high-K dielectric materials [16–18]. However, one of the major shortcomingsthat still remains in all these methodologies is the reduction of subthreshold swing within desirable limit and also nullifying the origin of undesirable parasitic capacitances. Though some literature has reported the better result for the first problem, simultaneous tackling of both still remains a dream, even in case of quantum-wire FET. Leakage current reduction causes havoc in chip-level performance, and therefore, scientists are looking into alternative approaches to pacify both the undesired effects. Precisely, in sub-22 nm era, controlling leakage current remains a dream by the existing architectures. This forms the backbone and dire need of unconventional and novel designs, which can be realized in the existing fablabs.

One of the solutions provided by researchers is FinFET architecture [19–21], where both previously mentioned problems can be tackled, but at the cost of complexity. In other words, beyond 22 nm, CMOS structure gets the most promising replacement by FinFET, as reported in the literature in the last decade [22]; though a lot of fabrication challenges make physical realization, and therefore, circuit implementation, difficult. Problems in patterning, doping, and higher access resistance make it extremely difficult for growth; however, various engineering ways are devised to reduce those factors so that novel memory can be proposed [23]. The prime advantage of replacing MOSFET by FinFET is its quasi-planar behavior with better noise control like substrate noise, cross-talk noise, charge sharing noise, ground noise, and leakage noise [24]. These together effectively make higher ON-to-OFF current ratio and lower average power consumption. It is already reported that FinFET overcomes existing short-channel effect along with exhibition of high driving current [25], which supports its candidacy in making high-speed switching device. Through channel stacking, area overhead problem can drastically be reduced in FinFET [26], which favors logic-based circuit design.

Results for delay and average power consumption for FinFET-based logic gates at different technology nodes [22] are computed, which states in favor of rapid reduction of power at lower gate length. However, delay increases, which demands optimization [23]. Here comes the importance of the predictive technology model, which has been very recently proposed [24] and has become quite acceptable among the FinFET research community. The role of predictive technology models are also emphasized in the literature where Fin height, pitch, and width are varied for optimum performance. The results are also verified using SPICE models.

From a geometrical pointofview, FinFET is a thin bodywith multiple-gate control. It supports double-gate, tri-gate, and various multi-gate configurations with varying geometries. FinFET also supports SOI substrates. It is a three-dimensional transistor that is used to make microchips, as reported by the premier industry, due to its improved gatecontrollability and quasi-planar configuration [25]. Among different FinFETs, vertical structures become popular owing to possible applications in high-power systems. III-V materials are considered due to switching applications. It is the best probable replacement of conventional MOSFET precisely because of its subthreshold characteristics. It belongs to the SOI family, as the conducting channel wraps around the Fin. The device can be made fully or partially depleted. For partially depleted structure, SOI layer width is much thicker than gate depletion width [26]. However, if fully depleted architecture is considered, then the top layer of the silicon is very thin compared to the insulator layer beneath. Multi-gate architectural concept is incorporated inside the conventional FinFET [27], for which double-gate and tri-gate FinFETs are realized in the last decades. Because of their quasi-planar structure, they exhibit greater control over barrier potential, and thereby effectively reduce the DIBL.

7.3 III-V MATERIAL-BASED MOSFETs

To use in high-performance digital logic applications, MOSFETs made of III-V group semiconductor materials are considered one of the capable device structures [28–31]. Currently, it is projected that III-V group materials-based MOSFETs may obtain more drive current and higher transconductance in comparison to Si-based similar devices [28].

Though silicon is the mostly used material in the electronics industry, other heterostructure materials are also used. A wide range of compound semiconductor materials can be obtained using elements of the periodic table's columns III and V, like GaAs, InP, $In_xGa(1\text{-}x)As$, etc. The main parameter that defines the important characteristics of these materials is the bandgap energy. Figure 7.1 shows the bandgap energy for different materials such as germanium, silicon, and other III-V compounds used in the semiconductor industry [29]. Desired bandgap energy as well as electron mobility of the channel may be obtained by using different mole fraction of the materials during fabrication. Lower effective mass (m^*) of InGaAs/InAs in comparison to silicon causes the high electron mobility and high injection velocity. Injection velocity and electron mobility are related to effective electron mass with the following relations as expressed in Equation 7.1.

$$\mu \approx \frac{1}{m^*} \text{ and } v_{inj} \approx \sqrt{\frac{1}{m^*}} \tag{7.1}$$

Few important material properties of III-V group materials along with Si are tabulated in Table 7.1. These smaller effective masses also provide higher tunneling probability, which ensures its better ON-state performance of the nanoscale heterostructure transistors [32–33].

Besides greater transport properties, compound semiconductors based on III-V group materials also enjoy the matured fabrication technology. Few challenges remain in terms of high-k gate dielectrics with lower interface trap density that is crucial for MOSFET operations. Another major challenge is the manufacturing scheme suitable to integrate III-V materials on Si wafers. However, it is already observed that InGAs is the intensively studied semiconducting material and front-runner to replace Si in the transistor channel in the 10-nm CMOS logic technology node or beyond [33–35]. Many researchers have already reported its digital, analog, and RF performance parameters [36–37]. Many heterostructure MOSFETs based on III-V group material have been projected and compared with their similar device made of silicon. From published results, it is implicit that the III-V group materials-based heterostructure device showed exceptional control of SCEs, higher on current (I_{ON}), lower delay than that of silicon-based device of the same dimension [37]. It is important to mention that the thickness of SiO_2 as gate dielectric material has been reduced to the value where the direct tunneling may happen and mostly increases the leakage current, affecting its utility in circuit operation. To overcome this problem, materials with high dielectric constant have been introduced [38]. Hafnium-based oxides (HfO_2) is one of the many contenders of potential high-k

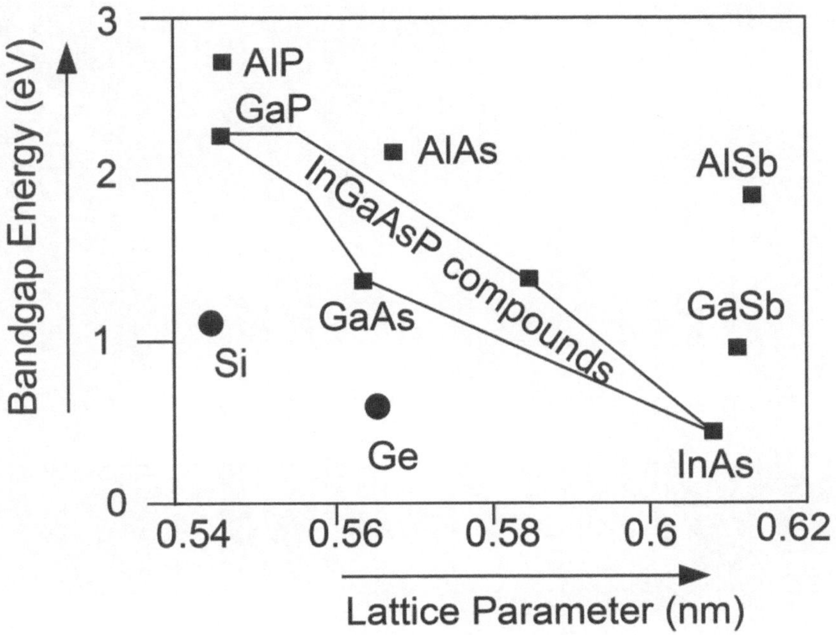

FIGURE 7.1 Bandgap Energy of Si, Ge, and III-V Compound Semiconductors

TABLE 7.1
Bulk Carrier Mobility of Different Channel Materials

Material	Relative permittivity	Mobility (cm^2/V.s)	
		Electron	Hole
Si	11.9	1,400	470
Ge	16.2	3,900	1,900
GaAs	12.9	8,500	400
InAs	15.2	40,000	500
InSb	16.8	77,000	850

gate dielectrics that have been recommended to replace SiO_2. In recent studies, the asymmetric behavior of the underlap DG heterostructure MOSFET was also analyzed and reported [39].

7.4 FinFETs Device Characteristics

Considering the structures, the most radical change in MOSFET structure happened with the invention of FinFET. Professor Chenming Hu of the University of California and his team were the first to use the term as the channel shape of the device in 3-D

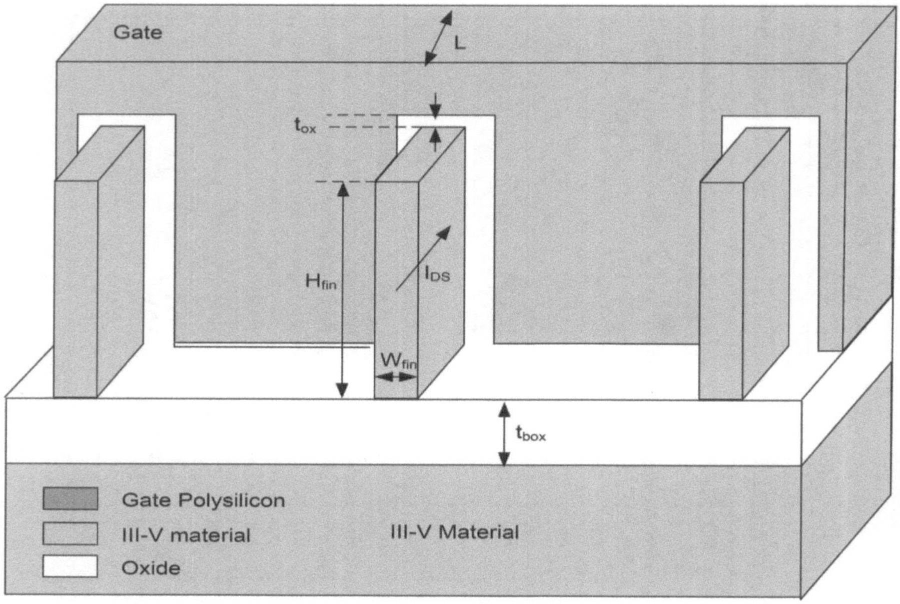

FIGURE 7.2 Structure of FinFET

structure [40] looks like a fin. The name FinFET technology was appropriate as the FET structure is like a set of fins when viewed from either source or drain side. A schematic view of the FinFET device structure is shown in Figure 7.2.

FinFET technology has been the natural choice as it helps in the persistent increase in the levels of integration [41]. Following "Moore's Law", to achieve the higher levels of integration, many device parameters have been changed [42–43]. Primarily to fabricate more devices in a given area, the feature sizes have been smaller and smaller. However, many important figures such as line voltage and power dissipation have been reduced along with an increase in frequency performance. But as the process technologies reached near 22 nm node, the appropriate scaling of several device parametersbecame unachievable. Power supply voltage, one of the main factors in decisive dynamic power, was predominantly affected. It is understood that optimizing for performance of the device resulted in unwanted compromise in other areas such as power. All these issues made it necessary to think beyond the traditional planar transistor and look at some more revolutionary options. In 2011, Intel was the first company to use FinFET in mass production. FinFET provided a significant performance improvement and also provided lower leakage current in comparison to a planar MOSFET as shown in Figure 7.3 [44].

The original structure of FinFET had thick oxide on Fin top of channel and used SOI for process simplicity. The triple-gate MOSFET consists of a thin-film, narrow silicon channel with gate on all the three sides [45]. Because of its 3-D structure, tri-gate FET forms a conducting channel on three sides of a vertical Fin configuration, providing "fully depleted" operation. To improve the device performance

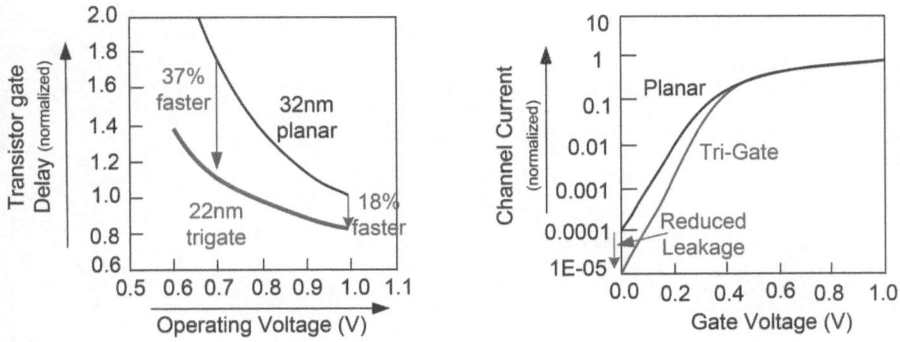

FIGURE 7.3 Performance and Leakage Current Comparison between Planar MOSFET and Fin-FET

(From Mark Bohr, Intel.)

further, multiple Fins may be connected together resulting in a higher drive current. Implementations of these related technologies include the quantum-wire silicon-on-insulator MOSFET [45–46] and the triple-gate MOSFET [47–48].

With the intention of improving the electrostatic integrity of the triple-gate MOSFETs, the sidewall regions of the gate are extended to some distance in the buried oxide situated below the channel region to form Π-gate device [49–50] and Ω-gate device [51–53]. These Π-gate device and Ω-gate device eventually increase the number of gates from 3 to 3.5. Other than only structural change, the use of strained silicon technology, a metal gate along with "high-k" dielectric as gate insulator helps in further boosting the drive current of the device [54–55]. From literature study, it is understood that FinFETs structure and related technology have already been considered as the number one solution at the technology node beyond 22nm because of its superior performance and improved gate controllability in comparison to its planar counterparts. They also provide superior I_{ON}/I_{OFF} ratio and display higher current density in comparison to planar structures. Owing to these advantages, FinFETs find a number of applications in areas like high-speed digital ICs, analog ICs, SRAMs, DRAMs, and flash memories. With so many recent researchers already proved that tri-gate transistors are significant innovation required to continue "Moore's Law".

More recently, it is established that "multiple surrounding-gate" channels may be stacked keeping gate, source, and drain terminals common to increase the further current drive per unit area. Other multi-gate MOSFET structures are available in literature. Cross-sections of the channel region for the different gate orientations available in literature are shown in Figure 7.4 [56].

Researchers have already revealed that performance parameters of FinFETs are greatly dependent on Fin geometry [57–58]. Recent studies on Fin shape focused on assessing the effect of Fin cross-sectional shape on SCEs, and findings on leakage current variation were found [58]. The effect of variability in metal grain work function and Fin-edge roughness variability affecting device characteristics are studied and

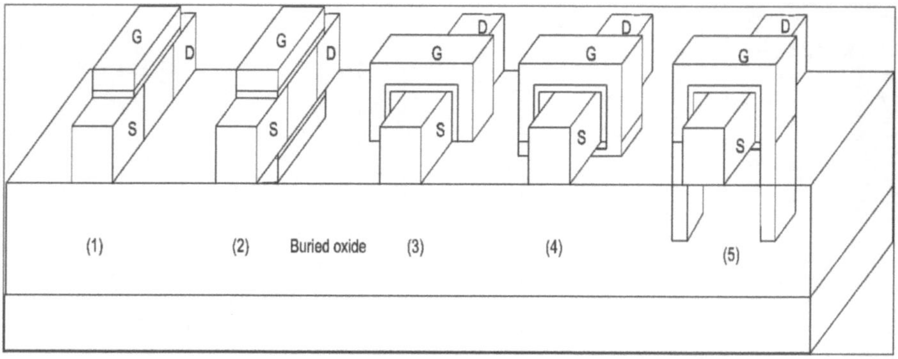

FIGURE 7.4 Cross-Sectional View of Different Gate Structures [56]

reported for FinFET [59]. Other than SOI wafers, multiple-gate devices can also be fabricated on bulk silicon wafers. Advantages of bulk wafer technology include lower wafer cost and better substrate heat dissipation rate [60]. However, in bulk FinFET fabrication processes, additional isolation steps are required that add cost to the device.

7.5 SWITCHING CHARACTERISTICS OF FinFETs

The switching characteristics of FinFET structure have not been explored thus far in great detail, precisely for III-V materials. Studies have reported mostly on the figure of merit, and a few on MHz region switching performance, related to the split-gate (SG) architecture. This has also been proposed to reduce resistance, and therefore improve switching characteristics. This feature basically reduces the gate power dissipation [61], as gate-to-drain capacitance is decreased. It also enhances the drift doping concentration, as a result of which transconductance is increased. High-frequency (~MHz) switching characteristics with superior figure of merit are exhibited by vertical GaN FinFET [62], which also demonstrates better reliability when integrated into power applications. It has also higher breakdown voltage for specified chip cross-section compared to other similar devices with alternate material system and provides better thermal management.

Recent works also suggested that device turn-off speed can be reduced by split-gate (SG) structure, and simultaneously switching loss is also decreased [63]. This enables the device for medium-power application. Authors also suggested that ON-to-OFF current ratio can greatly be enhanced through source-drain underlapping [64]. Research suggests that among other existing architectures, high-K stack is always considered for mobility enhancement [65], as it reduces dielectric capacitance. Vertical architectures are recently considered for improved electrical performance compared to conventional FinFET [63], precisely the parasitic capacitances, and therefore, switching characteristics are improved a lot. Less switching time and reduced power loss are reported for single-gate structure. In recent past, models are also presented, including fringing capacitance. This helps in near accurate modeling and determination of parasitic capacitances [66], which are extremely critical for analog/RF performance.

Major improvement of switching characteristics is obtained by introducing various layers in the conventional vertical FinFET and also by varying Fin aspect ratio [67]. Different oxide materials are also used to reduce the dielectric capacitances, which are suitable for low-waste power conversion applications.

This brief survey work justifies the background of research on FinFET and the corresponding significance of the mathematical modeling for near-accurate estimation of subthreshold swing. This will play a pivotal role in determining switching behavior. In the next section, we therefore express the review of III-V materials for FinFET fabrication and fabrication challenges inthe analysis in brief.

7.6 REVIEW ON III-V MATERIAL-BASED DEVICES

In this section, III-V materials for both n- and p-MOSFET devices and related advancements are reviewed, including development in gate stack technology and its allied reliability. Advantages of III-V materials in advanced CMOS technology nodes may be utilized with full potential, if it is integrated on 300 nm wafers, which can add the benefits of high-volume manufacturing technology of CMOS technology established by the big foundries. So, lots of attention is given on integration of III-V devices onto an Si platform. A cross-sectional view of a fabricated InGaAs FinFET is shown in Figure 7.5.

The planar complementary metaloxide semiconductor (CMOS) transistor is currently being changed by three-dimensional (3-D) Fin structures to allow better

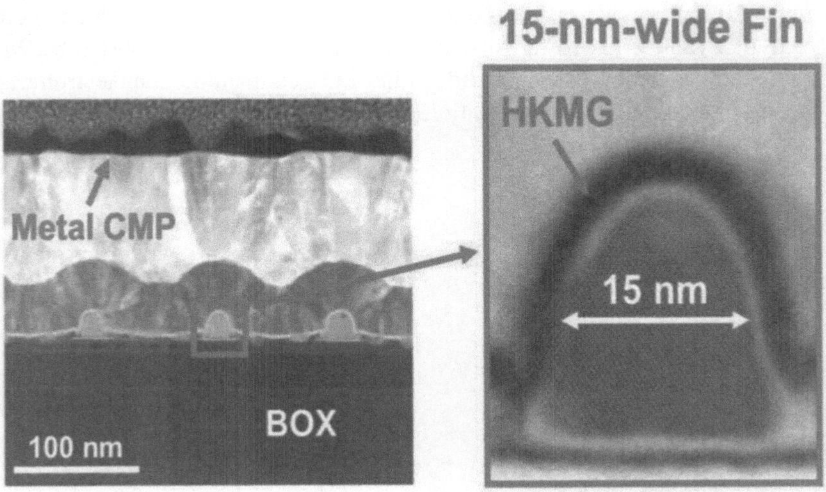

FIGURE 7.5 A Cross-Section of the Self-Aligned InGaAs-OI FinFET Architecture

(Figure courtesy of Henry H. Radamson, taken from the paper "Miniaturization of CMOS". Published in *Micromachines* 2019, 10, 293. https://doi.org/10.3390/mi10050293 [67].)

access for the gate electrode on three sides of the channel, and the nanowire transistors come in the next stage with the gate surrounded from all four sides. Intel is already manufacturing FinFETs made of silicon in some of its product line (since 2011). FinFETs may be viewed as intermediate between planar and gate-all-around nanowire transistors. These improvements for silicon have their counterparts in III-V research. In the recent past, various demonstrations of FinFETs using compound semiconductor of InGaAs materials have been reported. In almost all cases, Fins are fabricated using reactive ion etching (RIE) techniques.

To target sub-7 nm CMOS applications, research interest in InGaAs MOSFETs has shifted from planar to 3-D device architectures, including FinFET, tri-gate, and gate-all-around transistor designs. Scaling of Fin width for InGaAs FinFET is going on aggressively. Recently, the scaled InGaAs FinFETs are demonstrated with Fin widths down to 2.5 nm, combining atomic layer etching (ALE) with in situ ALD for the first time [68].

The first inversion-mode InGaAs FinFET with gate length down to 100 nm with ALD Al2O3 as gate dielectric was developed and demonstrated by Wuetal. [69]. Detailed analysis report along with obtained values of SS, DIBL, gate leakage current, and V_T roll-off was presented. A schematic view of the series of FinFETs along with a cross-sectional view is shown in Figure 7.6.

The researchers claim the first demonstration of an unstrained indium arsenide (InAs) Fin field-effect transistor (FinFET) with 20nm fin height (H_{fin}) [68].

InAs material-based FinFET devices having Fin widths of the range 25–35 nm were also reported [71]. Devices consists of 10 Fins, each having a height (H_{fin}) of 20 nm, and the length of gate, L_g = 1 μm. The source and drain regions are 3 μm wide, and the length of the un-gated Fin extension and source-drain extensions were both 50 nm, which means a total gate-to-source distance of 100 nm. For the high-k gate stack, the effective oxide thickness (EOT) is 1.2 nm. In that report, an InAs FinFET with H_{fin} = 20 nm provided peak transconductance of 1,420 μS/μm, which is comparable to that value attained for InAs channel planar devices as reported by Wang et al. [72] with identical gate length and similar EOT confirming that electron transport properties like field-effect mobility for the Fins having dimensions in the range

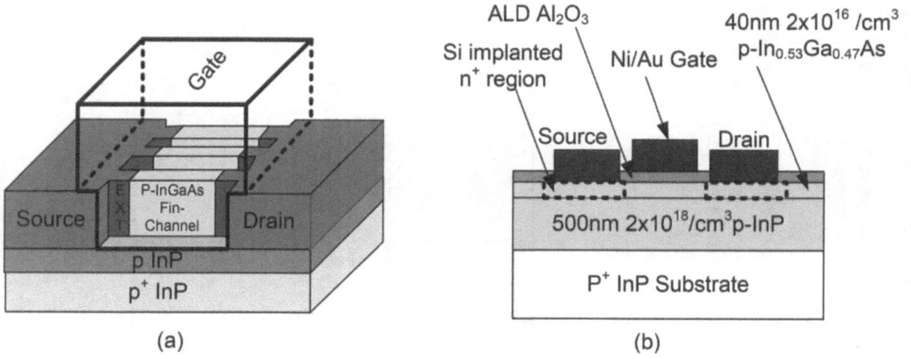

FIGURE 7.6 Schematic View of the Series of FinFETs along with Cross-Sectional View

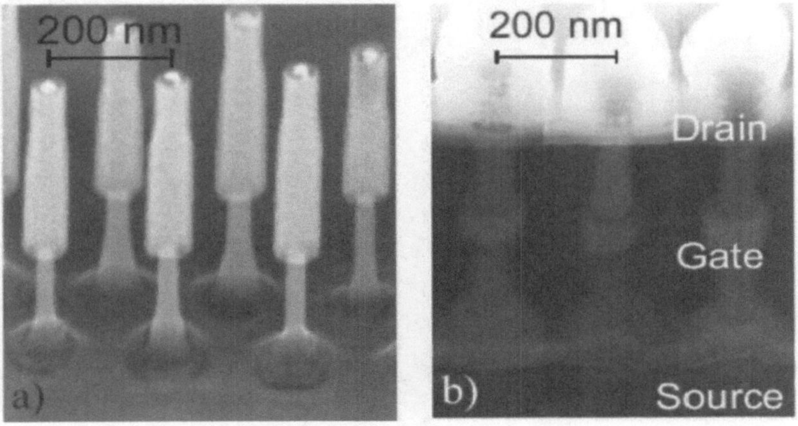

FIGURE 7.7 InAs Vertical Gate-All-Around FET (a) Nanowires after Thinning of Channel and (b) Final Device Structure

(Figure courtesy of Henry H. Radamson, taken from the paper "State of the Art and Future Perspectives in Advanced CMOS Technology", *Nanomaterials* 2020, 10, 1555. https://doi.org/10.3390/nano10081555 [70].)

of 25–35 nm closely match with planar InAs technology. Tilt views of the SEMs for InAs vertical gate-all-around FET are shown in Figure 7.7 [70].

An n-channel InGaAs on insulator FinFET process and the highest ON current to date for CMOS-compatible InGaAs devices integrated on silicon (Si) were reported by Djara et al. [70].

The CMOS-compatible fabrication flow and cross-section schematic of the self-aligned InGaAs-OI FinFET are presented in Figure 7.8(a) and Figure 7.8 (b), respectively. The gate leakage current for a 50nm gate-length (LG) transistor was below 400 pA/μm with 0.5 V and gate potential in the range of 0.2V to + 1V. The saturation transconductance peaked at ~615 μS/μm. The saturation subthreshold swing had a minimum of 92 mV/decade. The drain-induced barrier lowering (DIBL) was 57 mV/V. The threshold in saturation was 0.09V. The ON current (I_{ON}) was 156 μA/μm [70].

7.7 CHALLENGES IN FABRICATION

To continue the CMOS scaling path, several new materials such as SiGe stressors in source/drain and introduction of high-k dielectric, such as HfO_2 along with various new architectures like FD-SOI, FinFET, nanowire, etc., are considered. But in the case of Si technology, the major targets are downscaling dimensions to maintain better performance to have new functions on a single chip. To empower the chase of the scaling roadmap following Moore's law, new materials and new architectures need

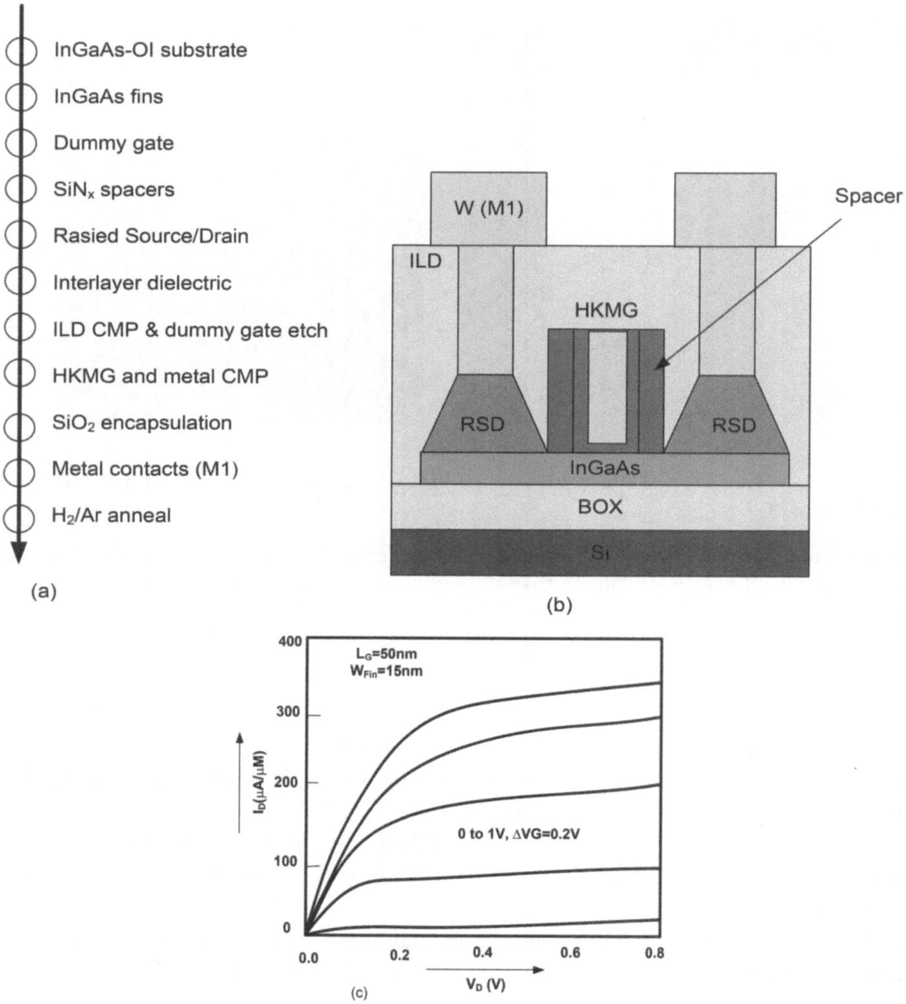

FIGURE 7.8 (a) CMOS-Compatible RMG Fabrication Process Flow, (b) Cross-section Schematic of the Self-Aligned InGaAs-OI FinFET Architecture, and (c) ID-VD Output Characteristics of an InGaAs-OI FinFET with L_g = 50 nm and W_{Fin} = 15 nm

to be integrated on silicon. From that point of view, III-V compounds are noteworthy semiconductor materials for advanced nanoelectronic devices because of their unique intrinsic properties that make them very attractive. Therefore, the key interest is on replacing classical Si channel of the MOSFET devices with III-V channel materials having higher mobility as shown in Figure 7.9. This makes them appropriate to be integrated with Si nanoelectronic devices easily.

However, the obvious material mismatch between InGaAs and Si considering its structures, lattice constants, thermal parameters, polarity, etc. becomes the main challenge in the heterogeneous integration of such radical devices on a substrate made on silicon. Therefore, in such hetero-epitaxial process, this material mismatch

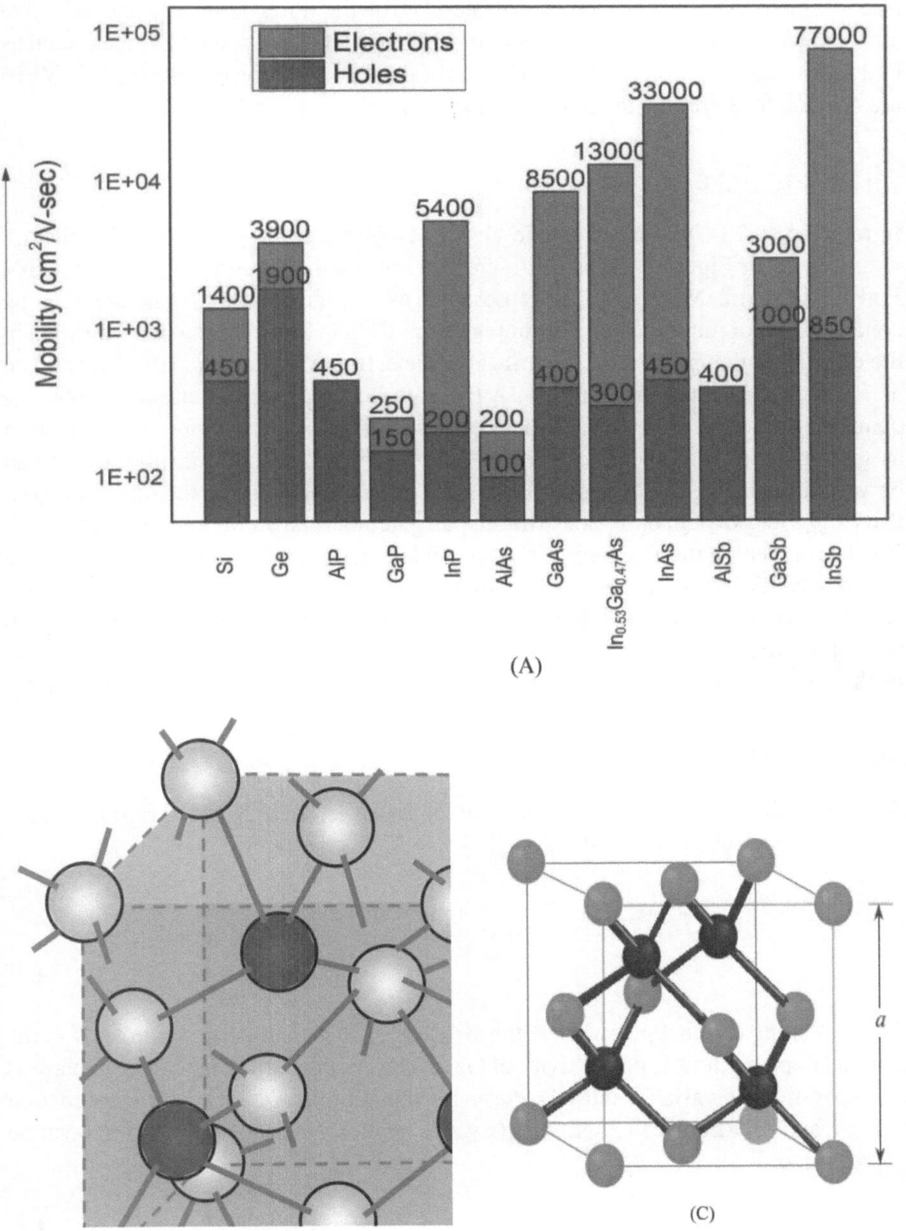

FIGURE 7.9 (a) Electron and Hole Mobility in III-V Semiconductors Compared with Si and Ge semiconductors, (b) Si Diamond Lattice Structure, and (c) III-V Zinc Blended Lattice Structure

will generate some crystalline defects in high density. These ensuing defects like misfit/threading dislocations, stacking faults, or antiphase boundaries and twins will intensely degrade the performance of the device. Several options have been well-thought-out to decrease the defect density in III-V group compounds on Si. Techniques like strain relaxed buffers, epitaxial lateral growth, wafer bonding, and selective area growth accompanied by the two most popular III-V growth techniques, such as molecular beam epitaxy (MBE) and metal-organic vapor-phase epitaxy, (MOVPE) are utilized to overcome the material mismatch-related issues.

7.8 DEVICE STRUCTURE

In references [73–78], the schematic view of III-V material-based FinFET structure is presented in Figure 7.10. W_{fin},H_{fin} are the Fin width and Fin height of the III-V FinFET structure. Moreover, L_g is represented as gate length of this structure. Doping concentration of the source and drain regions is $10^{21}/cm^3$, and doping concentration in the channel region is $10^{18}/cm^3$. High-K gate dielectric material (e.g., HfO_2) was taken as an insulating material to diminish the gate leakage current caused by obvious quantum tunneling effect across the gate oxide. Conservatively, polysilicon is used as gate material in conventional MOSFETs. The work function in polysilicon can be varied to change the flat band voltage resulting in a change in the threshold voltage (V_{th}). But polysilicon is not favorable as gate material for nanoscale MOSFETS because of its high thermal budget process and degradation due to the depletion of the doped polysilicon [79]. In this context, metal gate is preferred nowadays over polysilicon [80–81]. Therefore, molybdenum is considered a gate material. Moreover, SiO_2 is used as buried oxide (BOX), and silicon is used as substrate of this structure. The BOX layer thickness of 10 nm and substrate thickness of 10 nm are used in this work.

7.9 MATHEMATICAL MODELING

By using 3-D Poisson's equation, the channel potential of GAN-based FinFET structure can be expressed as [82–87]:

$$\frac{\partial^2\psi(x,y,z)}{\partial x^2}+\frac{\partial^2\psi(x,y,z)}{\partial y^2}+\frac{\partial^2\psi(x,y,z)}{\partial z^2}=\frac{qN}{\varepsilon} \tag{7.2}$$

where ψ is the channel potential of the structure, q is the electronic charge, N is the channel doping, and ε is permittivity of GaN. The channel potential can be explained by separation of variable intol-D Poisson's equation $\psi_1(x)$, 2-D Laplace equation $\psi_2(x,z)$, and 3-D Laplace equation $\psi_3(x,y,z)$. Therefore, the channel potential can be expressed as

$$\psi(x,y,z)=\psi_1(x)+\psi_2(x,z)+\psi_3(x,y,z) \tag{7.3}$$

This channel potential of the structure can be solved by the following boundary condition.

$$\psi(0,y,z)=V_{GS}-V_{fb};\qquad \psi(W_{fin},y,z)=V_{GS}-V_{fb} \tag{7.4}$$

$$\psi_2(0,z)=0,\qquad \psi_2(W_{fin},z)=0 \tag{7.5}$$

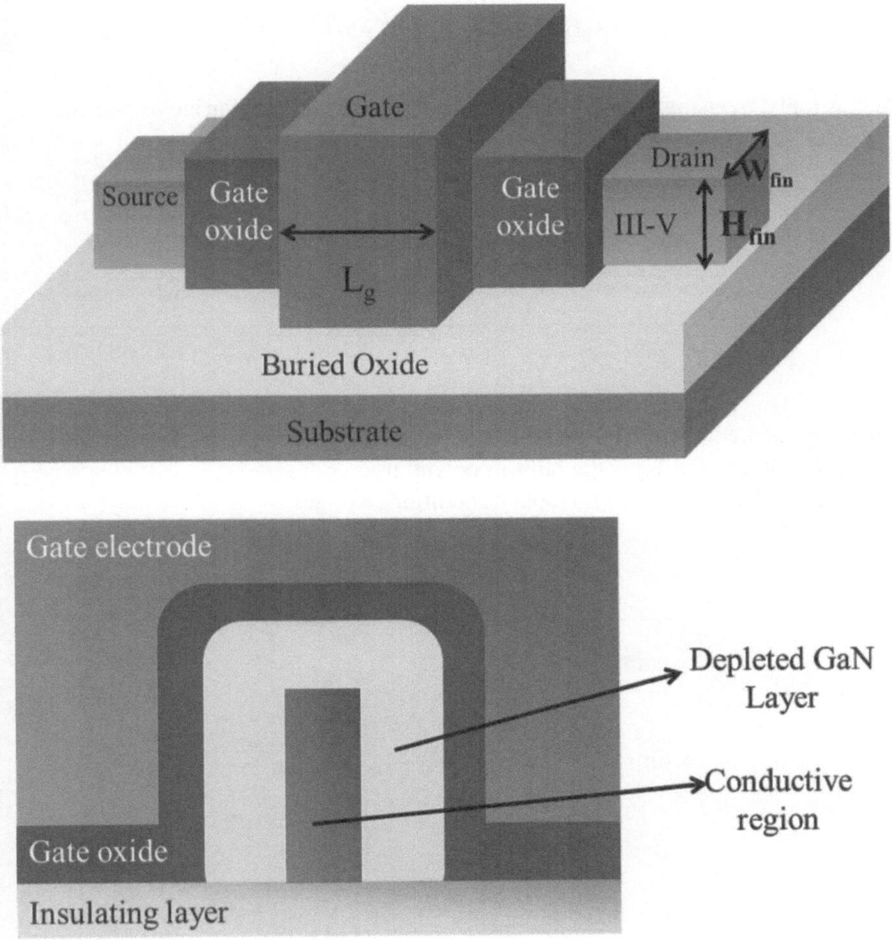

FIGURE 7.10 3-D Schematic View of III-V Material-based FinFET Structure with Wide Nano Channel (Wide Fin Shape)

$$\psi_2(\mathrm{x},0) = V_{bi} - \psi_1(x); \qquad \psi_2(\mathrm{x},\mathrm{L}_g) = V_{bi} + V_{DS} - \psi_1(x) \tag{7.6}$$

$$\psi_3(0,\mathrm{y},z) = 0; \qquad \psi_3(\mathrm{W}_{fin},\mathrm{y},z) = 0 \tag{7.7}$$

$$\frac{\partial \psi_3(x,0,z)}{\partial y} = -\psi_1(x) - \psi_2(\mathrm{x},\mathrm{z}) \tag{7.8}$$

$$\psi_3(\mathrm{x}, H_{fin}, z) = V_{GS} - V_{fb} - -\psi_1(x) - \psi_2(\mathrm{x},\mathrm{z})$$

$$\psi_3(x,y,0)=0; \qquad \psi_3(x,y,L_g)=0 \tag{7.9}$$

Accordingly, by using boundary conditions and separation of variables, we get:

$$\psi_1(x)=\frac{qNx^2}{2\varepsilon}-\frac{qNW_{fin}x}{2\varepsilon}+V_{GS}-V_{fb} \tag{7.10}$$

$$\psi_2(x,z)=\psi\big|_{z=0}+\psi\big|_{z=L_g} \tag{7.11}$$

$$\psi_3(x,y,z)=\psi\big|_{y=0}+\psi\big|_{y=H_{fin}} \tag{7.12}$$

where W_{fin} is the fin width of the structure, V_{GS} is the gate to source voltage, V_{fb} is the flat band voltage, and V_{bi} is the built-in potential.

Further, after solution of boundary conditions, we get

$$\psi\big|_{z=0}=\sum_{n=1}^{\infty}\left[\begin{array}{l}\dfrac{\frac{2}{(2n-1)}}{\pi\sinh\left(\frac{(2n-1)\pi}{W_{fin}}L_g\right)}\sinh\left(\dfrac{(2n-1)\pi}{W_{fin}}(L_g-z)\right)\\ \times\sin\left(\dfrac{(2n-1)\pi}{W_{fin}}x\right)\left\{2\left(V_{bi}-V_{GS}+V_{fb}\right)-\dfrac{2qNW_{fin}^{\;2}}{(2n-1)^2\pi^2\varepsilon}\right\}\end{array}\right] \tag{7.13}$$

$$\psi\big|_{z=L_g}=\sum_{n=1}^{\infty}\left[\begin{array}{l}\dfrac{\frac{2}{(2n-1)}}{\pi\sinh\left(\frac{(2n-1)\pi}{W_{fin}}L_g\right)}\sinh\left(\dfrac{(2n-1)\pi}{W_{fin}}(z)\right)\\ \times\sin\left(\dfrac{(2n-1)\pi}{W_{fin}}x\right)\left\{2\left(V_{bi}+V_{DS}-V_{GS}+V_{fb}\right)-\dfrac{2qNW_{fin}^{\;2}}{(2n-1)^2\pi^2\varepsilon}\right\}\end{array}\right] \tag{7.14}$$

$$\psi\big|_{y=0}=\sum_{n=1}^{\infty}\sum_{m=1}^{\infty}\left[\begin{array}{l}\left(\psi_1(x)-\psi_2(x,z)\right)\dfrac{16}{(2m-1)(2n-1)\pi^2}\left(\dfrac{-K_z\pi}{H_{fin}}\right)\\ \times\sin\left(\dfrac{(2n-1)\pi}{W_{fin}}x\right)\sin\left(\dfrac{(2m-1)\pi}{L_g}z\right)\dfrac{\cosh\left(\frac{K_z\pi}{H_{fin}}(H_{fin}-y)\right)}{\cosh(K_z\pi)}\end{array}\right] \tag{7.15}$$

$$\psi\big|_{y=H_{fin}}=\sum_{n=1}^{\infty}\sum_{m=1}^{\infty}\left[\begin{array}{l}\left(V_{GS}-V_{fb}-\psi_1(x)-\psi_2(\mathrm{x,z})\right)\dfrac{16}{(2n-1)\pi^2}\\ \times\sin\left(\dfrac{(2n-1)\pi}{W_{fin}}x\right)\sin\left(\dfrac{(2m-1)\pi}{L_g}z\right)\dfrac{\cosh\left(\dfrac{K_z\pi}{H_{fin}}(H_{fin}-y)\right)}{\cosh(K_z\pi)}\end{array}\right] \tag{7.16}$$

$$K_z=\sqrt{\left(\frac{2n-1}{W_{fin}}\right)^2+\left(\frac{2m-1}{L_g}\right)^2}H_{fin}$$

Moreover, the minimum point $z_{\min}$ along the length can be expressed as

$$z_{\min}=0.5L_g+\left(\frac{0.5W_{fin}}{\pi}\right)\ln\left(\frac{V_{bi}-V_{GS}+V_{fb}}{V_{bi}+V_{DS}-V_{GS}+V_{fb}}\right)$$

To calculate the threshold voltage V_{th} for $\psi(W_{fin}/2, H_{fin}, z_{\min})$, the value of V_{GS} is $2\Phi_F + \Phi_z$, where Φ_F is the Fermi potential and $\Phi_z = 5\Phi_t$, where Φ_t is the thermal voltage.

Therefore, we enlarge the sine hyperbolic, sine term, and $z_{\min}$ terms and retain only the first-order terms to obtain an approximate V_{th}, which can be expressed as

$$V_{th}=\frac{-u_2+\sqrt{u_2^{\,2}-4u_3u_4u_5}}{2u_1} \tag{7.17}$$

where,

$$u_1=\left[\begin{array}{l}\left(\dfrac{\left(W_{fin}\left(\dfrac{1}{V_{bi}+V_{fb}}\right)-\dfrac{1}{(V_{bi}+V_{fb}+V_{DS})}\right)}{2\pi}\right)\left(\dfrac{16}{(2n-1)\pi L_g}(\cosh(K_z\pi))^{-1}\right)\\ \left(2\left(\left(\dfrac{8}{\sinh\left(\dfrac{(2n-1)\pi}{W_{fin}}L_g\right)}\right)+ep_5\left(p_3+\left(\dfrac{16}{(2n-1)^2\pi^2}\right)\right)\right)-2+\left(\dfrac{1}{L_g}\right)\right)(a+1)\end{array}\right]$$

$$u_2 = \left[\begin{array}{l} W_{fin}\dfrac{L_g}{2}\left(\left(-\dfrac{K_z\pi}{H_{fin}}\right)+1\right)\left(\left(\left(\dfrac{8}{\sinh\left(\dfrac{(2n-1)\pi}{W_{fin}}L_g\right)}+e\left(\dfrac{8}{\sinh\left(\dfrac{(2n-1)\pi}{W_{fin}}L_g\right)}\right)\right)\left(p_3+\left(\dfrac{16}{(2n-1)^2\pi^2}\right)\right)\right)-1\right) \\ +1-\left(\dfrac{8}{\sinh\left(\dfrac{(2n-1)\pi}{W_{fin}}L_g\right)}+e\left(\dfrac{8}{\sinh\left(\dfrac{(2n-1)\pi}{W_{fin}}L_g\right)}\right)\right)\left(p_3+\left(\dfrac{16}{(2n-1)^2\pi^2}\right)\right) \\ +W_{fin}\left\{ \begin{array}{l} \left(\dfrac{1}{2}+\dfrac{W_{fin}}{2\pi}\left(\ln\left(V_{bi}+V_{fb}\right)-\ln\left(V_{bi}+V_{fb}+V_{DS}\right)\right)\right)\left(\left(\dfrac{1}{L_g}+\left(\begin{array}{l}(ek_5)\left(p_3+\left(\dfrac{16}{(2n-1)^2\pi^2}\right)\right)\\ +\dfrac{8}{\sinh\left(\dfrac{(2n-1)\pi}{W_{fin}}L_g\right)}\end{array}\right)\right)-1\right) \\ \left(\left(-\dfrac{K_z\pi}{H_{fin}}\right)+1\right)-\left(-\dfrac{K_z\pi}{H_{fin}}\right)ep_{11}-pk_{12} \end{array} \right\} \end{array} \right]$$

$$u_3 = \left(\left(-\frac{K_z\pi}{H_{fin}}\right)\left(W_{fin}\left(p_{11}+p_{12}\right)\right)\right)$$

$$u_4 = \left(\left(eW_{fin}\right)\left(2\left(\left(\left(\frac{8}{\sinh\left(\dfrac{(2n-1)\pi}{W_{fin}}L_g\right)}\right)+ep_5\left(p_3+\left(\frac{16}{(2n-1)^2\pi^2}\right)\right)\right)\right)-2+\left(\frac{1}{L_g}\right)\right)(a+1)\right)$$

$$u_5 = \left(p_{11}-\left(2\phi_F+\phi_z\right)-\left(\frac{L_g}{2}+\frac{W_{fin}}{2\pi}\left(\ln\left(V_{bi}+V_{fb}\right)-\ln\left(V_{bi}+V_{fb}+V_{DS}\right)\right)\right)\right)$$

$$p_{11} = \left(\left(-\frac{qNW_{fin}{}^2}{8\varepsilon}\right)-V_{fb}\right)+\left(p_7+\left(\frac{8L_g p_5}{(2n-1)^2\pi^2}\right)\right)$$

$$p_3 = 2V_{bi}+2V_{fb}+V_{DS}$$

$$p_{11} = \left(\left(-\frac{qNW_{fin}^{\ 2}}{8\varepsilon}\right) - V_{fb}\right) + \left(p_7 + \left(\frac{8L_g p_5}{(2n-1)^2 \pi^2}\right)\right)$$

$$p_1 = \left(\left(-\frac{qNW_{fin}^{\ 2}}{8\varepsilon}\right) - V_{fb}\right)$$

$$p_{10} = \left(\left(\frac{L_g}{2}\left(2V_{bi} + 2V_{fb} + V_{DS}\right)\right) + V_{DS}\right)\left(\frac{8}{\sin\left(\frac{(2n-1)\pi}{W_{fin}} L_g\right)}\right) + \left(\frac{8L_g p_5}{(2n-1)^2 \pi^2}\right)$$

$$p_7 = \left(\left(\frac{L_g}{2}\left(2V_{bi} + 2V_{fb} + V_{DS}\right)\right) + V_{DS}\right)\left(\frac{8}{\sin\left(\frac{(2n-1)\pi}{W_{fin}} L_g\right)}\right)$$

$$p_{12} = p_1 + p_{10} + V_{fb}$$

Moreover, the subthreshold swing of this structure can be developed by solving 3-D Poisson's equation in the channel region and can be expressed as

$$S = \left(\frac{q}{2.3KT}\frac{\partial\psi(x,y,z)}{\partial V_{GS}}\right)^{-1}$$

$$\Rightarrow S = \left[\frac{q}{2.3KT}\left(\frac{\partial\psi_1}{\partial V_{GS}} + \left.\frac{\partial\psi_2}{\partial V_{GS}}\right|_{z=0} + \left.\frac{\partial\psi_2}{\partial V_{GS}}\right|_{z=L_g} + \left.\frac{\partial\psi_3}{\partial V_{GS}}\right|_{y=0} + \left.\frac{\partial\psi_3}{\partial V_{GS}}\right|_{y=H_{fin}}\right)\right]^{-1} \tag{7.18}$$

Drain current of this structure in the linear region and saturation region can be achieved by the following equations [88–94]:

$$I_{D,lin} = \left((V_{GS} - V_{th})^{\delta}\, V_{DS} - \frac{1}{2}V_{DS}^2\right) \times \left(\frac{2W_{fin}\mu\varepsilon}{H_{fin}(L_g - l_d - \frac{V_{DS}}{E_C})} + \lambda_a \frac{2W_{fin}\varepsilon}{H_{fin}(L_g - l_d)^2}\right) \tag{7.19}$$

$$I_{d,sat}=\left(\beta(V_{GS}-V_{th})^{\delta}\,V_{Dsat}-\frac{1}{2}V_{Dsat}^{2}\right)\times\left(\frac{2W_{fin}\mu\varepsilon}{H_{fin}(L_g-l_d-\frac{V_{DS}}{E_C})}+\lambda_a\frac{2W_{fin}\varepsilon}{H_{fin}(L_g-l_d)^2}\right) \tag{7.20}$$

where μ is electrons mobility, L_g is the channel effective length, l_d is factor for channel length modulation [95], λ_a is the effect of velocity overshoot [92], and V_{Dsat} denotes the saturation voltage, which is specified by

$V_{Dsat}=\dfrac{V_{GS}-V_{th}}{1+\dfrac{V_{GS}-V_{th}}{L_gE_C}}$ where $E_C=\dfrac{2v_{sat}}{\mu}$, v_{sat} is the saturation velocity of electron [94].

β, δ are the parameters for smoothing the drain current [96].

Moreover, the following formula can be used for calculation of analog and RF performance

Transconductance: $g_m=\dfrac{\partial I_D}{\partial V_{GS}}$

Transconductance generation factor: $TGF=\dfrac{g_m}{I_D}$

Output conductance: $g_d=\dfrac{\partial I_D}{\partial V_{DS}}$

Intrinsic gain: $A_v=\dfrac{g_m}{g_d}$

Cutoff frequency: $f_T=\dfrac{g_m}{2\pi(C_{GD}+C_{GS})}$

where, C_{GS}, C_{GD} are gate-to-source capacitance and gate-to-drain capacitance.
Gain frequency product

$$GPF=\left(\frac{g_m}{g_d}\right)\times f_t=A_v\times f_t \tag{7.21}$$

Gain bandwidth product

$$GBW=\frac{g_m}{20\pi C_{GD}} \tag{7.22}$$

7.10 PERFORMANCE ANALYSIS

The electron mobility of higher values in III-V compound semiconductors generates from the lower effective mass of the carriers and that results in a solid boost of the device performance in terms of higher drive current (I_{ON}) and quicker switching speed at lower power supply voltage (V_{dd}), allowing a solid drop in leakage power dissipation.

Utilization of devices made of III-V materials for logic circuit applications were assessed by many researchers in recent times. They evaluated the suitability of these materials by studying the channel transport properties of the well-established HEMT devices. Many excellent planar devices considering quantum-well heterostructures such as InGaAs with a delta-doped InAlAs barrier layer at the bottom are also demonstrated utilizing the advancements of high-quality gate stack on III-V.

In this section, the different parameters important for digital and analog performance analysis such as the ON current, OFF current, I_{ON}/I_{Off} ratio, the transconductance (g_m), output resistance R_o, etc. are studied.

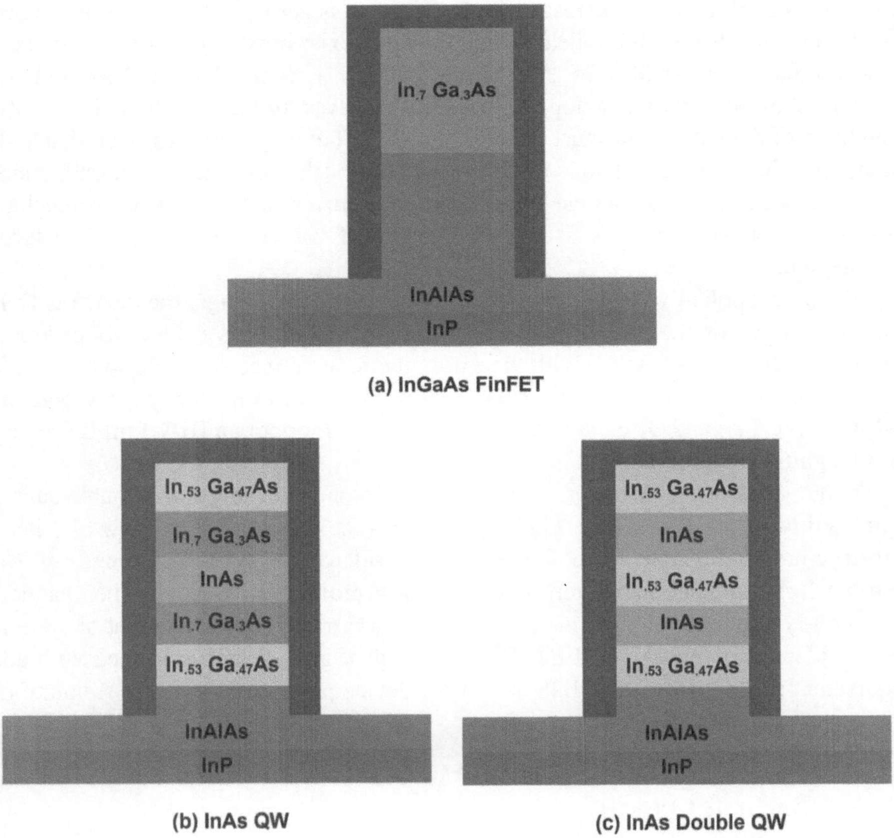

FIGURE 7.11 Schematic View of (a) InGaAs FinFET, (b) Indium Arsenide (InAs) Single Quantum-WellFinFETs, and (c) Dual Quantum-WellFinFETs

TABLE 7.2
Device Parameters Used for Simulation

Parameters	InGaAs FinFET	InAs Single QW	InAs Dual QW
Gate length	26 nm	26 nm	26 nm
Fin width	8 nm	8 nm	8 nm
V_D	0.5V	0.5V	0.5V
Gate oxide	HfO_2 (k=22)	HfO_2 (k=22)	HfO_2 (k=22)
Gate work function	4.65 eV	4.65 eV	4.65 eV

Any switch consisting of the electronic devices has two states, the OFF state and the ON-state. For an ideal switch, the resistance in OFF state is infinity, or the OFF current is equal to zero. The resistance in ON-state is zero and the switching times should be zero. But such an ideal switch does not exist in reality. All real switches have normally very small OFF state current, which is often called the leakage current. We may consider an analogy to mechanical valves for liquids where during off condition of the valve, flow must cut off completely, but it may not firmly do that and sometimes leaks small amounts of fluids. Because of the current, flow in OFF state of the device is called a leakage current. Leakage current may have various mechanisms for electronic devices. It may be because of volume conduction or surface conduction.

When the applied voltage is more than some threshold voltage, the device is ON and at this condition, current following through the device is called ON current. Normally, the load resistance with the external circuit connected to the switch must be limited as the switch has very small ON resistance. In this context, analysis of ON Current (I_{ON}) and OFF current (I_{OFF}) and I_{ON}/I_{OFF} ratio for an III-V FinFET is the most significant parameter.

A comparison of III-V material-based FinFET along with single and double quantum well (QW) FET is shown. Figure 7.11 shows the cross-sectional view of a heterostructure FinFET structure. The device considered having a gate length of 26 nm, an InAs channel, and 5-nm n-doped source/drain (S/D) regions. The channel has an n-type doping of 10^{19} cm^{-3}. An 8 nm-wide Fin and having a height of 10 nm is considered for InGaAs FinFET [97]. The high-κ gate dielectric is used with an equivalent oxide thickness of 1.25 nm. Other device parameters used for simulation are tabulated in Table 7.2.

7.10.1 ON State Current

The ON current of the device is the most important performance for its switching application. Figure 7.12shows the value of I_d (ON state) as a function of V_{gs} for the three different structures shown earlier. It is understood that InAs dual quantum-well FinFET possessesthe highest ON current among the structures considered.

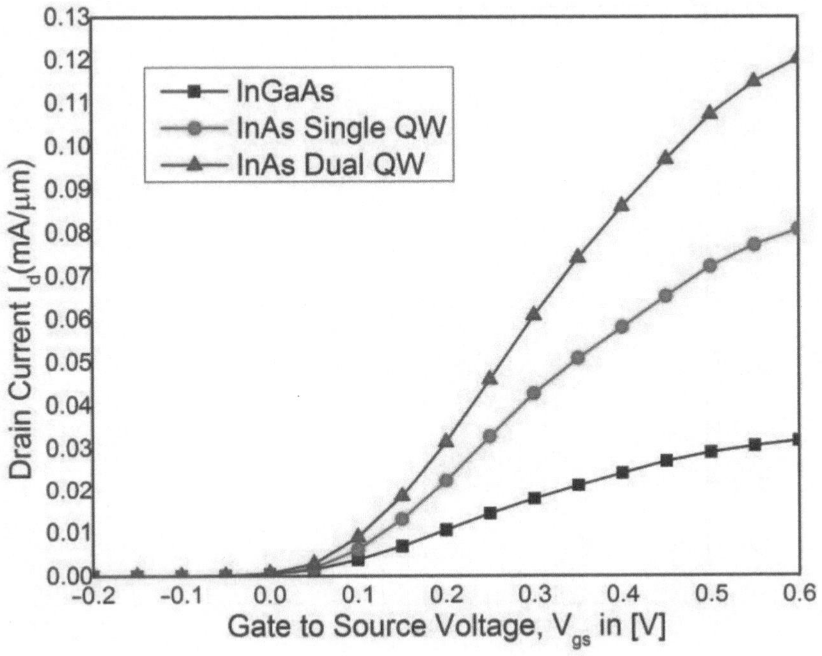

FIGURE 7.12 Variation of Drain Current I_d as a Function of Gate Voltage V_{gs} for Different Structures

7.10.2 ON/OFF Current Ratio

The ON/OFF current ratio of the device is shown in Figure 7.13. It is the most significant parameter to measure for a switch. From the figure, it is detected that I_{ON}/I_{OFF} ratio of drain current increases considerably for InAs-based FinFETs when compared to InGaAs FinFET. Though transfer characteristics show an improvement of I_{ON} for dual QW structures in comparison to single QW structures, its ON/OFF ratio decreases a little bit. This is due to the increase of I_{OFF} for the dual QW FET.

7.10.3 Transconductance

The variation of transconductance g_m as a function gate voltage (V_{gs}) for different device structure is shown in Figure 7.14. It is understood that a substantial improvement in g_m is observed for a device with quantum-well structures. Figure 7.14 also makes it clear that InAs-based dual QWFF offers the highest transconductance value.

7.10.4 Sheet Resistance

FinFET is the current state-of-the-art device architecture for logic technology. In the quest to demonstrate III-V technology for logic applications, numerous highly scaled InGaAs FinFETs have been made with dimensions approaching those of the

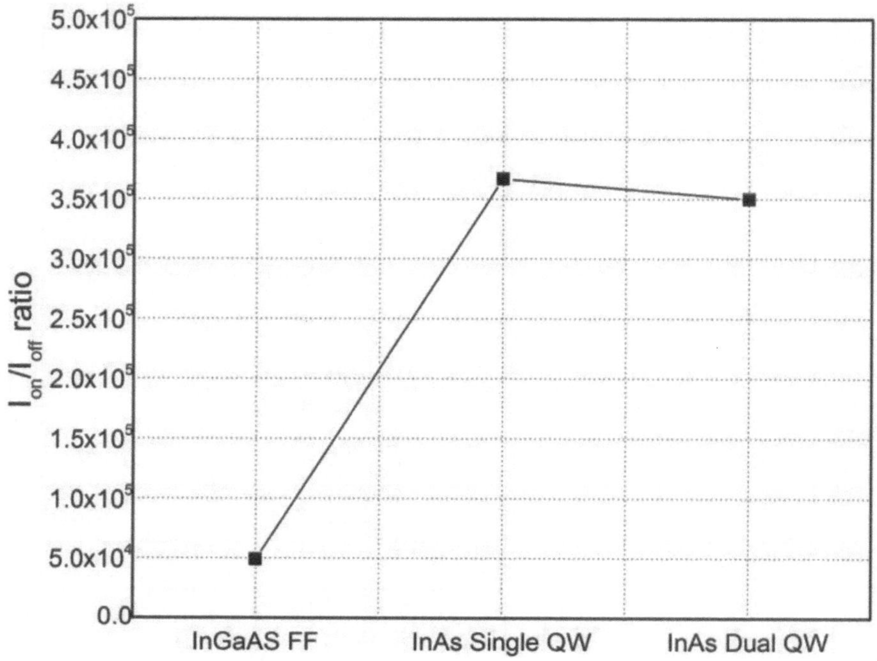

FIGURE 7.13 Ratio of I_{ON}/I_{OFF} for Different Structures

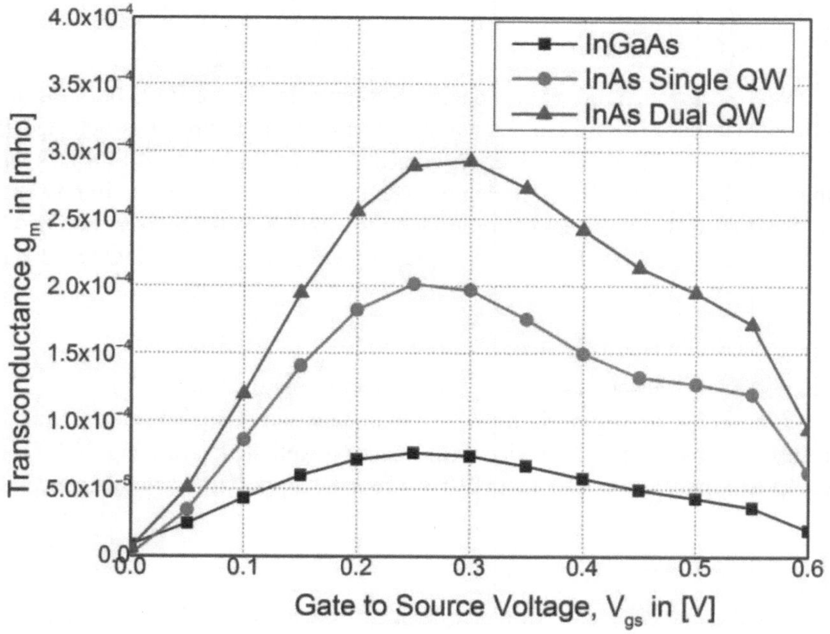

FIGURE 7.14 Transconductance (g_m) Variation with V_{gs} for Different Structures

state-of-the-art Si FinFETs. However, unlike planar InGaAs MOSFETs, performance of most InGaAs FinFETs is still lagging behind Si FinFETs. Rapid degradation of transconductance is also observed as Fin width scales down, raising concern over the scalability of InGaAs FinFETs technology down to 5 nm Fin width.

The damaged sidewalls with excessive roughness and defects may cause extensive oxide trapping during device operation. These kinds of variability in the device structure are also very important for device performance. Measured sheet resistance with different Fin width is reported earlier [98]. It is observed that the sheet resistance of the device increases with decrease in Fin width. In a vertical III-V nanowire MOSFET, it is demonstrated that excellent device performance may be achieved, including a high transconductance and high I_{ON}. Effect of gate length scaling on g_m and Ron is shown in Figure 7.15. Transconductance value as high as $g_m = 3.1$ mS/µm for any MOSFET on Si is reported [99].

A comparison of device performance with different indium mole fraction and device dimension is shown in Table 7.3.

At advanced sub-10nm process nodes, these devices with multiple gates should have thin channels. The quantization as a result of thin channels helps with suppression of the SCEs by widening the bandgap and enhancing the effective mass. However, the trade-off for the improved electrostatics causes a reduction in the ON-performance. Split C-V mobility data of QW planar devices show a rigorous drop in mobility as the channel is thinned from 15 to 3nm with a commensurate decrease in the drive current [100]. Variations of I_{ON}, I_{ON}/I_{OFF}, and SS with channel thickness are shown in Figure 7.16.

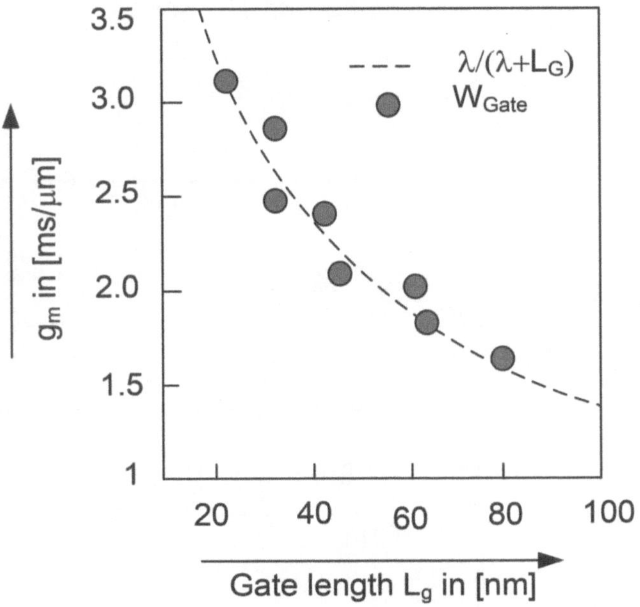

FIGURE 7.15 Variation of Transconductance (g_m) for Different Gate Lengths

TABLE 7.3
Benchmark of InGaAs FinFET Devices

Benchmark of III-V FinFET Devices					
InAs mole fraction	Fin height (nm)	Fin width (nm)	Gate length (nm)	Transconductance (μS/μm)	SS (mV/dec)
1	20	25	1,000	650	148
		35	1,000	1,430	310
0.7	25	50	100	280	190
0.53	40	40	60	1,100	95
0.53	20	30	80	1,800	82
0.7	10	20	120	1,620	114
0.53	9	40	30	1,640	84

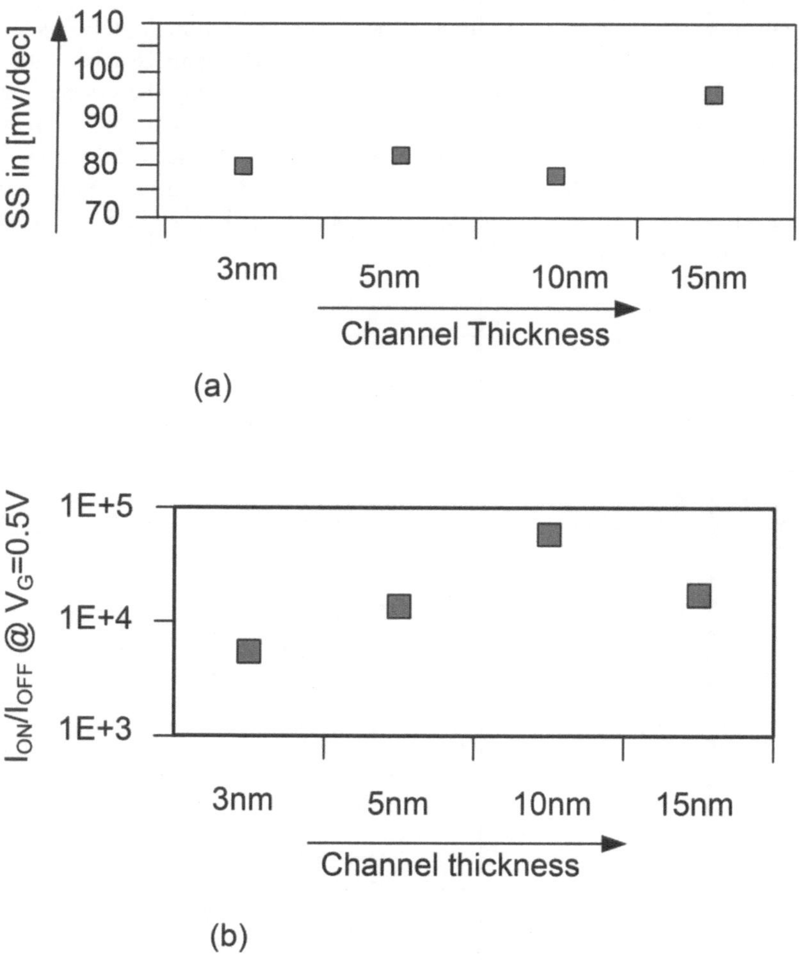

FIGURE 7.16 Variations of (a) SS and (b) I_{ON}/I_{OFF} with Channel Thickness

7.11 CONCLUSION

In this chapter, a revolution of III-V material-based FinFET structures is discussed. Advantages of III-V material-based devices in comparison to its Si counterpart are explained. ON state current, ON/OFF current ratio, and transconductance (g_m) of the device based on TCAD simulation results are presented. A comparison of device performance as a switch is made. These findings show great promise for InGaAs-based FinFETs and provide helpful insights for its design with enhanced characteristics.

REFERENCES

[1] Del Alamo, J. A. (2011) Nanometre-scale electronics with III–V compound semiconductors, *Nature*, 479(7373), 317–323.

[2] Lubow, A., Ismail-Beigi, S., Ma, T. P. (2010) Comparison of drive currents in metal-oxide-semiconductor field-effect transistors made of Si, Ge, GaAs, InGaAs, and InAs channels, *Applied Physics Letters*, 96, 122105.

[3] Xuan, Y., Wu, Y. Q., Ye, P. D. (2008) High-performance inversion-type enhancement-mode InGaAs MOSFET with maximum drain current exceeding 1 A/mm, *IEEE Electron Device Letters*, 29(4), 294–296.

[4] Agrawal, N., Thathachary, A. V., Mahapatra, S., Dutta, S. (2015) Impact of varying indium(x) concentration and quantum confinement on PBTI reliability in $In_xGa_{1-x}As$ FinFET, *IEEE Electron Device Letters*, 36(2), 120–122.

[5] Xue, F., Jiang, A., Zhao, H., Chen, Y.-T., Wang, Y., Zhou, F., Lee, J. (2012) Sub-50-nm $In_{0.7}Ga_{0.3}As$ MOSFETs with various barrier layer materials, *IEEE Electron Device Letters*, 33(1), 32–34.

[6] www.itrs2.net/

[7] Zeitzoff, P. M., Huff, H. R. (2005) MOSFET scaling trends, challenges, and key associated metrology issues through the end of the roadmap, *AIP Conference Proceedings*, 788, 203–213.

[8] Sabry, Y. M., Abdolkader, T. M., Farouk, W. F. (2011) Simulation of quantum transport in double-gate MOSFETs using the non-equilibrium Green's function formalism in real-space: A comparison of four methods, *International Journal of Numerical Modeling: Electronic Networks, Devices and Fields*, 24, 322–334.

[9] Deyasi, A., Sarkar, A., Roy, K., Chowdhury, A. R. (2019) Effect of high-K dielectric on differential conductance and transconductance of ID-DG MOSFET following Ortiz-Conde model, *Microsystem Technologies.*

[10] Agarwal, A., Pradhan, P. C., Swain, B. P. (2018) Triple gate SOI MOSFET. In: Kalam, A., Das, S., Sharma, K. (eds) *Advances in Electronics, Communication and Computing.* Lecture Notes in Electrical Engineering, vol 443. Springer, Singapore.

[11] Lemme, M., Mollenhauer, T., Henschel, W., Wahlbrink, T., Gottlob, H., Favi, J., Baus, M., Winkler, O., Spangenberg, B., Kurz, H. (2003) Subthreshold behavior of triple-gate MOSFETs on SOI Material, *Solid State Electronics Letters*, 48, 529–534.

[12] Deyasi, A., Sarkar, A. (2018) Analytical computation of electrical parameters in GAAQWT and CNTFET using NEGF method, accepted in *International Journal of Electronics*, 105(12), 2144–2159.

[13] Song, J. Y., Choi, W. Y., Park, J. H., Lee, J. D., Park, B-G. (2006) Design optimization of gate-all-around (GAA) MOSFETs, *IEEE Transactions on Nanotechnology*, 5(3), 186–191.

[14] Prakash, P., Sundaram, K. M., Bennet, M. A. (2018) A review on carbon nanotube field effect transistors (CNTFETs) for ultra-low power applications, *Renewable and Sustainable Energy Reviews*, 89, 204–213.

[15] Rani, S., Singh, B. (2021) CNTFET Based 4-Trit Hybrid Ternary Adder-Subtractor for low Power & High-Speed Applications, Silicon [online].

[16] Tsai, M. C., Wang, C. I., Chen, Y. C., Chen, Y. J., Li, K. S., Chen, M. C., Chen, M. J. (2018) Suppression of short channel effects in FinFETs using crystalline ZrO_2 high-K/ Al_2O_3 buffer layer gate stack for low power device applications, *Semiconductor Science and Technology*, 33(3), 035013.

[17] Suhag, A. K., Sharma, R. (2017) Design and simulation of nanoscale double gate MOSFET using high K material and ballistic transport method, *Materials Today: Proceedings*, 4(9), 10412–10416.

[18] Dhiman, G., Pourush, R., Ghosh, P. K. (2018) Performance analysis of High-K material gate stack based nanoscale junction less double gate MOSFET, *Materials Focus*, 7(2), 259–267.

[19] Maszara, W. P., Lin, M. R. (2013) FinFETs—Technology and circuit design challenges, *Proceedings of the ESSCIRC*, 16–20 Sept., Bucharest, Romania.

[20] Verma, S., Tripathi, S. L., Bassi, M. (2019) Performance analysis of FinFET device using qualitative approach for low-power applications, *IEEE Conference on Devices for Integrated Circuit*, 23–24 Mar., Kalyani, India.

[21] Pal, R. S., Sharma, S., Dasgupta, S. (2017) Recent trend of FinFET devices and its challenges: A review, *IEEE Conference on Emerging Devices and Smart Systems*, 3–4 Mar., Mallasamudram, India.

[22] Horiguchi, N., Parvais, B., Chiarella, T., Collaert, N., Veloso, A., Rooyackers, R., Verheyen, P., Witters, L., Redolfi, A., de Keersgieter, A., Brus, S., Zschaetzsch, G., Ercken, M., Altamirano, E., Locorotondo, S., Demand, M., Jurczak, M., Vandervorst, W., Hoffmann, T., Biesemans, S. (2011) FinFETs and their futures: Semiconductor-on-insulator materials for nanoelectronics applications, *Engineering Materials,* 7, 141–152.

[23] Kushwah, R. S., Chauhan, M., Shrivastava, P., Akashe, S. (2014) Modeling and simulation of FinFET circuits with predictive technology models, *Radioelectronics and Communications Systems*, 57(12), 553–558.

[24] Zhichao, L. (2007) Short channel effects in independent-gate FinFETs, *IEEE Electron Device Letters*, 28(2), 145–147.

[25] Hu, J., Zhang, Y., Han, C., Zhang, W. (2015) An investigation of super-threshold FinFET logic circuits operating on medium strong inversion regions, *The Open Electrical & Electronic Engineering Journal*, 9(1), 22–32.

[26] Musala, S., Srinivasulu, A. (2016) FinFET based 4-BIT input XOR/XNOR logic circuit, *IEEE International Conference on applied Electronics*, 6–7 Sept., Pilsen, Czech Republic.

[27] Hajare, R., Lakshminarayana, C., Sumanth, S. C., Anish, A. R. (2015) Design and evaluation of FinFET based digital circuits for high speed ICs, *IEEE International Conference on Emerging Research in Electronics, Computer Science and Technology*, 17–19 Dec., Mandya, India.

[28] Bhattacharjee, A. B. (2009) *Compact MOSFET Models for VLSI Design*, New York, John Wiley & Sons.

[29] Sze, S. M. (1981) *Physics of Semiconductor Devices*, 2nd edition, New York, John Wiley & Sons.

[30] Hill, R. J. W., et al. (2007) Enhancement-mode GaAs MOSFETs with an In0.3Ga0.7As channel, a mobility of over 5000 cm2/V.s, and transconductance of over 475 μS/μm, *IEEE Electron Device Letters*, 28(12), 1082–1083.

[31] Nishida, A., Hasegawa, K., Ohama, R., Fujikawa, S., Hara, S., Fujishiro, H. I. (2013) Comparative study on nano-scale III-V double-gate MOSFETs with various channel materials, *Physica Status Solidi C*, 10(11), 1413–1416.

[32] Baraskar, A., Gossard, A. C., Rodwell, M. J. W. (2013) Lower limits to metal-semiconductor contact resistance: Theoretical models and experimental data, *Journal of Applied Physics,* 114(15), 154516.

[33] Sun, Y., Kiewra, E. W., De Souza, J. P., Bucchignano, J. J., Fogel, K. E., Sadana, D. K., Shahidi, G. G. (2008) Scaling of $In_{0.7}Ga_{0.3}As$ Buried-Channel MOSFETs, *IEDM Technical Digest*, 1–4.

[34] Deyasi, A., Chowdhury, A. R., Roy, K., Sarkar, A. (2018) Effect of high-K dielectric on drain current of ID-DG MOSFET using Ortiz-Conde model, *IEEE EDKCON*, 176–181.

[35] Morassi, L., Verzellesi, G., Zhao, H., Lee, J. C., Veksler, D., Bersuker, G. (2012) Errors limiting split-CV mobility extraction accuracy in buried-channel InGaAs MOSFETs, *IEEE Transactions on Electron Devices*, 59, 1068–1075.

[36] Murmann, P., Nikaeen, D., Connelly, J., Dutton, R. W. (2006) Impact of scaling on analog performance and associated modeling needs, *IEEE Transactions on Electron Devices*, 53(9), 2160–2167.

[37] Tewari, S., Biswas, A., Mallik, A. (2012) Study of InGaAs-Channel MOSFETs for analog/mixed-signal system-on-chip applications, *IEEE Electron Device Letters*, 33(3), 372–374.

[38] Pradhan, K. P., Mohapatra, S. K., Sahu, P. K., Behera, D. K. (2014) Impact of high-k gate dielectric on analog and RF performance of nanoscale DG-MOSFET, *Microelectronics Journal*, 45, 144–151.

[39] Pardeshi, H., Pati, S. K., Raj, G., Mohankumar, N., Sarkar, C. K. (2012) Investigation of asymmetric effects due to gate misalignment, gate bias and underlap length in III–V heterostructure underlap DG MOSFET, *Physica E*, 46, 61–67.

[40] Hu, C., et al. (2000) FinFET-a self-aligned double-gate MOSFET scalable to 20 nm, *IEEE Transactions on Electron Devices*, 47(12), 2320–2325.

[41] Huang, X., Lee, W. C., Kuo, C. (2001) Sub-50 nm P-channel FinFET, *IEEE Transactions on Electron Devices*, 48(5), 880–886.

[42] Mohapatra, S. K., Pradhan, K. P., Singh, D., Sahu, P. K. (2015) The role of geometry parameters and fin aspect ratio of sub-20nm SOI-FinFET: An analysis towards analog and RF circuit design, *IEEE Transactions on Nanotechnology*, 14(3), 546–554.

[43] Zhao, H., Yeo, Y-C., Rustagi, S. C., Samudra, G. S. (2008) Analysis of the effects of fringing electric field on FinFET device performance and structural optimization using 3-D simulation, *IEEE Transactions on Electron Devices*, 55(5), 1177–1184.

[44] https://newsroom.intel.com/press-kits/leading-edge-intel-technology-manufacturing/

[45] Lemme, M. C., Mollenhauer, T., Henschel, W., Wahlbrink, T., Baus, M., Winkler, O., Granzner, R., Schwierz, F., Spangenberg, B., Kurz, H. (2004) Subthreshold behaviour of triple-gate MOSFETs on SOI material, *Solid State Electronics*, 48(4), 529.

[46] Colinge, J. P., Baie, X., Bayot, V., Grivei, E. (1996) A silicon-on-insulator quantum wire, *Solid State Electronics*, 39, 49.

[47] Chau, R., Doyle, B., Kavalieros, J., Barlage, D., Murthy, A., Doczy, M., Arghavani, R., Datta, S. (2002) Advanced depleted-substrate transistors: Single-gate, The SOI MOSFET: From Single Gate to Multigate double-gate and tri-gate, *Extended Abstracts of the International Conference on Solid State Devices and Materials (SSDM)*, 68.

[48] Doyle, B. S., Datta, S., Doczy, M., Jin, B., Kavalieros, J., Linton, T., Murthy, A., Rios, R., Chau, R. (2003) High performance fully-depleted tri-gate CMOS transistors, *IEEE Electron Device Letters*, 24(4), 263.

[49] Park, J. T., Colinge, J. P., Diaz, C. H. (2001) Pi-gate SOI MOSFET, *IEEE Electron Device Letters*, 22, 405.

[50] Park, J. T., Colinge, J. P. (2002) Multiple-gate SOI MOSFETs: Device design guidelines, *IEEE Transactions on Electron Devices*, 49(12), 2222.

[51] Yang, F. L., Chen, H. Y., Cheng, F. C., Huang, C. C., Chang, C. Y., Chiu, H. K., Lee, C. C., Chen, C. C., Huang, H. T., Chen, C. J., Tao, H. J., Yeo, Y. C., Liang, M. S., Hu, C. (2002) 25 nm CMOS Omega FETs, *Technical Digest of IEDM*, 255.

[52] Yang, Fu-Liang, et al. (2004) 5nm-gate nanowire FinFET, *Symposium on VLSI Technology*, 196.

[53] Ritzenthaler, R., Dupré, C., Mescot, X., Faynot, O., Ernst, T., Barbé, J. C., Jahan, C., Brévard, L., Andrieu, F., Deleonibus, S., Cristoloveanu, S. (2006), Mobility behavior in narrow Ω-gate FET devices, *Proceedings IEEE International SOI Conference*, 77.

[54] Krivokapic, Z., Tabery, C., Maszara, W., Xiang, Q., Lin, M. R. (2003), High-performance 45-nm CMOS technology with 20-nm multi-gate devices, *Extended Abstracts of the International Conference on Solid State Devices and Materials (SSDM)*, 760.

[55] Kavalieros, J., Doyle, B., Datta, S., Dewey, G., Doczy, M., Jin, B., Lionberger, D., Metz, M., Rachmady, W., Radosavljevic, M., Shah, U., Zelick, N., Chau, R. (2006) Tri-gate transistor architecture with High-k gate dielectrics, *Metal Gates and Strain Engineering. Symposium on VLSI Technology*, paper 7.1.

[56] Colinge, J. P (2008) *FinFETs and Other Multi-Gate Transistors*, New York, Springer.

[57] Subramanian, V., Mercha, A., Parvais, B., Loo, J., Gustin, C., Dehan, M., Collaert, N., Jurczak, M., Groeseneken, G., Sansen, W., Decoutere, S. (2007) Impact of fin width on digital and analog performances of n-FinFETs, *Solid-State Electronics*, 51(4), 551–559, Apr.

[58] Gaynor, B. D., Hassoun, S. (2014) Fin shape impact on FinFET leakage with application to multithreshold and ultralow-leakage FinFET design, *IEEE Transactions on Electron Devices*, 61(8), 2738–2744, Aug.

[59] Seoane, N., et al. (2016) Comparison of Fin-Edge roughness and metal grain work function variability in InGaAs and Si FinFETs, *IEEE Transactions on Electron Devices*, 63(3), 1209–1216,.

[60] Park, T. S., Choi, S., Lee, D. H., et al. (2003) Fabrication of body-tied FinFETs (Omega MOSFETs) using bulk Si wafers, *IEEE Symposium on VLSI Technology*, 135–136.

[61] Luo, X., Ma, D., Tan, Q., Wei, J., Wu, J., Zhou, K., Sun, T., Liu, Q., Zhang, B., Li, Z. (2016) A split gate power FINFET With improved ON-resistance and switching performance, *IEEE Electron Device Letters,* 37(9), 1185–1188.

[62] Zhang, Y., Sun, M., Perozek, J., Liu, Z., Zubair, A., Piedra, D., Chowdhury, N., Gao, X., Shepard, K., Palacios, T. (2019) Large-area 1.2-kV GaN vertical power FinFETs with a record switching figure of merit, *IEEEElectron Device Letters*, 40(1), 75–78.

[63] Wang, H., Xiao, M., Sheng, K., Palacios, T., Zhang, Y. (2021) Switching performance analysis of vertical GaNFinFETs: Impact of Interfin designs, *IEEE Journal of Emerging and Selected Topics in Power Electronics*, 9(2), 2235–2246.

[64] Pandey, A., Raycha, S., Maheshwaram, S., Manhas, S. K., Dasgupta, S., Saxena, A. K., Anand, B. (2013) FinFET device Capacitances: Impact of input transition time and output load, *IEEE 5th International Nanoelectronics Conference*, 2–4 Jan., Singapore.

[65] Rudenko, T., Kilchytska, V., Collaert, N., De Gendt, S., Rooyackers, R., Jurczak, M., Flandre, D. (2005) Specific features of the capacitance and mobility behaviors in FinFET structures, *Proceedings of 35th European Solid-State Device Research Conference*,16 Sept., Grenoble, France.

[66] Rodriguez, S. S., Tinoco, J., Martinez-Lopez, A. G., Alvarado, J. (2013) Parasitic gate capacitance model for triple-gate FinFETs, *IEEE Transactions on Electron Devices*, 60(11), 3710–3717.

[67] Jian, Z. A., Mohanty, S., Ahmadi, E. (2021) Switching performance analysis of 3.5 kV Ga_2O_3 Power FinFETs, *IEEE Transactions on Electron Devices*, 68(2), 672–678.

[68] Lu, W., et al. (2018) First transistor demonstration of thermal atomic layer etching: InGaAs FinFETs with sub-5 nm Fin-width featuring in situ ALE-ALD, *IEDM Technical Digest*, 39.1.1–39.1.4, 2018.

[69] Wu, Y. Q., Wang, R. S., Shen, T., Gu, J. J. and Ye, P. D. (2009) First experimental demonstration of 100 nm inversion-mode InGaAs FinFET through damage-free sidewall etching, *2009 IEEE International Electron Devices Meeting (IEDM)*, Baltimore, MD, USA, 2009, pp. 1–4, doi: 10.1109/IEDM.2009.5424356.

[70] Djara, V., Deshpande, V., Sousa, M., Caimi, D., Czornomaz, L., Fompeyrine, J. (2016) CMOS-compatible replacement metal GatemInGaAs-OI FinFET with ION = 156 μA/μm at VDD = 0.5 V and IOFF = 100 nA/μm, *IEEE Electron Device Letters*, 37(2), 169–172, Feb.

[71] Oxland, R., et al. (2016) InAs FinFETs With Hfin = 20 nm FabricatedUsing a Top–Down etch process, *IEEE Electron Device Letters*, 37(3), 261–264, Mar. 2016, doi: 10.1109/LED.2016.2521001.

[72] Wang, S.-W., Vasen, T., Doornbos, G., Oxland, R., Chang, S.-W., Li, X., Contreras-Guerrero, R., Holland, M., Wang, C.-H., Edirisooriya, M., Rojas-Ramirez, J. S., Ramvall, P., Thoms, S., MacIntyre, D. S., Vellianitis, G., Hsieh, C. H., Chang, Y.-S., Yin, K. M., Yeo, Y.-C., Diaz, C. H., Droopad, R., Thayne, I. G., Passlack, M. (2015) Field effect mobility of InAs surface channel N-MOSFET with low Dit scaled gate stack, *IEEE Transactions on Electron Devices*, 62(8), 2429–2436.

[73] Boukortt, N. E. I., Hadri, B., Caddemi, A., Crupi, G., Patane, S. (2015) 3-D simulation of nanoscale SOI n-FinFET at a gate length of 8 nm using ATLAS SILVACO, *Transactions on Electrical and Electronic Materials*, 16(3), 156–161.

[74] Buryk, I. P., Golovnia, A. O., Ivashchenko, M. M., Odnodvorets, L. V. (2020) Numerical simulation of FinFET transistors parameters, *Journal of Nano- and Electronic Physics*, 12(3), 03005.

[75] Im, K. S., et al. (2013) High-performance GaN-based nanochannel FinFETs with/without AlGaN/GaN heterostructure, *IEEE Transactions on Electron Devices*, 60(10), 3012–3018.

[76] Yadav, C., Kushwaha, P., Khandelwal, S., Duarte, J. P., Chauhan, Y. S., Hu, C. (2014) Modeling of GaN-based normally-off FinFET, *IEEE Electron Device Letters*, 35(6), 612–614.

[77] Biswas, K., Sarkar, A., Sarkar, C. K. (2018) Fin shape influence on analog and RF performance of junctionless accumulation-mode bulk FinFETs, *Microsystem Technologies*, 24, 2317–2324.

[78] Biswas, K., Sarkar, A., Sarkar, C. K. (2017) Spacer engineering for performance enhancement of junctionless accumulation-mode bulk FinFETs, *IET Circuits, Devices & Systems*, 11(1), 80–88.

[79] Wong, B., Mittal, A. Cao, Y., Starr, G. W. (2004) *Nano-CMOS Circuit and Physical Design*, Hoboken, NJ, Wiley-IEEE Press.

[80] Ha, D., Takeuchi, H., Choi, Y-K., King, T-J. (2004) Molybdenum gate technology for ultrathin-body MOSFETs and FinFETs, *IEEE Transactions on Electron Devices*, 51(12), 1989–2004.

[81] Hussain, M. M., Smith, C. E., Harris, H. R., Young, C. D., Tseng, H-H., Jammy, R. (2010) Gate-first integration of tunable work function metal gates of different thicknesses into high-k/metal gates CMOS FinFETs for multi-V_{Th} engineering, *IEEE Transactions on Electron Devices*, 57(3), 626–631.

[82] Jung, H. K. (2010) Threshold voltage dependence on bias for FinFET using analytical potential model, *Journal of Information and Communication Convergence Engineering*, 8, 107–111.

[83] Fasarakis, N., Tsormpatzoglou, A., TaSis, D. H., Dimitriadis, C. A., Papathanasiou, K., Jomaah, J., Ghibaudo, G. (2011) Analytical unified threshold voltage model of short-channel FinFETs and implementation, *Solid State Electronics*, 64, 34–41.

[84] Katti, G., DasGupta, N., DasGupta, A. (2004) Threshold Voltage model for mesa-isolated small geometry fully depleted SOI MOSFETs based on analytical solution of 3-D Poisson's equation, *IEEE Transactions on Electron Devices*, 51, 1169–1177.

[85] Sharma, A., Goud, A. A., Roy, K. (2015) Sub-10 nm FinFETs and tunnel- FETs: From devices to systems, *Proceeding of 2015 Design, Automation and Test in Europe Conference and Exhibition (DATE)*, Grenoble, 1443–1448.

[86] Sinha, S., Yeric, G., Chandra, V., Cline, B., Cao, Y. (2012) Exploring sub-20nm FinFET design with predictive technology models, *Proceeding of DAC Design Automation Conference*, San Francisco, 283–288.

[87] Das, R., Baishya, S. (2019) Analytical modeling of threshold voltage and subthreshold swing in Si/Ge Heterojunction FinFET, *Applied Physics A*, 125, 682.

[88] Suzuki, K., Sugii, T. (1995) Analytical models for n+-p+ double gate SOI MOSFET's, *IEEE Transactions on Electron Devices*, 42(11), 1940–1948.

[89] Arora, N. D., Rios, R., Huang, C. L., Raol, K. (1994) PCIM: A physically based continuous short-channel IGFET model for circuit simulation, *IEEE Transactions on Electron Devices*, 41(6), 988–997.

[90] Roldan, J. B., Gamiz, F., Lopez-Villanueva, J. A., Carceller, J. E. (1997) Modeling effects of electron velocity overshoot in a MOSFET, *IEEE Transactions on Electron Devices*, 44(5), 841–846.

[91] Roldan, J. B., Gamiz, F., Lopez-Villanueva, J. A., Cartujo, P., Carceller, J. E. (1998) A model for the drain current of deep submicrometer MOSFET's including electron-velocity overshoot, *IEEE Transactions on Electron Devices*, 45(10), 2249–2251.

[92] Reddy, G. V., Kumar, M. J. (2005) A new dual material double gate (DMDG) nanoscale SOI MOSFET: Two dimensional analytical modelling and simulation, *IEEE Transactions on Electron Devices*, 4(2), 260–268.

[93] Basak, A., Sarkar, A. (2020) Drain current modelling of asymmetric junctionless dual material double gate MOSFET with high K gate stack for analog and RF performance, *Silicon* [online].

[94] Basak, A., Chanda, M., Sarkar, A. (2019) Drain current modeling of unipolar junction dual material double-gate MOSFET (UJDMDG) for SoC applications, *Microsystem Technologies* [online].

[95] Chen, Y. G., Kuo, J. B., Yu, Z., Dutton, R. W. (1995) An analytical drain current model for short-channel fully-depleted ultrathin silicon-on-insulator NMOSdevices, *Solid State Electronics*, 38(12), 2051–2057.

[96] Sakurai, T., Newton, A. R. (1990) Alpha-power law MOSFET model and its applications to CMOS inverter delay and other formulas, *IEEE Journal of Solid-State Circuits*, 25(2), 584–594.

[97] Biswas, K., Sarkar, A., Sarkar, C. K. (2016) Performance assessment of Indium Arsenide (InAs) Single and Dual quantum-well heterostructure FinFETs, *2016 3rd International Conference on Devices, Circuits and Systems (ICDCS)*, Coimbatore, India, 2016, pp. 208–211.

[98] Zhao, X., Vardi, A., del Alamo, J. A. (2019) Fin-width scaling of highly doped InGaAs fins, *IEEE Transactions on Electron Devices*, 66(6), 2563–2568, June, doi: 10.1109/TED.2019.2912618.

[99] Kilpi, O., et al. (2020) High-performance vertical III-V nanowire MOSFETs on Si with gm > 3 mS/μm, *IEEE Electron Device Letters*, 41(8), 1161–1164, Aug., doi: 10.1109/LED.2020.3004716.

[100] Alian, Alireza, Pourghaderi, Mohammad, Mols, Yves, Cantoro, Mirco, Ivanov, Tsvetan, Collaert, Nadine, Thean, Aaron V-Y. (2013) Impact of the channel thickness on the performance of ultrathin InGaAs. *2013 IEEE International Electron Devices Meeting*, pp. 16.6.1–16.6.4.

8 Negative Capacitance Field-Effect Transistors to Address the Fundamental Limitations in Technology Scaling

Harsupreet Kaur

CONTENTS

8.1 INTRODUCTION

Over the past few decades, complementary metal oxide semiconductor (CMOS) technology has evolved at a remarkable rate and has witnessed the emergence of denser and faster integrated circuits with high yield and reliability. Starting with a few transistors on a single chip in the early 1970s, it has increased to a billion transistors in a span of 40 years. The demands for greater integration and higher performance are the key factors driving the CMOS technology, and the result has been a tremendous expansion in technology and communication markets, including the market associated with high-performance microprocessors as well as low static power applications such as wireless systems.

It all started when the first transistor was invented in 1947 by Bardeen, Brattain, and Shockley at the Bell Laboratory. It was followed by the introduction of the bipolar junction transistor (BJT) in 1949 by Shockley. The biggest revolution happened

DOI: 10.1201/9781003200987-8

when Jack Kilby at Texas Instruments first made the monolithic integrated circuit in 1958. This was a significant breakthrough in the semiconductor technology, and Kilby was later awarded the Nobel Prize in 2000 for what proved to be a watershed development for the microelectronics industry. In 1960, the first commercial IC was introduced by Fairchild Corporation, followed by TTL IC in 1962. Since then, technology has progressed swiftly and steadily at a remarkable pace to pave way for ever more powerful, cost-effective, and energy-efficient products.

CMOS technology has been the enabler for advancement of the present electronics and communication age, and silicon has been a natural choice of material for IC manufacturing because of its abundance in nature, and more importantly its native oxide, which is most suitable for IC fabrication. Almost 90% of the electronic circuits fabricated worldwide are made of silicon using CMOS technology. CMOS technology is the most popular technology because of its low power and less area requirement. For several decades, the continuous performance improvement has essentially relied on the dimension downsizing of the basic component of integrated circuits, the metal oxide semiconductor field-effect transistor (MOSFET).

In 1965, Gordon Moore predicted that number of components per chip would increase by a factor of two every year and then second update of Moore's law in 1975 predicted that number of components per chip would increase by a factor of two every two years. In view of this, in 1974 Robert Dennard [1] described scaling techniques for MOSFETs that result in reduced transistor area, high-drive current (high performance), low active power, etc. Dennard's scaling method is also known as "traditional scaling" and was successful in the industry up to 130-nm generation in the early 2000s. However, as the gate length of a MOSFET decreases, capacitive control of the channel potential by the gate becomes more difficult. This is because the drain potential can significantly influence the channel potential resulting in severe short-channel effects (SCEs). The reason behind various SCEs is reduced gate control over the channel region as a result of increased charge sharing from source/drain as dimensions are scaled down. This consequently reduces the threshold voltage with the reduction of channel length. Some other SCEs that degrade the device performance are higher subthreshold conduction, drain-induced barrier lowering (DIBL) and corresponding threshold voltage reduction, punch-through, channel length modulation and hot carrier effects, random dopant fluctuation, etc. Scaling also necessitates reduction of gate oxide thickness, but when gate oxide thickness is reduced, tunneling probability increases and results in an increase in oxide leakage current. Using a high-*k* dielectric can overcome the problem to some extent, as a high-*k* dielectric can provide a similar gate electric field even with a physically thick high-*k* gate dielectric. This can reduce the direct tunneling leakage.

Moreover, increased OFF current, I_{OFF}, is a severe concern that may limit CMOS scaling because of significant passive power consumption. Furthermore, the requirement for improved performance and greater functionality leads to an increase in both static and dynamic power consumption, and hence power management is one of the most critical issues in very large-scale integration circuit and systems [2]. To have sustained growth of CMOS technology, it is extremely important to design low power and energy-efficient circuits and systems. Improved energy efficiency can be achieved by lowering both dynamic (switching) and leakage power consumption. Lower power also results in lower operating temperatures of electronic devices, which consequently leads to better performance and also improves reliability. However, lowering of power using traditional techniques becomes increasingly difficult beyond 22 nm technology node. This is because power supply voltage cannot be scaled as swiftly due to performance degradation caused by simultaneous reduction in threshold voltage. The simultaneous scaling of threshold voltage, which is essential for maintaining a requisite I_{ON}/I_{OFF} ratio leads to a substantial increase in the subthreshold leakage (OFF state) current. Another metric that is used to describe the abruptness of switching behavior is known as the subthreshold swing (SS), inverse of the subthreshold slope, which has a fundamental lower limit of 60mV/decade. Practical nanoscale CMOS devices have much higher value of subthreshold swings. Although lowering of the subthreshold swing below 60 mV/decade is a nontrivial task, it is extremely critical for achieving any significant improvements in the energy efficiency of ultra large-scale integration.

Therefore, to overcome the aforementioned issues, research studies [3–5] focused on innovations in transistor materials and structures to continue scaling with improved performance and reduced cost.

8.2 IMPROVEMENTS AND ALTERNATIVES TO OVERCOME THE CHALLENGES

In 2003, Intel's 90-nm technology introduced strained silicon transistors to increase electron and hole mobility, which eventually increased drive current without further scaling down SiO_2 layer [2]. In 2007, Intel introduced the 45-nm technology, and the use of high-k metal gate transistor was demonstrated by replacing SiO_2 layer with hafnium oxide. The use of high-k material significantly reduced leakage current and improved drive current, and hence provided substantial improvement in transistor performance. However, the performance of devices with high-k gate insulator was also found to degrade due to fringing field effects and interfacial layer defects [6].

Hence, gate stack design [7–8] was proposed in which a thin interfacial layer of SiO_2 was used between high-*k* dielectric and silicon channel to improve the interface state properties. Furthermore, various multi-gate geometry MOSFETs (double gate, surrounding gate, FinFET, etc.) have been explored and it has been reported that they exhibited improved gate control and steep subthreshold slope

However, in spite of the alternatives adopted, there are various issues that still demand attention, such as difficulty in fabrication of advanced nonplanar multi-gate and nanowire MOSFETs below 10 nm gate length, limitation on V_{dd} scaling, and difficulty in overcoming the fundamental limit of subthreshold swing, SS ~60 mV/dec. In addition, with the increase in the number of components per chip, the total power dissipated (dynamic power and static power) increases to critically high levels, thereby resulting in degraded performance and reliability of ICs. However, over the years, it has been observed that static power dissipation forms a major component of total power as can be noted, and hence it is very crucial to explore new technologies that can overcome the lower limit of V_{dd} and are suitable for ultra low voltage/power applications.

8.3 FERROELECTRIC MATERIAL-BASED FETs OR NEGATIVE CAPACITANCE FETs

Various alternatives have been proposed to address the challenges associated with miniaturization. Negative capacitance (NC) FETs or ferroelectric (FE) FETs that employ a ferroelectric material in the gate stack have also attracted immense attention. Research groups have integrated ferroelectric material with novel device designs to counter different issues, the most important being that of static power dissipation. Apart from this, these devices are also reported to result in better current drivability, lower OFF current, and very interestingly, the drain-induced barrier rising (DIBR) effect.

FE materials have an interesting property of being spontaneously polarized in the absence of an external electric field, and the direction of polarization can be reversed by application of an external field larger than the coercive field [9]. This property of spontaneous polarization is attributed to the non-centrosymmetricity of FE material crystal. Ferroelectric material-based FETs or negative capacitance FETs are based on incorporation of FE material as gate insulator and have been researched extensively by many research groups for various ultra low voltage/power applications [10–11]. As mentioned earlier, among other reasons, scaling of conventional FETs is also limited by the inability to remove heat generated in the switching process, which results in degraded energy efficiency [10]. Furthermore, with an increase in packing density, large number of devices are integrated on a chip. The increase in subthreshold current results in increase in static power dissipation, which consequently leads to overheating and significant deterioration in performance and reliability of chips. To address this issue, Salahuddin et al. [10] proposed that by replacing the standard insulator in a MOSFET by ferroelectric material as insulator of appropriate thickness, voltage amplification can be achieved. It was suggested that this would also result in sub-60 mV/dec operation, which implies steep and faster switching characteristics. The voltage amplification achieved is primarily because of the negative capacitance effect of ferroelectric material, which eventually amplifies total capacitance of the device and results in steep switching characteristics.

This is because SS of conventional MOSFET is given as:

$$SS = \frac{\partial V_{gs}}{\partial\left(\log_{10} I_{ds}\right)} = \frac{\partial V_{gs}}{\partial\left(\psi_s\right)} \frac{\partial \psi_s}{\partial\left(\log_{10} I_{ds}\right)}$$

where ψ_s, V_{gs}, I_{ds} denote surface potential, gate voltage, and drain current, respectively.

To achieve SS < 60 mV/dec, $\frac{\partial V_{gs}}{\partial\left(\psi_s\right)}$ must be > 1, and this can be achieved by voltage amplification capability of FE insulator.

The negative capacitance (NC) phenomenon in ferroelectrics can be explained by energy landscape of a ferroelectric material [13].

It can be observed that negative curvature in the energy landscape corresponds to negative capacitance region since the capacitance (C) in terms of free energy (U) is

$$C = \left[\frac{d^2U}{dQ^2}\right]^{-1}$$

The energy landscape of a ferroelectric material is described by Landau-Khalatnikov (LK) equation, according to which the free energy of the system can be represented as a Taylor series expansion of an order parameter [12].

For a ferroelectric material, free energy *U(P)* is expressed as an even order polynomial of polarization as follows:

$$U(P) = \alpha P^2 + \beta P^2 + \gamma P^6 - E_{ext} P$$

where $E_{ext} = V_{fe} / t_{fe}$ denotes external field, and α, β and γ are material-dependent Landau parameters. The voltage applied across FE layer is denoted by V_{fe}, and t_{fe} is the thickness of FE layer. The state of the system is obtained by minimizing *U(P)* w. r. t. *P* to obtain spontaneous polarization as follows:

To find points of stable equilibrium in the absence of applied field, the following conditions are imposed:

$$\frac{dU(P)}{dP} = P(2\alpha + 4\beta P^2 + 6\gamma P^4) = 0$$

$$\frac{d^2U(P)}{dP^2} > 0$$

When $\alpha < 0$, the material exhibits ferroelectricity and α is temperature-dependent [9, 12] and is expressed as $\alpha(T) = \alpha_0 (T - T_c)$, where T_c is Curie temperature or transition temperature; α_o is always positive and $\alpha(T)$ is negative below T_c. Further, depending on signs of β and γ, two types of transitions are defined [9]. *First-order transition* is when $\alpha < 0$, $\beta < 0$ and $\gamma > 0$, while *second-order transition* holds for materials with $\alpha < 0$, $\beta > 0$ and $\gamma > 0$.

Furthermore, it has been reported that negative capacitance state is unstable and to employ the ferroelectric materials for device operation, it is necessary to stabilize it. To stabilize negative capacitance, a linear positive capacitor is added in series with the ferroelectric layer [10–11]. This positive series capacitor can be a semiconductor channel capacitor or any dielectric capacitor. When a positive series capacitor is added with ferroelectric negative capacitance, the total energy of the system stabilizes. Furthermore, to stabilize NC region and to ensure optimal gain characteristics, it is important to optimize ferroelectric layer parameters, such as thickness of ferroelectric layer and coercive field and remanent polarization to achieve hysteresis-free device operation [13–14]. Furthermore, optimization can also be achieved by improving capacitance matching between ferroelectric capacitance (C_{fe}) and positive capacitance of underlying MOSFET.

8.4 PROPERTIES OF DIFFERENT FERROELECTRIC MATERIALS

Over the last few years, various materials have been reported to exhibit ferroelectric behavior. In some materials such as Rochelle salt ($NaKC_4H_4O_6 \cdot 4H_2O$), barium titanate ($BaTiO_3$), lead zirconium titanate (PZT, $PbZr_{1-x}$, Ti_xO_3), and strontium bismuth tantalate (SBT, $SrBi_2Ta_2O_9$), ferroelectricity was reported very early. Rochelle salt has a peculiar property that it is ferroelectric between two temperatures, 255 K and 296 K [15], while barium titanate exhibits three ferroelectric phases in temperature ranges: 278 K–393 K, 193 K–278 K, and T<193 K, respectively. The other two most widely studied perovskite ferroelectric materials are lead zirconium titanate (PZT) and strontium bismuth tantalate (SBT). These materials were among the prominent ones that were integrated with FETs to realize improvement in device operation. However, PZT exhibits high spontaneous polarization (20–40 $\mu C/cm^2$) and high transition temperature (370°C) [16–18]. The ferroelectric properties of PZT depend on composition of titanium (Ti) and zirconium (Zr), and it has been reported that for Ti/Zr ratios of 60/40 and 70/30, high switchable polarization can be achieved [18]. Further, SBT is a layered perovskite structure and exhibits lower spontaneous polarization (~10 $\mu C/cm^2$) as compared to PZT. However, these materials have high dielectric constant; SBT has dielectric constant in the range of 150–250 whereas PZT has ~1,300 [17], due to which it is difficult to scale down these materials and hence, it is difficult to integrate them in current technology node. Therefore, research groups across the world realized that ferroelectric materials with lower dielectric constants would be more suitable to integrate with current technology node devices.

In this direction, ferroelectric properties have also been discovered in some polymers [19] and polyvinylidene fluoride (PVDF), and its copolymers are among the most important organic ferroelectrics since they exhibit low dielectric constant (~13). Polyvinylidene fluoride-trifluoroethylene (PVDF-TrFE) copolymer has been used in many experimental studies in 70/30 percentage of VDF/TrFE, and it has been reported that PVDF-TrFE possesses excellent ferroelectric properties [20].

Recently, ferroelectricity was also reported in doped hafnium oxide, and this has paved the way for integration of sub-10 nm ferroelectric gate insulator in submicron devices [17, 21–24]. Doped HfO_2 offers several advantages, such as high bandgap, excellent thermodynamic stability, and fabrication feasibility with Si [21–22] among others. FE-HfO_2 exhibits stable ferroelectric properties even below 10 nm thickness along with high coercive field, due to which it has emerged as an ideal material for integration of FE-based devices into current CMOS technology nodes [17–24]. It has been reported that ferroelectric properties of doped HfO_2 depend on several controllable process parameters, such as deposition techniques, dopant atomic radius, dopant percentage concentration, annealing temperature, modulation by capping layer, and thickness and choice of substrate [22, 24–26]. Hence, to critically analyze the effect of these factors, various studies have focused on exploring the impact of different dopants on the switching properties of FE-HfO_2.

In addition, experimental studies have been carried out to examine the performance of silicon doped hafnium oxide [22, 24, 27–30]. It has been reported that when Si:HfO_2-based films are crystallized in the presence of a cap, orthorhombic phase is observed, which exhibits ferroelectric behavior and quantitative measurements have shown remanent polarization (P_r) ~ 10 $\mu C/cm^2$ and coercive field (E_c) ~ 1–2 MV/cm [22]. Also, Curie temperature of approximately 400°C has been reported by linearly extrapolating P_r values in temperature range 150°C–205°C [29]. In addition, high P_r ~ 24 $\mu C/cm^2$ and E_c ~ 1.2 MV/cm have been reported for Y-doped HfO_2 with concentration 2.3–5.2% along with Curie temperature 450°C [31]. Recently, ferroelectric properties in Y-doped HfO_2 have been experimentally demonstrated for thickness down to 3 nm, and switchable polarization (P_{sw}) ~ 10 μC/cm^2 was demonstrated for 3 nm thick Y-doped HfO_2 [32].

8.5 NOVEL DEVICE DESIGNS BASED ON FERROELECTRIC MATERIALS

It is possible to achieve voltage amplification and sub-60 mV/dec operation by using FE layer of appropriate thickness in gate stack configuration with an interfacial layer. This voltage amplification is attributed to negative capacitance (NC) phenomenon of FE layer, which has been experimentally demonstrated as well [33]. Furthermore, experimental demonstration *SS* values less than 60 mV/dec have been experimentally reported in metal-ferroelectric metal oxide semiconductor FET [34–35]. As discussed in the previous section, capacitance matching is extremely crucial in designing NCFETs, and many studies [11, 36–37] have been reported that have focused on stabilization of NC effect by optimizing FE layer thickness and other device parameters to achieve voltage amplification. Moreover, the concept of negative capacitance exhibited by FE materials has also been investigated [38] in three different configurations, that is, stand-alone FE, FE in series with dielectric capacitor, and an FE in series with a semiconductor. It has been demonstrated that in all the configurations, it is important to minimize total Gibbs free energy of complete system and not just FE. Hence, it has been demonstrated both theoretically and experimentally that by properly optimizing FE parameters, it is possible to achieve steep slope transistors suitable for ultra low voltage/power applications.

Various groups have also carried out theoretical studies on NCFETs. Jiménez et al. [39] developed a physics-based surface potential and drain current model for long channel double gate negative capacitance FET and considered SBT as the ferroelectric material to explore the impact of NC phenomenon on device performance. Analytical models of long channel gate-all-around NC transistors have also been reported [40], and it was demonstrated that device exhibits $SS < 60$ mV/dec along with high value of I_{ON} in comparison to conventional gate-all-around FET.

Furthermore, metal-ferroelectric-semiconductor (MFS) structure is difficult to fabricate and results in various interface defects at ferroelectric and silicon interface. In view of this, many research groups [41–43] have demonstrated NCFETs with structure similar to MFIS (metal-ferroelectric-insulator-semiconductor). In such devices, the insulator layer between FE and semiconductor aids in stabilization of the NC effect, thereby, making MFIS structures more promising for future transistor applications.

Moreover, FE materials exhibit ferroelectricity up to Curie temperature, and NC effect is strongly temperature-dependent. It has been reported that with an increase in temperature, the step-up conversion capability of NCFETs gets suppressed and vanishes completely above Curie temperature [44–45]. However, by optimizing FE layer parameters such as layer thickness, coercive field, and remanent polarization, performance of NCFETs at high temperatures can be enhanced. NCFETs are also suitable for various digital and analog circuit applications such as inverter [46–47] and ring oscillator [48], which have been extensively explored, and it has been reported that

FeFETs result in performance enhancement in digital and analog circuits at reduced power consumption.

Most of the early studies on ferroelectric FETs considered ferroelectric materials such as SBT, PZT, $BaTiO_3$, PVDF, etc. However, these materials have high dielectric constant and low E_c due to which it is a challenge to scale down such devices according to current technology nodes. Therefore, the discovery of ferroelectricity in doped HfO_2 has paved the way for integration of ferroelectric materials in nanoscale devices since it has low dielectric constant (~25), high coercive field (1–2 MV/cm), and remanent polarization (1–40 μC/cm^2) [12, 23]. Many experimental studies also successfully integrated ferroelectric HfO_2 in nanoscale devices for memory applications, and the operation capability has been proven for devices with gate length of 28 nm.

In view of various advantages of doped HfO_2 ferroelectric materials [49], some studies analyzed the performance of NCFETs by considering ferroelectric layer of doped HfO_2. It was reported that energy efficiency improved by 2.5 times for NCFETs with doped HfO_2 as ferroelectric layer in comparison to conventional MOSFETs.

However, with continued miniaturization, some detrimental effects crop up and several studies have reported some novel devices to integrate the advantages of advanced device designs such as channel engineering and gate electrode engineering techniques along with ferroelectric materials.

The efficacy of integrating graded channel profile along with ferroelectric insulator has also been explored for improved short-channel immunity, current drivability, and reduced power dissipation [50]. Furthermore, many experimental studies have reported fabrication steps for step linear doping profile, super steep retrograde profile, halo profile, etc. in source, channel, and drain for devices in nanometer regime. It has been demonstrated that by using techniques such as beamline ion implantation and integrated divergent beam (IDB) technology, the doping can be controlled precisely for devices with sub-22 nm dimensions so these devices have the potential to emerge as strong candidates for ultra voltage/low power applications. The study on GCNCSOI device also explored the impact of parasitic capacitance at high temperatures. The device characteristics were obtained by employing baseline approach and the device considered was metal-ferroelectric-metal-insulator-semiconductor (MFMIS) structure. The impact of GC design was extensively studied by obtaining results for a wide range of parameters, such as different values of doping concentration at source end and for different length ratios of high-doped region and low-doped region, and the results were compared with uniformly doped device. In addition, to assess the suitability of device for high-temperature analog applications, key figures of merit such as transconductance, transconductance efficiency, output conductance, and unity gain frequency were studied for optimum values of parasitic capacitance at 300, 340, and 380 K. It was demonstrated that GCNCSOI FET demonstrated superior performance at elevated temperatures.

Another novel device design, dual-material double gate ferroelectric field-effect transistor (DMGFeFET) [51], was explored, and this design integrated the advantages of gate electrode engineering and negative capacitance property to improve electrostatic integrity and transport efficiency and to result in efficient steep switching characteristics in nanoscale devices. An analytical model was formulated for short-channel DMGFeFET to explore the subthreshold characteristics of DMGFeFET by using Landau-Khalatnikov equation along with Poisson's equation to obtain variation of potential profile distribution, threshold voltage, and subthreshold current. A comprehensive comparative analysis of the proposed device with equivalent DMGFET demonstrated that DMGFeFET effectively suppressed short-channel effects and exhibited improved electrostatic integrity in terms of reduced threshold voltage roll-off, DIBL, and substantially reduced subthreshold current. Furthermore, the proposed device exhibited reduction in subthreshold current by three orders compared to conventional DMGFET, thereby implying reduced static power dissipation and efficient device operation.

8.6 JUNCTIONLESS FERROELECTRIC DEVICES

Conventional MOSFETs are approaching their scaling limits due to short-channel effects along with complex fabrication issues related to high thermal budget required for the formation of super steep doping profiles. In recent years, junctionless (JL) transistors [52] have also been widely researched, and various studies have reported that JL devices can overcome the requirement of complex fabrication procedures as they have the same doping type in source, drain, and channel, because of which they can be easily fabricated [52]. Moreover, these devices use bulk conduction instead of surface conduction, on account of which the carriers move through the channel with bulk mobility and experience less electric field in direction perpendicular to current flow, which results in less degradation in carrier mobility. JL transistors are also reported to exhibit near ideal subthreshold swing values, extremely low leakage currents, and high-mobility.

Along with complex fabrication procedures, increased power dissipation is also a challenge. To overcome these issues and to improve device performance, JL ferroelectric devices have been researched extensively. Some of the earliest work on such devices was a double gate ferroelectric junctionless transistor [53], which integrated the advantages of JL device and the negative capacitance property of ferroelectric materials. The use of silicon doped HfO2 ferroelectric in JL devices was also investigated for the first time.

The proposed device could be a strong contender for future CMOS technology as it can overcome the critical constraints of complex fabrication procedures and can also achieve the lower limit of voltage operation, thus, resolving the issue of severe heating at chip level. A study focused on developing an analytical model by employing Landau-Khalatnikov equation and parabolic potential approximation to analyze device behavior in different regions of operation. Various parameters such as potential, mobile charge density, and threshold voltage were studied and the analysis also focused on drain current characteristics of the proposed device. An exhaustive comparative analysis with the conventional double gate junctionless device demonstrated that the integration of ferroelectric layer led to significant improvement in terms of gain, minimum subthreshold values, subthreshold current, and improved current drivability.

It is also important to explore such devices for high-temperature applications such as temperature sensors, optical temperature sensing, etc. The high-temperature performance of double gate ferroelectric junctionless transistor was also analyzed [54], and a temperature-dependent model was developed to study the impact of temperature variations on various electrical parameters. The work demonstrated that double gate ferroelectric junctionless transistor exhibited superior performance even at elevated temperatures. Very recently, experimental studies have been carried out to demonstrate advantages of incorporating FE layer in conventional FETs, and hysteresis free sub-60 mV/dec switching has been successfully demonstrated, clearly indicating the potential of NCFETs for ultra low power/voltage switching application. In addition, recently, Seo et al. [55] have reported fabrication of JL FE FinFET for neuromorphic applications and the fabricated synapse showed ~80% pattern recognition accuracy.

It must be pointed out that junctionless transistors have high doping in the silicon body, due to which it becomes difficult to deplete the channel. Furthermore, because of heavy doping in channel, JL transistors exhibit high leakage current, and hence are not very feasible for low standby power applications. To exploit the advantages of these devices and to integrate these in the present CMOS design space, it is important to overcome these limitations. In view of this, junctionless cylindrical surrounding gate transistor (JLCSG) with ferroelectric layer as gate insulator [56] was proposed. Due to the improved gate controllability offered by surrounding gate geometry and NC effect of Si:HfO_2 ferroelectric (FE) insulator, the proposed device aids in depletion of channel and hence leads to significant suppression of short-channel effects along with sub-60 mV/dec device operation.

Furthermore, because heavy doping is employed in JL devices, it is crucial to optimize metal work function in order to fully deplete the channel. Hence, the impact of metal work function on the performance of JLCSG with ferroelectric insulator was also investigated [56]. The work demonstrated that optimization of metal work function and interfacial layer can provide better gate controllability, improved gain, and low point subthreshold slope values. Furthermore, substantial improvement in subthreshold current, saturation current, and I_{ON}/I_{OFF} ratio was also observed.

Another method that can aid in the depletion of channel in junctionless devices is the use of nonuniform doping profiles, which can also result in significant improvement in subthreshold characteristics as well as ON state drive current and I_{ON}/I_{OFF} ratio. Hence, the impact of vertical Gaussian-doped (GD) profile and NC phenomenon on short-channel double gate junctionless transistors (DGJLTs) [57] was analyzed by combining Landau-Khalatnikov equation with TCAD simulations according to the baseline approach. Device characteristics were studied for a wide range of parameters such as projected range (R_p), straggle (σ), and peak doping (N_{pk}) values. Also, to obtain optimum ferroelectric parameters, that is, coercive field and remanent polarization, charge density versus FE capacitance and MOS capacitance characteristics were also analyzed for double gate Gaussian doped negative capacitance junctionless transistor (DGGDNCJLT) for devices with different R_p and σ values. Furthermore, device design guidelines for short-channel DGGDNCJLTs were also presented to demonstrate the advantages of ferroelectric gate insulator and Gaussian doping profile in channel to achieve improved electrostatic integrity and non-hysteretic operation along with substantial gain in DGJLT.

8.7 ADVANCED CHANNEL MATERIAL-BASED FERROELECTRIC FETS

The introduction of advanced high-mobility channel materials has been proposed in the last few years to improve the carrier mobility and current drivability in FETs. Among various semiconductor materials, germanium (Ge) possesses the highest hole mobility, which makes it suitable for high-performance *p*-channel MOSFETs. In addition, germanium devices are easy to fabricate with the existing silicon technology process flow since Ge resembles silicon in structure as they both are group IV elements. This has made germanium a more favorable contender to replace Si in future technology nodes. To address the viability of Ge as the future channel material, there have been many reports on the experimental and fabrication techniques of Ge devices.

However, the most critical challenge in the fabrication of germanium devices is to grow a stable oxide (GeO_2) on Ge surface [58–59]. Therefore, high-*k* materials are used with Ge-based devices, which results in surface roughness and causes a large number of interface states and fixed oxide trap charges. These trap charges significantly affect the performance of Ge-based FETs, and hence it is important to study their impact on device operation.

To enhance the performance of *n*-channel MOSFET, silicon-germanium (SiGe) is also being extensively explored as a suitable channel material owing to its significantly higher electron mobility than silicon. The growing interest in SiGe has led to many experimental and theoretical studies in which SiGe has been integrated with novel device designs.

Among other novel channel materials with high carrier mobility and low bandgap than silicon, germanium-tin (GeSn) alloy has managed to draw the attention of researchers because of its spectacularly high carrier mobility. The lattice mismatch of Ge and Sn is about 15%, which causes the strain in GeSn alloy and is responsible for high carrier mobilities [60]. A small amount of Sn is required to obtain 60–85% enhancement in carrier mobility, implying that the bandgap of GeSn is not compromised much [60].

In recent years, there has been a concerted effort to explore devices with advanced channel materials along with ferroelectric materials as insulator layer to achieve improved device performance and to extend the CMOS design space. Few studies have also demonstrated the feasibility of fabricating Ge devices with these ferroelectric materials as gate insulator [61–63]. Many theoretical studies have also been carried out for planar ferroelectric GeFETs to understand the impact of negative capacitance on Ge-based devices. A capacitance model for planar negative capacitance Ge-nFET was proposed [64] and Fan et al. [65] studied the device characteristics of planar Ge ferroelectric nFET by using the baseline approach. Later, a comprehensive analytical drain current model was developed for long channel double gate germanium ferroelectric FET (DGGeFeFET) by using Poisson's equation and Pao-Sah's integral with Landau-Khalatnikov equation to obtain surface potential and drain current. The effect of interface trap states present within the bandgap of Ge was taken into account, and comparative analysis of the proposed device DGGeFeFET with conventional device showed that the presence of ferroelectric layer as gate insulator in DGGeFeFET effectively subdues the degradation in device characteristics caused by interface trap charges compared to conventional device [66].

Some studies have also focused on multi-geometry GeFeFETs. The effect of negative capacitance on Ge FinFET with fixed trap charges present at oxide/Ge-channel interface was studied by using 3-D Poisson's equation with Landau-Khalatnikov equation. The device characteristics of NCGeFinFET were studied using optimized values of P_r and E_c with and without the presence of fixed trap charges and were compared with conventional GeFinFET. It was demonstrated that degradation in device characteristics in the presence of negative trap charges could be minimized by further optimizing P_r and E_c in the calibrated range. The suitability of NCGeFinFET for digital circuit applications was also studied [67].

Since it is important to analyze the performance of FeFETs at high temperatures, the performance of NCGeFinFET for a wide temperature range was also investigated. Various parameters of NCGeFinFET such as surface potential, threshold voltage, gain, subthreshold swing, gate capacitance, drain current, and transconductance were studied for a wide range of temperature (300 K–380 K) and the characteristics have been exhaustively compared with the conventional Ge FinFET device. The study demonstrated that integration of ferroelectric layer not only led to a substantial reduction in OFF current but also provided immunity to trap charges.

Apart from these devices, the suitability of negative capacitance-based devices is also being explored for various other applications. Ferroelectric materials have also been integrated with 2-D materials, reconfigurable devices, HEMTs, etc. to explore their efficacy for both advanced ultra low voltage/low power applications as well as high-power application areas. These devices hold great promise, and it remains to be seen how suitably they can be integrated in future technology nodes.

REFERENCES

[1] Dennard, R., et al., "Design of Ion-Implanted MOSFETs with Very Small Physical Dimensions," *IEEE Journal of Solid State Circuits*, vol. 9, no. 5, pp. 256–268, 1974.

[2] Bohr, M. T., and I. A. Young, "CMOS Scaling Trends and Beyond," *IEEE Micro*, vol. 37, no. 6, pp. 20–29, 2017.

[3] Chen, Q., B. Agrawal, and J. D. Meindl, "A Comprehensive Analytical Subthreshold Swing (S) Model for Double-Gate MOSFETs," *IEEE Transactions on Electron Devices*, vol. 49, no. 6, pp. 1086–1090, June 2002.

[4] Auth, C. P., and J. D. Plummer, "Scaling Theory for Cylindrical, Fully-Depleted, Surrounding-Gate MOSFET's," *IEEE Electron Device Letters*, vol. 18, no. 2, pp. 74–76, Feb. 1997.

[5] Cheng, B., M. Cao, R. Rao, A. Inani, P. V. Voorde, W. M. Greene, J. M. C. Stork, Z. Yu, P. M. Zeitzoff, and J. C. S. Woo, "The Impact of High-Gate Dielectrics and Metal Gate Electrodes on Sub-100 nm MOSFETs," *IEEE Transactions on Electron Devices*, vol. 46, no. 7, pp. 1537–1544, July 1999.

[6] Inani, A., R. V. Rao, B. Cheng, and J. Woo, "Gate Stack Architecture Analysis and Channel Engineering in Deep Sub-Micron MOSFETs," *Japanese Journal of Applied Physics*, vol. 38, Part 1, No. 4B, pp. 2266–2271, Apr. 1999.

[7] Cheng, B., M. Cao, R. Rao, A. Inani, P. V. Voorde, W. M. Greene, J. M. C. Stork, Z. Yu, P. M. Zeitzoff, and J. C. S. Woo, "The Impact of High-Gate Dielectrics and Metal Gate Electrodes on Sub-100 nm MOSFETs," *IEEE Transactions on Electron Devices*, vol. 46, no. 7, pp. 1537–1544, July 1999.

[8] Djeffal, F., T. Bentrcia, M. A. Abdi, and T. Bendi, "Drain Current Model for Undoped Gate Stack Double Gate (GSDG) MOSFETs Including the Hot-carrier Degradation Effects," *Microelectronics Reliability*, vol. 51, no. 3, pp. 550–555, 2011.

[9] Kittel, C., "Dielectrics and Ferroelectrics," in *Introduction to Solid State Physics*, 5th ed., Wiley Eastern Reprint, New York, 1976, chapter 13, pp. 399–432.

[10] Salahuddin, S., and S. Datta, "Use of Negative Capacitance to Provide Voltage Amplification for Low Power Nanoscale Devices," *Nano Letters*, vol. 8, no. 2, pp. 405–410, Feb. 2008.

[11] Khan, A. I., C. W. Yeung, C. Hu, and S. Salahuddin, "Ferroelectric Negative Capacitance MOSFET: Capacitance Tuning & Antiferroelectric Operation," in *Proceedings of the IEDM Technical Digest*, pp. 11.3.1–11.3.4, Washington, DC, USA, Dec. 2011.

[12] Jimenez, D., E. Miranda, and A. Godoy, "Analytic Model for the Surface Potential and Drain Current in Negative Capacitance Field-effect Transistors," *IEEE Transactions on Electron Devices*, vol. 57, no. 10, pp. 2405–2409, Oct. 2010.

[13] Rabe, K. M., C. H. Ahn, and J. M. Triscone, *Physics of Ferroelectrics: A Modern Perspective*, Springer, Berlin, 2007.

[14] Xiao, Y., B. Zhang, H. Lou, L. Zhang, and X. Lin, "A Compact Model of Subthreshold Current with Source/Drain Depletion Effect for the Short-Channel Junctionless Cylindrical Surrounding-Gate MOSFETs," *IEEE Transactions on Electron Devices*, vol. 63, no. 5, pp. 2176–2181, 2016.

[15] Dekker, A. J., *Solid State Physics*, Macmillan & Co Ltd., London, 1957.

[16] Dawber, M., K. M. Rabe, and J. F. Scott, "Physics of Thin-film Ferroelectric Oxides," *Reviews of Modern Physics*, vol. 77, pp. 1083–1130, Oct. 2005.

[17] Muller, J., P. Polakowski, S. Mueller, and T. Mikolajick, "Ferroelectric Hafnium Oxide Based Materials and Devices: Assessment of Current Status and Future Prospects," *ECS Journal of Solid State Science and Technology*, vol. 4, no. 5, pp. N30–N35, Feb. 2015.

[18] Izyumskaya, N., Y. -I. Alivov, S. -J. Cho, H. Morkoç, H. Lee, and Y. -S. Kang, "Processing, Structure, Properties, and Applications of PZT Thin Films," *Critical Reviews in Solid State and Materials Sciences,* vol. 32, no. 3–4, pp. 111–202, 2007.

[19] Nakamura, K., and Y. Wada, "Piezoelectricity, Pyroelectricity, and the Electrostriction Constant of Poly(vinylidene fluoride)," *Journal of Polymer Science Part A-2: Polymer Physics*, vol. 9, no. 1, pp. 161–173, 1971.

[20] Chen, S., K. Yao, F. E. H. Tay, and L. L. S. Chew, "Comparative Investigation of the Structure and Properties of Ferroelectric Poly(vinylidene fluoride) and Poly(vinylidene fluoride–trifluoroethylene) Thin Films Crystallized on Substrates," *Journal of Applied Polymer Science*, vol. 116, pp. 3331–3337, 2010.

[21] Böscke, T. S., J. Müller, D. Bräuhaus, U. Schröder, and U. Böttger, "Ferroelectricity in Hafnium Oxide Thin Films," *Applied Physics Letters*, vol. 99, no. 10, p. 102903, Sept. 2011.

[22] Böscke, T. S., J. Müller, D. Bräuhaus, U. Schröder, and U. Böttger, "Ferroelectricity in Hafnium Oxide: CMOS Compatible Ferroelectric Field Effect Transistors," in *IEDM Technical Digest*, Washington, DC, USA, Dec. 2011, pp. 24.5.1–24.5.4.

[23] Mueller, S., J. Muller, U. Schroeder, and T. Mikolajick, "Reliability Characteristics of Ferroelectric Si:HfO_2 Thin Films for Memory Applications," *IEEE Transactions on Device and Materials Reliability*, vol. 13, no. 1, pp. 93–97, 2013.

[24] Park, M. H., Y. H. Lee, H. J. Kim, Y. J. Kim, T. Moon, K. D. Kim, J. Müller, A. Kersch, U. Schroeder, T. Mikolajick, and C. S. Hwang, "Ferroelectricity and Antiferroelectricity of Doped Thin HfO2-Based Films," *Advanced Materials*, vol. 27, no. 11, pp. 1811–1831, Feb. 2015.

[25] Clima, S., D. J. Wouters, C. Adelmann, T. Schenk, U. Schroeder, M. Jurczak, and G. Pourtois, "Identification of the Ferroelectric Switching Process and Dopant-dependent Switching Properties in Orthorhombic HfO_2: A first principles insight," *Applied Physics Letters*, vol. 104, no. 10, pp. 092906-1-092906-4, Mar. 2014.
[26] Schroeder, U., E. Yurchuk, J. Müller, D. Martin, T. Schenk, P. Polakowski, C. Adelmann, M. I. Popovici, S. V. Kalinin, and T. Mikolajick, "Impact of Different Dopants on the Switching Properties of Ferroelectric Hafnium Oxide," *Japanese Journal of Applied Physics*, vol. 53, no. 8S1, p. 08LE02, July 2014.
[27] Böscke, T. S., St. Teichert, D. Bräuhaus, J. Müller, U. Schröder, U. Böttger, and T. Mikolajick, "Phase Transitions in Ferroelectric Silicon Doped Hafnium Oxide," *Applied Physics Letters*, vol. 99, no. 11, pp. 112904, Sept. 2011.
[28] Martin, D., E. Yurchuk, S. Müller, J. Müller, J. Paul, J. Sundquist, S. Slesazeck, T. Schloesser, R. V. Bentum, M. Trentzsch, U. Schroeder, and T. Mikojajick, "Downscaling Ferroelectric Field Effect Transistors by Using Ferroelectric Si-doped HfO_2," in *Proceedings of the 13th International Conference on Ultimate Integration on Silicon (ULIS)*, Grenoble, France, 2012.
[29] Mueller, S., S. R. Summerfelt, J. Muller, U. Schroeder, and T. Mikolajick, "Ten-nanometer Ferroelectric Si:HfO_2 films for next-generation FRAM capacitors," *IEEE Electron Devices Letters*, vol. 33, no. 9, pp. 1300–1302, 2012.
[30] Yurchuk, E., J. Müller, R. Hoffmann, J. Paul, D. Martin, R. Boschke, T. Schl€osser, S. Müller, S. Slesazeck, R. Bentum, M. Trentzsch, U. Schr€oder, and T. Mikolajick, "HfO_2-based Ferroelectric Field-Effect Transistors with 260 nm Channel Length and Long Data Retention," in *Proceedings of the 4th IEEE International Memory Workshop (IMW)*, Milan, Italy, 2012.
[31] Shimizu, T., K. Katayama, T. Kiguchi, A. Akama, T. J. Konno, O. Sakata, and H. Funakubo, "The Demonstration of Significant Ferroelectricity in Epitaxial Y-doped HfO_2 film," *Scientific Reports*, vol. 6, pp. 32931, Sept. 2016.
[32] Tian, X., S. Shibayama, T. Nishimura, T. Yajima, S. Migita, and A. Toriumi, "Evolution of Ferroelectric HfO_2 in ultrathin region down to 3 nm," *Applied Physics Letters*, vol. 112, no. 10, pp. 102902, 2018.
[33] Khan, A. I., K. Chatterjee, B. Wang, S. Drapcho, L. You, C. Serrao, S. R. Bakaul, R. Ramesh, and S. Salahuddin, "Negative Capacitance in a Ferroelectric Capacitor," *Nature Materials,* vol. 14, no. 2, pp. 182–186, Feb. 2015.
[34] Rusu, A., G. A. Salvatore, D. Jiminez, and A. M. Ionescu, "Metal-Ferroelectric-Metal-Oxide-Semiconductor Field Effect Transistor with Sub-60mV/decade Subthreshold Swing and Internal Voltage Amplification," in *Proceedings of the IEDM*, pp. 16.3.1–16.3.4, San Francisco, CA, USA, Dec. 2010.
[35] Salvatore, G. A., D. Bouvet, and A. M. Ionescu, "Demonstration of Subthreshold Swing Smaller Than 60mV/decade in Fe-FET with P(VDFTrFE)/ SiO2 Gate Stack," in *Proceedings of the IEDM*, San Francisco, CA, USA, Dec. 2008.
[36] Liu, C., et al., "Simulation-based Study of Negative-capacitance Double-gate Tunnel Field-Effect Transistor with Ferroelectric Gate Stack," *Japanese Journal of Applied Physics*, vol. 55, no. 4S, p. 04EB08, 2016.
[37] Rusu, A., A. Saeidi, and A. M. Ionescu, "Condition for the Negative Capacitance Effect in Metal–Ferroelectric–Insulator–Semiconductor Devices," *Nanotechnology*, vol. 27, no. 11, p. 115201, 2016.
[38] Majumdar, K., S. Datta, and S. P. Rao, "Revisiting the Theory of Ferroelectric Negative Capacitance," *IEEE Transactions on Electron Devices*, vol. 63, no. 5, pp. 2043–2049, 2016.
[39] Jimenez, D., E. Miranda, and A. Godoy, "Analytic Model for the Surface Potential and Drain Current in Negative Capacitance Field-effect Transistors," *IEEE Transactions on Electron Devices*, vol. 57, no. 10, pp. 2405–2409, Oct. 2010.
[40] Xiao, Y. G., Z. J. Chen, M. H. Tang, Z. H. Tang, S. A. Yan, J. C. Li, X. C. Gu, Y. C. Zhou, and X. P. Ouyang, "Simulation of Electrical Characteristics in Negative Capacitance

Surrounding-gate Ferroelectric Field-effect Transistors," *Applied Physics Letters*, vol. 101, pp. 253511, Dec. 2012.

[41] Li, Y., Y. Lian, and G. S. Samudra, "Quantitative Analysis and Prediction of Experimental Observations on Quasi-static Hysteretic Metal-ferroelectric-metal-insulator-semiconductor FET and its Dynamic Behavior Based on Landau Theory," *Semiconductor Science and Technology*, vol. 30, no. 4, pp. 045011, Mar. 2015

[42] Xiao, Y., M. Tang, J. Li, B. Jiang, and J. He, "The Influence of Ferroelectric—electrode Interface Layer on the Electrical Characteristics of Negative-Capacitance Ferroelectric Double-Gate Field-Effect Transistors," *Microelectronics Reliability*, vol. 52, no. 4, pp. 757–760, 2012.

[43] Xiao, Y. G., M. H. Tang, Y. Xiaong, J. C. Li, C. P. Cheng, B. Jiang, H. Q. Cai, Z. H. Tang, X. S. Lv, X. C. Gy, and Y. C. Zhou, "Use of Negative Capacitance to Simulate the Electrical Characteristics in Double Gate-Ferroelectric Field-Effect Transistors," *Current Applied Physics*, vol. 12, no. 6, pp. 1591–1595, Nov. 2012.

[44] Salvatore, G. A., A. Rusu, and A. M. Ionescu, "Experimental Confirmation of Temperature Dependent Negative Capacitance in Ferroelectric Field Effect Transistor," *Applied Physics Letters*, vol. 100, pp. 163504, 2012.

[45] Xiao, Y. G., M. H. Tang, J. C. Li, C. P. Cheng, B. Jiang, H. Q. Cai, Z. H. Tang, X. S. Lv, and X. C. Gu, "Temperature Effect on Electrical Characteristics of Negative Capacitance Ferroelectric Field Effect Transistors," *Applied Physics Letters*, vol. 100, no. 8, pp. 083508, Feb. 2012.

[46] Li, Y., Y. Lian, K. Yao, and G. S. Samudra, "Evaluation and Optimization of Short Channel Ferroelectric MOSFET for Low Power Circuit Application with BSIM4 and Landau Theory," *Solid-State Electronics*, vol. 114, pp. 17–22, 2015.

[47] Li, Y., K. Yao, and G. S. Samudra, "Effect of Ferroelectric Damping on Dynamic Characteristics of Negative Capacitance Ferroelectric MOSFET," *IEEE Transactions on Electron Devices*, vol. 63, no. 9, pp. 3636–3641, Sept. 2016.

[48] Aziz, A., S. Ghosh, S. Datta, and S. K. Gupta, "Physics-Based Circuit-Compatible SPICE Model for Ferroelectric Transistors," *IEEE Electron Device Letters*, vol. 37, no. 6, pp. 805–808, 2016.

[49] Kobayashi, M., and T. Hiramoto, "On Device Design for Steep-Slope Negative-Capacitance Field-Effect-Transistor Operating at Sub 0.2V Supply Voltage with Ferroelectric HfO2 Thin Film," *AIP Advances*, vol. 6, no. 2, pp. 025113, Feb. 2016.

[50] Mehta, H., and H. Kaur, "Study on Impact of Parasitic Capacitance on Performance of Graded Channel Negative Capacitance SOI FET at High Temperature," *IEEE Transactions on Electron Devices*, vol. 66, no 7, pp. 2904–2909, July 2019.

[51] Mehta, H., and H. Kaur, "Subthreshold Analytical Model for Dual-Material Double Gate Ferroelectric Field Effect Transistor (DMGFeFET)," *Semiconductor Science and Technology*, vol. 34, no. 6, pp. 065008, 21st May 2019.

[52] Colinge, J. P., C. W. Lee, A. Afzalian, N. D. Akhavan, R. Yan, I. Ferain, P. Razavi, O'Neill, A. Blake, M. White, A. M. Kelleher, B. McCarthy, and R. Murphy, "Nanowire Transistors without Junctions," *Nature Nanotechnology*, vol. 5, no. 3, pp. 225–229, Mar. 2010.

[53] Mehta, H., and H. Kaur, "Modeling and Simulation Study of novel Double Gate Ferroelectric Junctionless (DGFJL) Transistor," *Superlattices and Microstructures*, vol. 97, pp. 536–547, Sept. 2016.

[54] Mehta, H., and H. Kaur, "High Temperature Performance of Si:HfO2 Based Long Channel Double Gate Ferroelectric Junctionless Transistors," *Superlattices and Microstructures*, vol. 103, pp. 78–84, Mar. 2017.

[55] Seo, M., M. H. Kang, S. B. Jeon, H. Bae, J. Hur, B. C. Jang, S. Yun, S. Cho, W. K. Kim, M. S. Kim, K. M. Hwang, S. Hong, S. Y. Choi, and Y. K. Choi, "First Demonstration of a Logic-Process Compatible Junctionless Ferroelectric FinFET Synapse for Neuromorphic Applications," *IEEE Electron Device Letters*, vol. 39, no. 9, Sept. 2018.

[56] Mehta, H., and H. Kaur, "Impact of Interface Layer and Metal Workfunction on Device Performance of Ferroelectric Junctionless Cylindrical Surrounding Gate Transistors," *Superlattices and Microstructures*, vol. 111, pp. 194–205, Nov. 2017.

[57] Mehta, H., and H. Kaur, "Impact of Gaussian Doping Profile and Negative Capacitance Effect on Double-Gate Junctionless Transistors (DGJLTs)," *IEEE Transactions on Electron Devices*, vol. 65, no. 7, pp. 2699–2706, July 2018.

[58] Saraswat, K., C. O. Chui, T. Krishnamohan, D. Kim, A. Nayfeh, and A. Pethe, "High Performance Germanium MOSFETs," *Materials Science and Engineering: B*, vol. 135, no. 3, pp. 242–249, Dec. 2006.

[59] Simoen, E., et al., "Challenges and Opportunities in Advanced Ge pMOSFETs," *Materials Science in Semiconductor Processing*, vol. 15, no. 6, pp. 588–600, 2012.

[60] Han, Genquan, et al., "High-mobility Germanium-Tin (GeSn) Pchannel MOSFETs Featuring Metallic Source/drain and sub-370°C Process Modules," in *Proceedings of the International Electron Devices Meeting*, Washington, DC, USA, 2011.

[61] Zhou, J., et al., "Frequency Dependence of Performance in Ge negative Capacitance PFETs Achieving Sub-30 mV/decade Swing and 110 mV Hysteresis at MHz," in *Proceedings of the IEEE International Electron Devices Meeting (IEDM),* San Francisco, CA, USA, 2017.

[62] Lomenzo, P. D., Q. Takmeel, C. M. Fancher, C. Zhou, N. G Rudawski, S. Moghaddam, J. L. Jones, and T. Nishida, "Ferroelectric Si-doped HfO2 Device Properties on Highly Doped Germanium," *IEEE Electron Device Letters*, vol. 36, no. 8, pp. 766–768, Aug. 2015.

[63] Li, J., et al., "Correlation of Gate Capacitance with Drive Current and Transconductance in Negative Capacitance Ge PFETs," *IEEE Electron Device Letters*, vol. 38, no. 10, pp. 1500–1503, Oct. 2017.

[64] Liao, Y. H., S. T. Fan, and C. W. Liu, "Modeling and Simulation of Negative Capacitance Gate on Ge FETs," *ECS Transactions*, vol. 75, no. 8, pp. 461–467, 2016.

[65] Fan, S. T., J. Y. Yan, D. C. Lai, and C. W. Liu, "The Hysteresis-free Negative Capacitance Field Effect Transistors Using Non-linear Polycapacitance," *Solid-State Electronics*, vol. 122, pp. 13–17, 2016.

[66] Bansal, M., and H. Kaur, "Impact of Negative Capacitance Effect on Germanium Double Gate pFET for Enhanced Immunity to Interface Trap Charges," *Superlattices and Microstructures*, vol. 117, pp. 189–199, 2018.

[67] Bansal, M., and H. Kaur, "Analysis of Negative Capacitance Germanium FinFET (NCGeFinFET) with the Presence of Fixed Trap Charges," *IEEE Transactions on Electron Devices*, vol. 66, no. 4, pp. 1979–1984, Apr. 2019.

9 Recent Trends in Compact Modeling of Negative Capacitance Field-Effect Transistors

Shubham Tayal, Shiromani Balmukund Rahi, Jay Prakash Srivastava, and Sandip Bhattacharya

CONTENTS

9.1 INTRODUCTION

The semiconductor industry has devoted much attention to the downscaling of field-effect transistors in the last few decades. The supply voltage (V_{DD}) reduction has become the vital tool to minimize the power density in integrated circuits (ICs). However, the continuous scaling of supply voltage (V_{DD}) has been the major challenge in current CMOS (complementary MOS) technology for ultra low power applications owing to thermal management issue [1–5]. It is usually believed that the continuing scaling of conventional FETs and its fleaveours like FinFET is limited by the sub-threshold swing (SS) that has approached the primordial cap of 60 mV/dec at 300 K

DOI: 10.1201/9781003200987-9

due to "Boltzmann Tyranny", that is, the incapability to eliminate the heat generated in the switching activity. Consequently, the devices with steep subthreshold slope are vital to attain energy-effective switching and low power as V_{DD} scales [1–6]. The several novel super steep-slope FETs, like impact ionization MOS (I-MOS) [2], Z^2-FET [7], tunnel FETs (TFETs), CNFETs, and negative capacitance (NC) FET, have been explored to overcome the "Boltzmann Tyranny" [1–8]. The TFET and NCFET are the two most common super steep slope device in research communities.

Tunnel FET is a quantum technical device following band-to-band tunneling as a transport phenomenon, resulting in limited current, suffering ambipolar current, and having a compatibility issue with conventional CMOS circuits and systems. On the other hand, the NCFET integrates a ferroelectric film inside the gate stack of an FET that behaves as a negative capacitance. The coupling of the ferroelectric materials directly to the channel in field-effect transistors provides a unique compatibility with conventional FET devices. An NCFET has the special characteristic that its gate stack isn't passive and comprises a process for amplification of the surface potential. In this chapter, we present the basics of NCFET along with the current advancement in the compact modeling field of the NCFET devices in various configurations [1–10].

9.2 STATE-OF-THE-ART STEEP-SLOPE DEVICES

The vital cap of SS (i.e., 60 mV/dec) on a FET device is the prime reason behind the limitation on scaling of supply voltage. It is the origin of on-chip power density issues in integrated circuits (ICs) and systems. At the same time, the steep subthreshold slope (SS) FET devices are simultaneously expanding the frontiers of viable ultra low power circuit and system at the energy scale. The research in steep-slope devices have been driven by the fundamental power issues encountered by conventional MOSFET as rapid device scaling followed the popular mathematical formula (Equation 9.1) commonly used for power calculation in the CMOS industry. Equation 9.2 and Equation 9.3 show the importance of power supply (V_{DD}). The rapid scaling of conventional FET device design parameters for maintaining performance and requirements cause an exponential increase in leakage current (I_{OFF}) [11–15].

$$P_{total} = P_{static} + P_{dynamic} \tag{9.1}$$

$$P_{static} = N_g \times I_{off} \times V_{DD} \tag{9.2}$$

$$P_{dynamic} = \alpha \times C_{totl} \times V_{DD}^2 \times f \tag{9.3}$$

Ng is the number of gates, α: the fraction of active gates, C_{total}: the total load capacitance of all gate, and f: switching frequency. Figure 9.1 illustrates the leakage current in conventional MOSFET and its dependency on device design matrix component. The depth and detailed investigation regarding the same are out of context here. However, the readers can refer to suitable articles and books.

As shown in Figure 9.1, the factor m known as body factor is measured by C_D and C_{OX}. The design parameters C_D and C_{OX} are becoming a critical factor in ultra-scaled

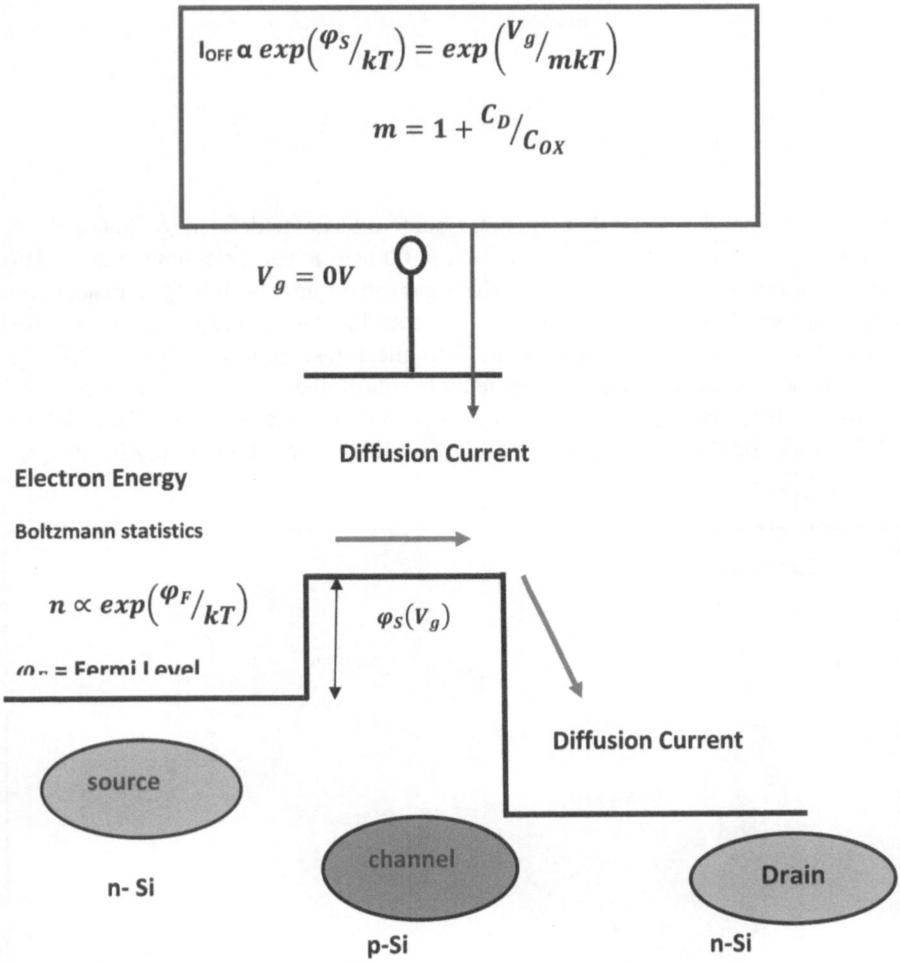

FIGURE 9.1 Basic Concept of Off-State Leakage Current

MOSFETs and its advanced FET devices like FinFET, SOI FET, etc. In the negative capacitance FET devices, the factor "m" is controlled by adding a thin additional layer of ferrodielectric materials in conventional FET and in its twin structures such as FinFET, SOI FET, and TFET [1–10]. Figure 9.2 shows the most popular developed field-effect devices used for various applications in the semiconductor industry.

On the basis of subthreshold slope (SS) defined by Equation 9.4, semiconductor devices are roughly classified into two categories. The first category contains MOS devices having SS value larger than 60 mV/decade measured by Equation 9.1, such as bulk MOSFET, FinFET, FDSOI, and NWFET [12–20]. For a typical Si-based FET, the ideal SS can be given as Equation 9.4 that can be further investigated as

$$SS = \left(\frac{d(log_{10}I_{DS})}{dV_G}\right)^{-1} = \frac{dV_G}{d\varphi_G} \times \frac{d\Psi_G}{d(log_{10}I_{DS})} = \left(1 + \frac{C_d}{C_{ox}}\right) \times \frac{d\Psi_G}{d(log_{10}I_{DS})}$$

$$= m \times n \qquad (9.4)$$

$$n = \frac{d\Psi_S}{d\left(log_{10} I_{DS}\right)} = log(10) \times \frac{k_B T}{q} = 2.3 \times 25.8 \frac{mV}{dec} \sim 60 \frac{mV}{dec} \qquad (9.5)$$

Here, the factor 'n' is known as transport factor and can be defined by Equation 9.5.

Ideally, the transport factor 'n' is stuck at 60 mV/dec at room temperature. This ideal condition in conventional FET structure created a great wall in front of semiconductor players. Some scientists suggested various FET architectures to get rid of this issue. However, practically only two FET architectures, namely, TFET and NCFET, have gained popularity and experimentally fulfill the requirements. Figure 9.3 shows the development flow of the two most commonly used super FET technology. TFET works on the principle of improving the 'm' factor whereas NCFET uses the

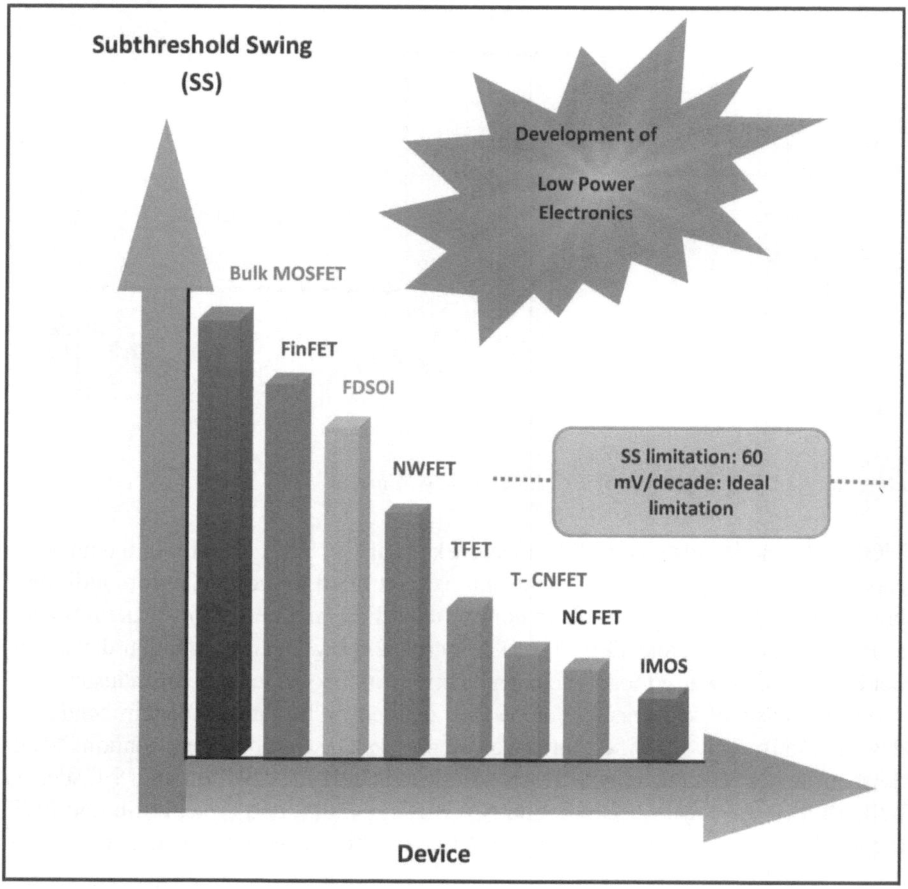

FIGURE 9.2 Development of Semiconductor Field-Effect Transistor

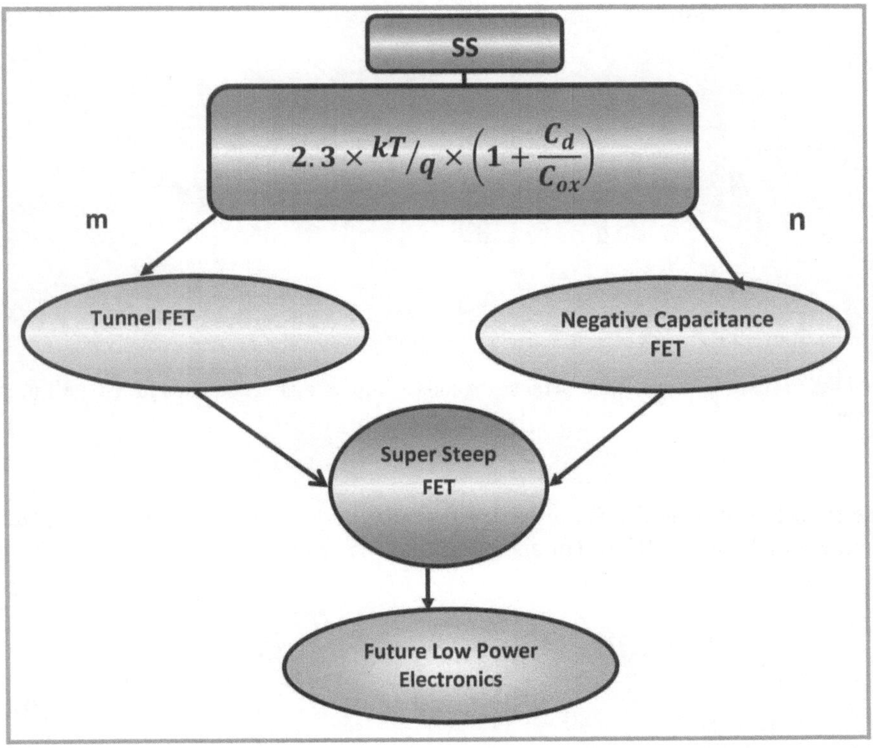

FIGURE 9.3 Development Flow Chart of Super FET Technology

phenomenon of improving the 'n' factor to improve the SS. In this chapter, we will focus briefly on TFET and the main focus will be on NCFET [11–22].

9.2.1 Tunnel FETs

Tunnel FETs (TFETs) are treated as one of the viable candidates for low-power operations owing to their ability to provide a sub-thermal subthreshold slope. This is possible because TFET is a quantum mechanical device that optimizes the transport factor "n" by using tunneling transport phenomenon instead of conventional drift-diffusion, which is temperature-independent [23–27].

TFETs are p-i-n gated junction devices that work in reverse regime. Figure 9.4 shows a pictorial representation of n-Tunnel FET. For an n-type TFET, p^+/n^+ doping is used for source/drain, while for p-type TFET, n^+/p^+ doping is used for source/drain [29–31].

As shown in Equation 9.6, transport current in tunnel devices is directly related to transport charges carriers during tunneling that occur in ON-state. Basically, ON-state in TFET is best energy band alignment in the device that offers minimum tunnel window for charge carriers. It is controlled by applied gate voltage on the device.

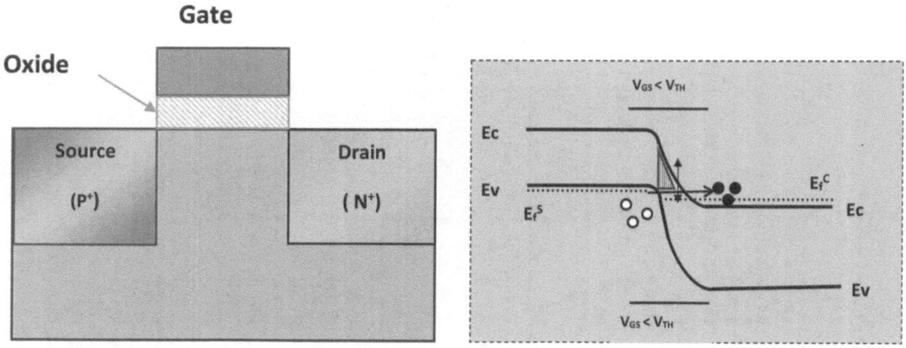

FIGURE 9.4 (A) N-type Tunnel Transistor and (b) Tunneling Phenomenon in N-type Tunnel Transistor

The tunneling probability and effective resultant current are governed by Wentzel-Kramers-Brillouin (WKB) quantum theory and can be given by Equation 9.6: [23–31]

$$I_{DS} \sim T(E) \propto \left(-\frac{4\sqrt{2m^*}E_g^{\frac{3}{2}}}{3|q|\bar{h}\left(E_g + \Delta\varnothing\right)} \sqrt{\frac{\varepsilon_{si}}{\varepsilon_{ox}} t_{ox} t_{si}} \right) \Delta\varnothing \tag{9.6}$$

In Equation 9.6, the term m* is the effective mass of charge particle, E_g is effective bandgap, $\Delta\Phi$ is the energy range for band-to-band tunneling (BTBT) window through which the carriers tunnel from one side to the other. The TFET device design matrix variables t_{ox}, t_{si}, ε_{ox}, and ε_{si} are the oxide and silicon films' thickness and dielectric constants, respectively. The remaining constant such as "ħ" is called the reduced Planck's constant and "|q|" is the electronic charge. The tunneling window ($\Delta\varnothing$) in the tunneling probability can be also expressed as $\Delta\varnothing = E_V^{ch} - E_C^S$. Thus, TFETs are excellent devices that provide very low SS and hence have very fast switching. However, unfortunately, TFETs are very subtle to defects present at the interface and in the semiconductor region. The temperature-dependent trap-assisted tunneling (TAT) process results in BTBT even when the FET is supposed to be in off state, which in turn degrades the subthreshold slope. That is why very few experimental TFETs reach an SS less than 60 mV/dec [24–31].

9.2.2 Negative Capacitance FETs

Further enhancement in FET operation and reliability necessitates new technology and device structures. Negative capacitance ferroelectric MOSFET (FeFET) [Figure 9.5 (a)] is a potential low-power transistor beyond conventional MOSFET. NCFET fundamentally changes the features of FETs and hence also changes the performance and power issues of upcoming circuit and system. When ferroelectric is in a series connection with a positive-capacitance material, its negative-capacitance range in the S-shape polarization-voltage curve becomes accessible due to

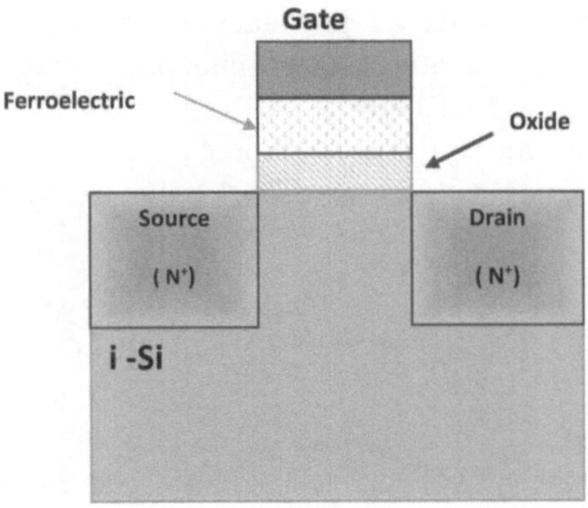

FIGURE 9.5 (A) In a Prototype N-Type NCFET

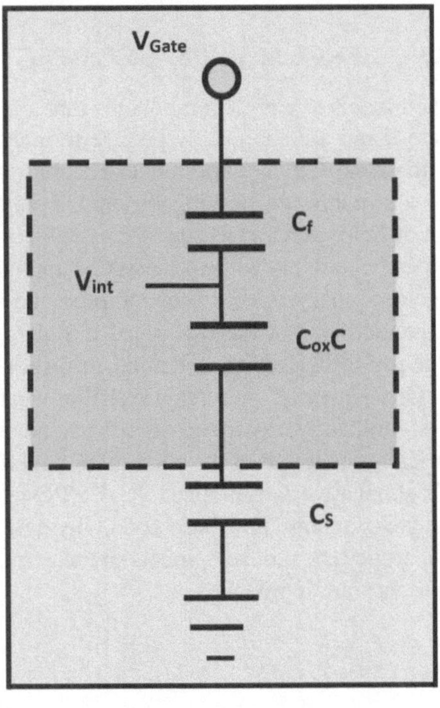

FIGURE 9.5 (B) Equivalent Capacitance NCFET Model

achievement of stability criterion of the whole device, that is, minimum total energy. This brings a beneficial effect of internal voltage amplification to FeFET [32–39].

The NCFETs aims to achieve a steep SS by enhancing the surface potential "φ_s", instead of changing the mechanism of carrier injection. In conventional high-*k* gate dielectric-based FETs, oxide capacitance (C_{ox}) is always positive, and thus it is impossible to attain $m < 1$. However, to achieve the same, gate dielectric material should provide a "negative capacitance", which can be attained by means of some ferroelectric materials. Using such a gate stack, the surface potential can increase faster than the gate voltage, creating a large amount of charge and a higher current compared to a conventional FET [1–6].

Figure 9.5 (b) presents the equivalent circuit of a typical NCFET with a connected NC capacitor (C_f) in-series with the gate stack followed by internal amplification factor β.

$$\beta = \frac{\partial V_{int}}{\partial V_g} = \frac{C_f}{\left(C_f + C_{int}\right)} \tag{9.7}$$

Here, C_f and C_{int} are the ferroelectric capacitance and equivalent capacitance of the base transistor from the gate terminal, respectively. Further, to get a significant increase in FET performance owing to the NC effect, the absolute value of the $|C_f|$ and C_{int} needs to be relatively close [39–52].

9.3 MODELING AND BENCHMARKING COMPACT MODELING

Compact models are the models of circuit components that are adequately easy to be fused in circuit simulators and are adequately precise to make the result helpful to circuit designers. The trade-off between the model accuracy and simplicity makes the compact modeling a stirring and defying research field for device physicists. Incessant scaling down of field-effect transistor (FET) devices has made it essential to integrate new physical phenomena, while widespread applications have prompted the incorporation of the secondary effects to accomplish the vital model accuracy. Simultaneously, the expanded size of the integrated circuits, which would now be able to be exposed to the full SPICE analysis, denied corresponding expansion in the model execution time. Consequently, significant exertion went into compact model reformulation so that significantly expanded accuracy and model refinement are cultivated without restrictive reduction in the computational efficiency [53].

The compact models of negative capacitance (NC) FET experienced revolutionary changes in the past few years and are now based on many new physical phenomena. In this section, we present the recent developments in the compact modeling field of the NCFET devices with various approaches.

9.3.1 Pahwa's Model

9.3.1.1 MFMIS Structure

The negative capacitance (NC) FET in metal-ferroelectric-metal-insulator-semiconductor (MFMIS) configuration is analyzed quantitatively by Pahwa et al. [54] in 2016. A physics-based compact model is presented by considering the transient as well as temperature effect for circuit designs. The schematic view of an NCFET along with its equivalent capacitor divider circuit model is elucidated in Figure 9.6. The ferroelectric material used for investigation is $Pb(Zr_{0.2}Ti_{0.8})O_3$ (PZT). The sandwiching of the metal layer between ferroelectric and FET gate made it simple to consider them as two different circuit units joined by a wire.

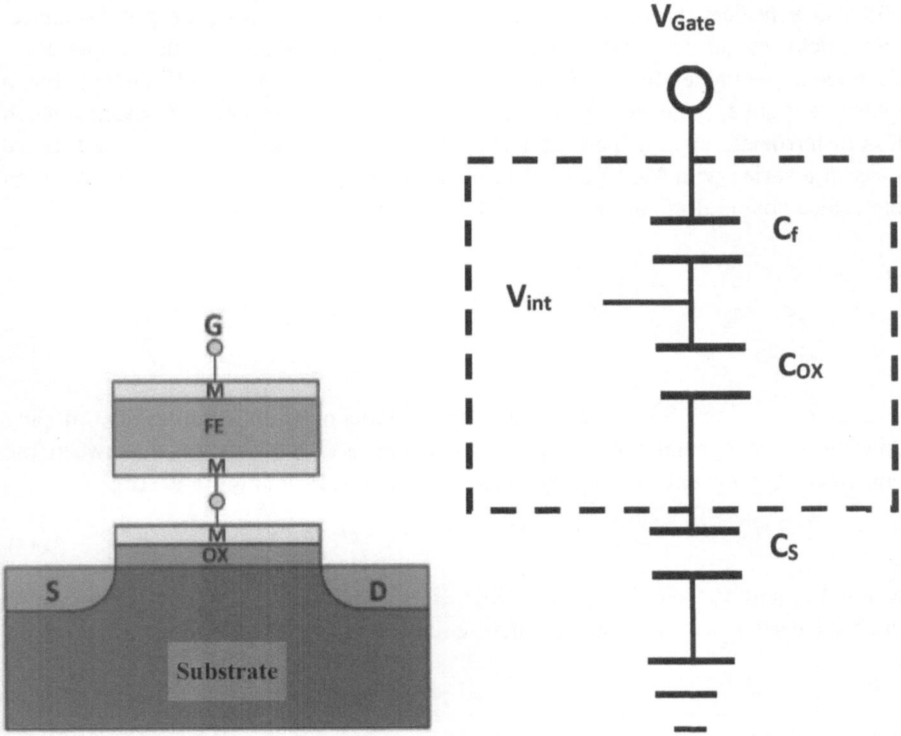

FIGURE 9.6 (a) Schematic View and (b) Equivalent Capacitor Divider Circuit Model of a NCFET [54]

The ferroelectric capacitance (C_f) model is developed using Landau-Khalatnikov (LK) theory and equation:

$$\delta \frac{dP}{dT} = -\frac{\partial G}{\partial P} \tag{9.8}$$

where G is Gibb's free energy given as $\alpha P^2 + \beta P^4 + \gamma P^6$—EP. P and E represent the polarization and electric field applied to the ferroelectric, respectively, that can be calculated as:

$$E = \frac{V_f}{t_f} = 2\alpha P + 4\beta P^2 + 6\gamma P^5 + \delta \frac{dP}{dT} \tag{9.9}$$

$$P = Q - \epsilon_0 \frac{V_f}{t_f} \tag{9.10}$$

$$C_f = \frac{\partial Q}{\partial V_f} = \frac{1}{t_f\left(2\alpha + 12\beta Q^2 + 30\gamma Q^4\right)} \tag{9.11}$$

where Q is the gate charge density. V_f, C_f, and t_f denote the potential drop, capacitance, and thickness of the ferroelectric, respectively. Further, α is the temperature-dependent parameter that makes ferroelectric also temperature-dependent. For a quasi-static state, Q can be approximated equal to P as electric field ($E = V_f/t_f$) is much less in ferroelectrics. Further, the term dP/dT term may also be overlooked. Since C_f is in a series with MOSFET, which makes the Q to be expressed as a function of surface potential ψ_s of the MOSFET [55]

$$Q = \pm\sqrt{2q\epsilon_s N_A}\left[\varnothing_t e^{-\psi_s/\phi_t} + \psi_s - \phi_t + e^{\frac{-2\phi_f + V_C}{\phi_t}}\left(\phi_t e^{\psi_s/\phi_t} - \psi_s - \phi_t\right)\right]^{1/2} \tag{9.12}$$

where $\epsilon_s, N_A, \phi_t, \phi_f, \& V_C$ represents the semiconductor permittivity, doping concentration, thermal voltage, Fermi potential, and the difference between the quasi-Fermi levels of electron and holes, respectively. From Figure 9.6 (b):

$$V_G = V_{FB} + V_f + V_{Ox} + \psi_s \tag{9.13}$$

where V_{FB} and V_{Ox} specify the flat-band voltage and oxide potential, respectively. From the earlier two equations, Q can be calculated in terms of V_f for a particular V_G.

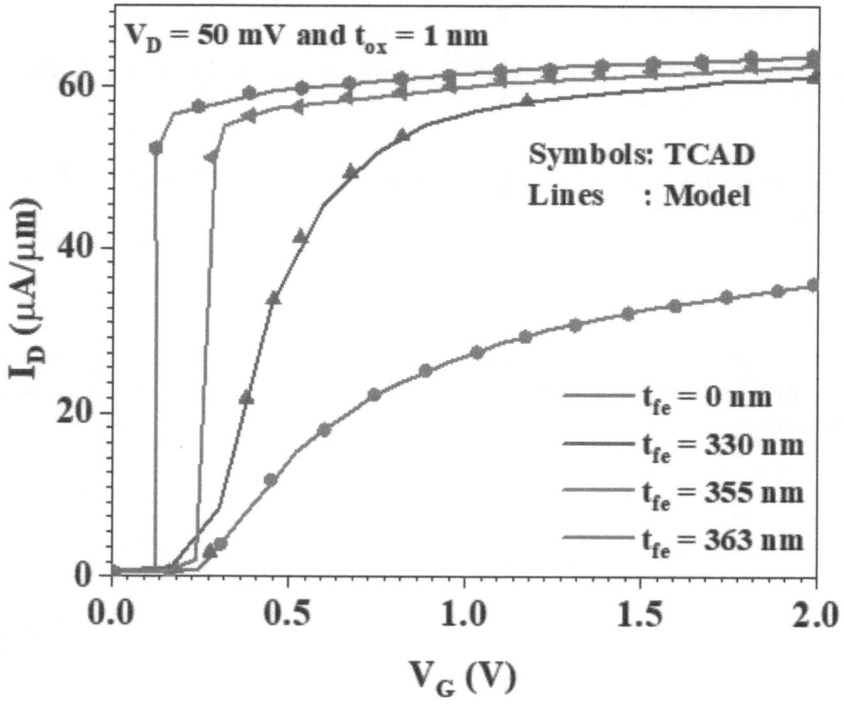

FIGURE 9.7 Model Validation with TCAD [56]

Thus, by expressing V_f in terms of Q and using BSIM6 MOSFET model by the simulator, temperature and transient effects of a NCFET device can be extracted.

Further, the internal voltage gain (A_V) owing to negative capacitance is obtained from Figure 9.6 (b) and is given as

$$A_V = \frac{\partial V_{int}}{V_G} = \frac{|C_f|}{|C_f| - C_{int}} \tag{9.14}$$

Pahwa et al. thoroughly validated the aforesaid model of NCFET using TCAD simulation as shown in Figure 9.7 [56]. It is found that large thickness of PZT layer results in negative output differential resistance in addition to high ON current.

9.3.1.2 MFIS Structure

Khan et al. [57] reported that the effect of negative capacitance is unstable in steady state for any MFMIS structure that is having gate leakage. To overcome this issue, MFIS structure as shown in Figure 9.8 has been proposed in which negative capacitance effect is not affected by the gate leakage. They reported that MFIS structure outperforms the MFMIS structure in terms of ON current (I_{on}) and subthreshold swing (SS).

The modeling for MFIS structure has been reported by Jimenez et al. [52] and Chen et al. [58] for double gate and bulk MOSFET as the baseline FET, respectively. However, their modeling technique is time consuming owing to the involvement of

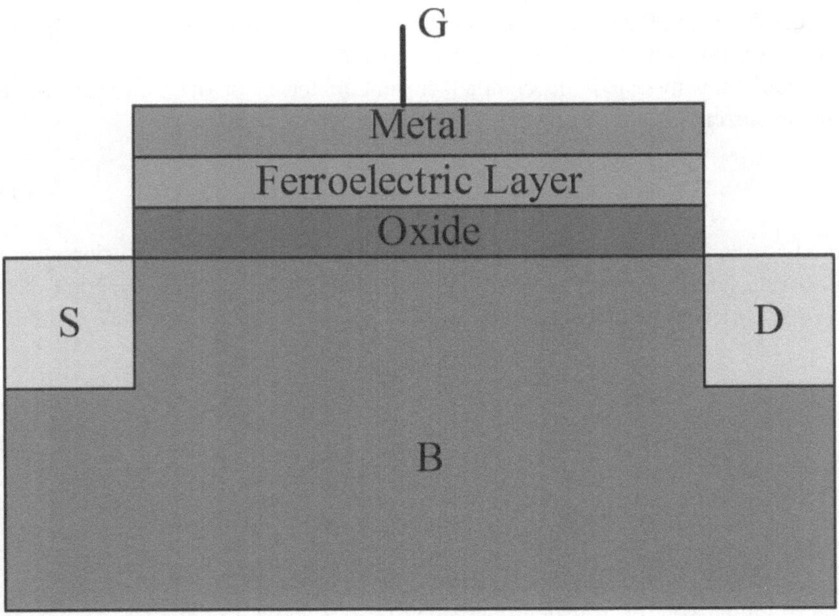

FIGURE 9.8 Schematic View of the NCFET in MFIS Configuration [57]

implicit equations. To get rid of this issue, a new compact model has been proposed by Pahwa et al. [59] in 2017, which uses explicit equations. Landau-Devonshire theory [60] relates the voltage across the ferroelectric (V_f) as a function of gate charge density (Q) as

$$V_f = Et_f = aQ + bQ^3 \tag{9.15}$$

$$a = -\frac{3\sqrt{3}}{2}\frac{E_C t_f}{P_r},\quad b = \frac{3\sqrt{3}}{2}\frac{E_C t_f}{P_r^3}$$

where E_c and P_r denote the ferroelectric properties and are the remnant polarization and coercive field, respectively. From voltage balance condition

$$V_G - V_{FB} = \left(a + 1/C_{OX}\right)Q + bQ^3 + \psi_s \tag{9.16}$$

Further, Q as a function of ψ_s [55]

$$Q = sign(\psi_s)\gamma C_{ox}\left[\psi_s + V_t\left(e^{-\psi_s/V_t} - 1\right) + e^{-\frac{2\varnothing_f + V_c}{V_t}}\left(V_t e^{\psi_s/V_t} - \psi_s - V_t\right)\right]^{1/2} \tag{9.17}$$

An implicit equation in terms of "Q" will result by placing the value of ψ_s from equation into equation. Pahwa et al. [59] reported an efficient computational technique in which they derived a competent initial guess of Q for weak inversion, strong inversion, and the transition region separately.

Further, they modeled the channel current in terms of drift current (I_{dr}) and diffusion current (I_{df}).

$$I = \overline{I_{dr}} + \overline{I_{df}} \tag{9.18}$$

where $\overline{I_{dr}}$ and $\overline{I_{df}}$ represent the average values of drain and diffusion current, respectively, and are given in terms of gate charge ($Q_{GD(S)}$) and inversion charge ($Q_{ID(S)}$) densities at the drain (source) end.

$$\overline{I_{dr}} = \mu\frac{W}{L}\int_{Q_{GS}}^{Q_{GD}} Q_I\left(a + \frac{1}{C_{ox}} + 3bQ^2\right)dQ \tag{9.19}$$

$$\overline{I_{df}} = \mu\frac{W}{L}V_t\left(Q_{ID} - Q_{IS}\right) \tag{9.20}$$

The comparison of transfer characteristics of MFMIS–NCFET and MFIS–NCFET for Vds of 50mV and 1V has been presented in Figure 9.9 (a) and 9.9 (b),

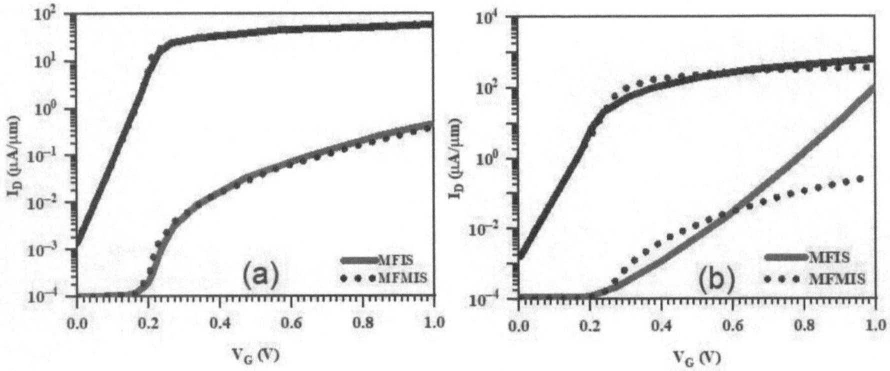

FIGURE 9.9 Comparison of Transfer Characteristics of MFMIS–NCFET and MFIS–NCFET for $V_D$of (a) 50mV and (b) 1V [57]

respectively. At 50mV, there is no substantial difference observed between the two. However, at 1V, the MFIS structure is found to be superior to the MFMIS structure.

9.3.2 Liang's Model

Aziz et al. [61] succeeded in modeling the dynamic character of NCFET by considering the ferroelectric layer as a nonlinear voltage-dependent capacitor. However, this model was unsupportive to the dc simulations. This problem has been solved in Xiao's model [62] in which the law of conservation of charge is used to model the NCFET, but Xiao's model is potent only for certain fitting conditions. Further, the parasitic components were not involved in the model proposed by Chenming et al. [63]. To improve the accuracy, Liang et al. [64] proposed a more effective compact model for NCFET by considering the short-channel effects (SCEs) where the parasitic elements of the ferroelectric layer have also been considered.

Figure 9.10 depicts the NCFET structure with its equivalent model taken into consideration. The parallel capacitor (C_0) and series resistor (R_f) are the parasitic elements associated with ferroelectric layer. To support the dc simulation, the capacitor has been modeled by using a unity-gain voltage-controlled voltage source (X_f). The gate charge (Q_G) of the baseline MOSFET can be calculated using BSIM4 SPICE model [63] as:

$$Q_G = W_{eff} \int_0^{L_{eff}} \left(V_G' - V_{FB} - \psi_S - V_y \right) dy \tag{9.21}$$

where L_{eff} and W_{eff} are the effective channel length and width, respectively. Further V_G' is modeled as:

$$V_G' = 0.9nV_t \ln[1 + e^{\left(\frac{V_{G,eff} - V_{th} - 0.02}{0.9nV_t}\right)} \tag{9.22}$$

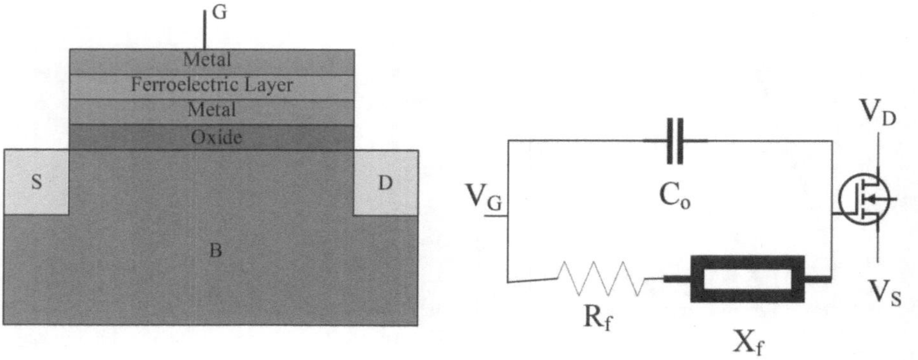

FIGURE 9.10 NCFET Structure with Its Equivalent Model Used in Liang's Model [64]

where n represents the subthreshold swing and $V_{G,eff}$ represents the effective gate-source voltage, owing to the existence of gate-oxide depletion. To include the SCEs and DIBL effects, threshold voltage (V_{th}) has also been included. Using the time-dependent LK equation and from [61],

$$V_f = R_f \frac{dQ_f}{dt} + t_f \left(\frac{2\alpha Q_f}{A_f} + \frac{4\beta Q_f^3}{A_f^3} + \frac{6\gamma Q_f^5}{A_f^5} \right) \tag{9.23}$$

$$R_f = \rho \frac{t_f}{A_f}$$

$$C_0 = \varepsilon_0 \frac{A_f}{t_f}$$

where V_f and Q_f are potential across X_f and charge stored by X_f, respectively. Further, the area and thickness of ferroelectric material are given by A_f and t_f, respectively. ρ represents the kinetic coefficient factor. So, as per law of conservation of charge,

$$Q_f = Q_G - V_f C_0 \tag{9.24}$$

By using aforesaid equations, drain-source current (I_{ds}) can be calculated as per the biasing potential of the MOSFET. Even the dependency of I_{ds} on terminal voltages can also be obtained. The authors have verified the model by comparing their simulation results with those from models in [61] and [62].

9.3.3 Dabhi's Model

The compact model for negative capacitance (NC) FDSOI FET with MFIS structure as shown in Figure 9.11 for the first time was proposed by Dabhi et al. [65] in 2020. The model has been developed by utilizing the BSIM-IMG framework.

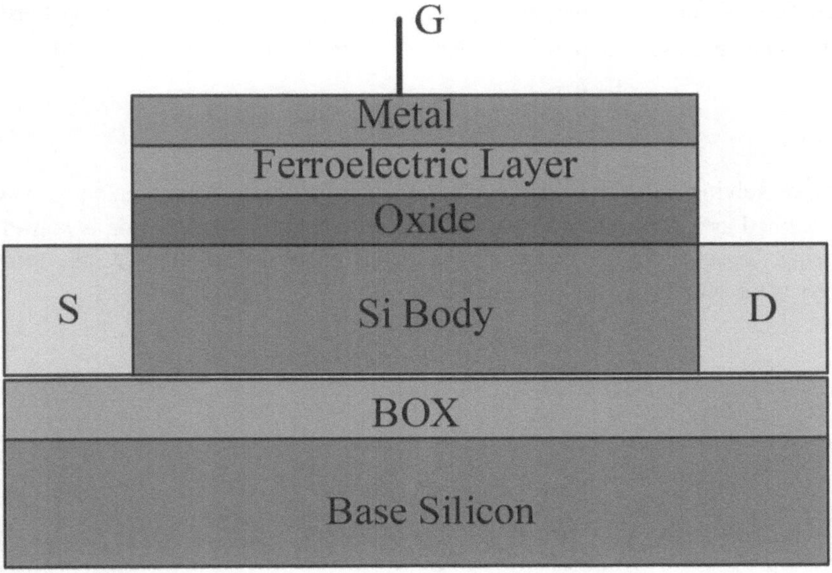

FIGURE 9.11 NC-FDSOI FET with MFIS Structure [65]

The Poisson equation across silicon body has been solved as [65–67]:

$$E_{s1}^2 - E_{s2}^2 = \frac{2qN_cV_t}{\varepsilon_{si}}\left[e^{\frac{\psi_{s1}-V_{ch}}{V_t}} - e^{\frac{\psi_{s2}-V_{ch}}{V_t}}\right] \tag{9.25}$$

As per Gauss's law,

$$E_{s1} = C_{ox1}\left(\frac{V_{fg} - \Delta\varnothing_{m1} - V_f - \psi_{s1}}{\varepsilon_{si}}\right) \tag{9.26}$$

$$E_{s2} = C_{ox2}\left(\frac{V_{bg} - \Delta\varnothing_{m2} - \psi_{s2}}{\varepsilon_{si}}\right) \tag{9.27}$$

where $E_{s1(s2)}$, $\psi_{s1(s2)}$, $\Delta\varnothing_{m1(m2)}$, $V_{fg\,(bg)}$, $T_{ox1\,(ox2)}$, and $C_{ox1\,(ox2)}$ denote the front (back) gate electric field, gate surface potential, gate work function, gate voltage, gate oxide thickness, and gate oxide capacitance, respectively. The gate charge density (Q) is obtained as in [68]

$$Q = \varepsilon_{si}\left[\frac{2qN_cV_t}{\varepsilon_{si}}e^{\frac{\psi_{s1}-V_{ch}}{V_t}} \times \left[1 - e^{\frac{(K_{si}-1)\psi_{s1} + K_{ox}\left(V_{bg}-\Delta\varnothing_{m2}\right)}{V_t}}\right] + \left(C_{ox2}\frac{V_{bg} - \Delta\varnothing_{m2} - \psi_{s1}}{T_{si} + \frac{\varepsilon_{si}}{\varepsilon_{ox}}T_{ox2}}\right)^2\right]^{\frac{1}{2}} \tag{9.28}$$

where C_{si} is silicon capacitance and $K_{si} = C_{si}/(C_{si} + C_{ox2})$, $K_{ox} = C_{ox2}/(C_{si} + C_{ox2})$. Further, the front and back surface potentials, that is, ψ_{s1} & ψ_{s2} are coupled as in [69].

$$\psi_{s2} = K_{si}\psi_{s1} + K_{ox}\left(V_{bg} - \Delta\varnothing_{m2}\right) \tag{9.29}$$

Thus, by solving the aforesaid equations, the implicit form of Poisson equation can be obtained as a function of single variable ψ_{s1}. This implicit relation is solved by utilizing the second-order householder's algorithm in six iterations. Consequently, the inversion charge density (Q_{inv}) and drain current (I_{ds}) are obtained as

$$Q_{inv} = \frac{2qN_cV_t}{E_{s1} + E_{s2}} e^{\frac{\psi_{s1} - V_{ch}}{V_t}} \times \left[1 - e^{\frac{(K_{si}-1)\psi_{s1} + K_{ox}\left(V_{bg} - \Delta\varnothing_{m2}\right)}{V_t}}\right] \tag{9.30}$$

$$I_{ds} = \mu\frac{W}{L}\left[\frac{Q_{inv,s} + Q_{inv,d}}{2}\left(\psi_{s1,s} - \psi_{s1,d}\right) + \eta V_t\left(Q_{inv,s} - Q_{inv,d}\right)\right] \tag{9.31}$$

where $\psi_{s1,s}\left(Q_{inv,s}\right)$ and $\psi_{s1,d}\left(Q_{inv,d}\right)$ are the surface potential (inversion charge density) at source and drain ends, respectively. Further, as in BSIM-IMG model [66], all the second-order effects have been considered. This model has been validated by comparing it with the experimental results of [70] as shown in Figure 9.12.

9.3.4 Gaidhane's Model

Beyond 7 nm technology node, the FinFET devices need to be replaced by the advanced FETs with better gate controllability such as nanosheet and nano-wire FETs due to reduced short-channel effects (SCEs). At such deep technology node, the quasi-ballistic transport must be considered. Gaidhane et al. [71] in 2020 proposed a compact model for NCFET in nanosheet configuration as shown in Figure 9.13.

The ballistic transport along with drift-diffusion transport has been included in developing the model using the LK equation for ferroelectric layer. A 2-D Poisson equation in its reduced form as given in [11] is used for NCEFT as:

$$\frac{1}{k}\frac{\partial^2\psi_C}{\partial^2 x} + \frac{V_G + V_{FB} + V_f - \psi_C}{\zeta} = \frac{q}{\varepsilon_s}n(x) \tag{9.32}$$

where ψ_C is the central potential and k is an intermediate parameter and is equal to $(1 + \varepsilon_{ins}T_{NS} / 4\varepsilon_s T_{ins})$. T_{NS} and T_{ins} are nanosheet thickness and thickness of interfacial oxide layer, respectively. ζ is given as $\varepsilon_s T_{NS} T_{ins} k / 4\varepsilon_s T_{ins}$. $n(x)$ is the charge density and may be given as:

$$n(x) = n_{s-d}(x) + n_{d-s}(x) \tag{9.33}$$

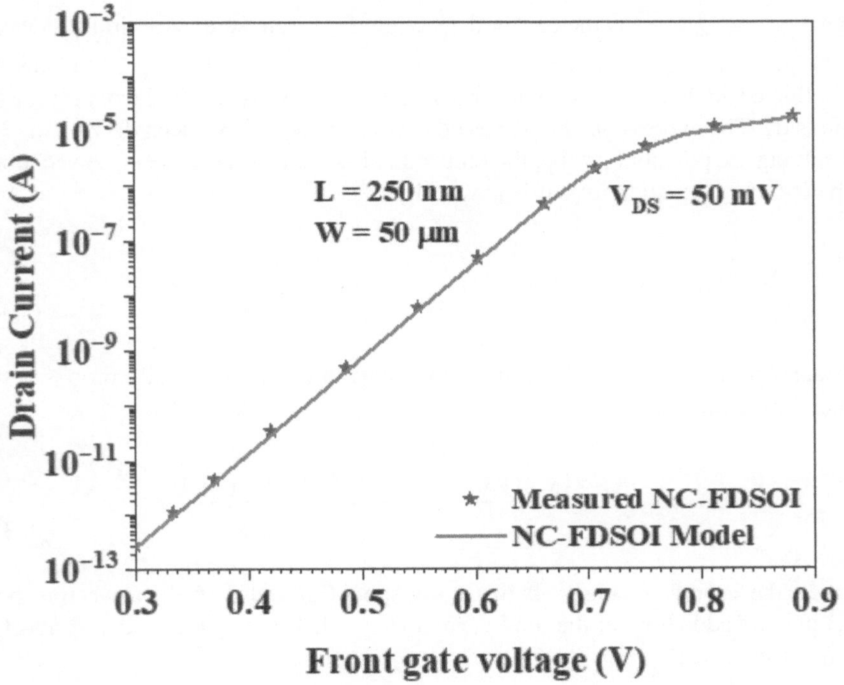

FIGURE 9.12 Model Validation with TCAD [65]

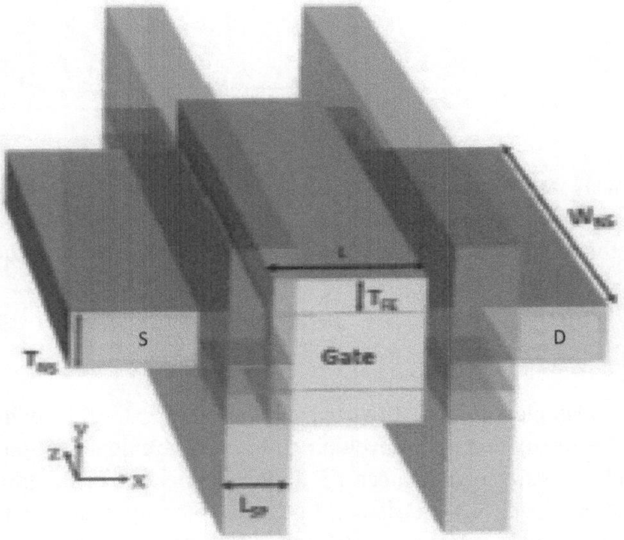

FIGURE 9.13 NCFET in Nanosheet Configuration [71]

where $n_{s(or\,d)-d(or\,s)}(x)$ is the carrier density injected from source (or drain) side and can be given as $[n_{b-s(or\,d)}(x)+n_{dd-s(or\,d)}(x)]$. $n_{b-s(or\,d)}(x)$ and $n_{dd-s(or\,d)}(x)$ denote the ballistic and drift-diffusive carrier densities respectively injected from source (or drain) side. The expressions for aforesaid carrier densities have been derived in [71]. For solving the potential profile, the reduced Poisson equation has been solved along with continuity equation for ballistic velocity (v_b) :

$$[n_{b-s(or\,d)}]v_b + \mu[n_{dd-s(or\,d)}]\frac{\partial \psi_C}{\partial x} + \mu\varnothing_t \frac{\partial n_{dd-s(or\,d)}}{\partial x} = Const. \tag{9.34}$$

Gaidhane et al. [71] have solved these equations in detail and finally proposed the model for surface potential as:

$$\psi_s = \frac{\psi_C + (k-1)(V_G - V_{FB} + \alpha_f n(x) - \beta_f n(x)^3)}{k} \tag{9.35}$$

Further, the current owing to ballistic transport (I_b) and drift-diffusive transport (I_{diff}) may be added to get the total current (I_{total}), that is, $I_{total} = I_b + I_{diff}$. I_b and I_{diff} can be given as [71]:

$$I_b = qT_{NS}W_{NS}\left(n_{b-s}v_{b(x=0)} - n_{b-d}v_{b(x=L)}\right)\left[\frac{1-e^{P\frac{V_{ds}}{\varnothing_t}}}{1+e^{-P\frac{V_{ds}}{\varnothing_t}}}\right] \tag{9.36}$$

$$I_{dd} = \frac{q\mu_{eff}T_{NS}W}{L}\int_0^L \{n_{dd-S}(x) - n_{dd-d}(x)\}\frac{d\psi(x)}{dx}dx \tag{9.37}$$

The previously mentioned model has been validated with the TCAD data for different thickness (T_f) of ferroelectric material as shown in Figure 9.14. It has been demonstrated that voltage amplification effect can be reduced by using quasi-ballistic transport.

9.4 CONCLUSION

In this chapter, we explored the negative capacitance FET (NCFET) technology, basic structure, and its impact on newly developing ultra low circuit and system in terms of compact models. Negative capacitance FET (NCFET) is one of the most promising steep-subthreshold slope (SS) transistors. In this chapter, operation and modeling of negative capacitance field-effect transistor (NCFET) are described. Various compact models have been reviewed for NCFET in different configurations. Pahwa's models

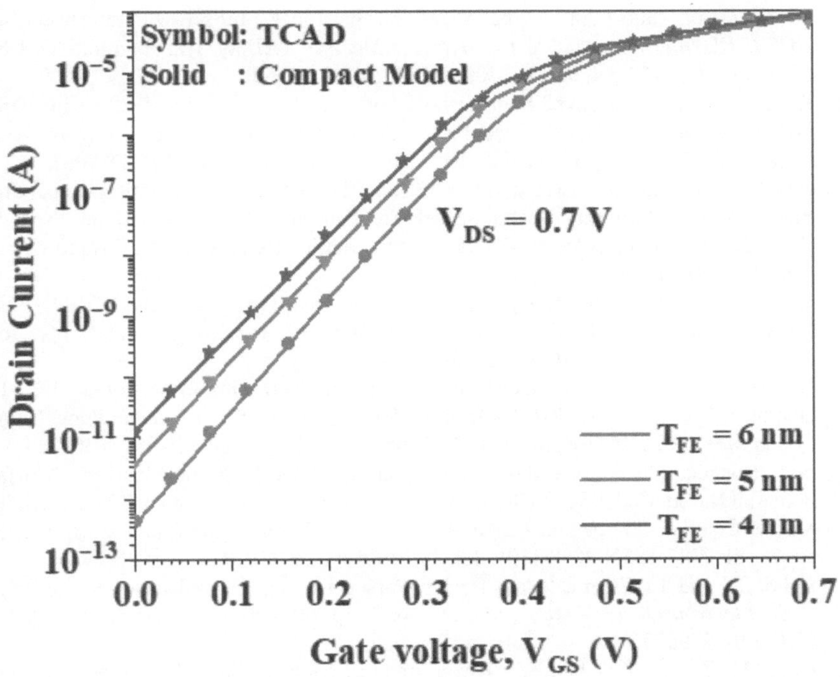

FIGURE 9.14 Model Validation with TCAD [71]

explain the MFMIS and MFIS configurations. Further, Liang's model proposed a more effective compact model for NCFET by considering the short-channel effects (SCEs) where the parasitic elements of the ferroelectric layer have also been considered. Dabhi's model gives a framework for negative capacitance (NC) FDSOI FET with MFIS structure for the first time in 2020. The model has been developed by utilizing the BSIM-IMG framework. Gaidhane et al. proposed a compact model for NCFET in nanosheet configuration for future technology nodes. Finally, this chapter concludes that NCFET may be exploited to reduce subthreshold swing for ultra low power MOSFET applications. Consequently, the modeling of NCFET may be considered a broad field of research for further improving its performance at a circuit level.

REFERENCES

[1] Seabaugh, A. C., and Q. Zhang, "Low-Voltage Tunnel Transistors for Beyond CMOS Logic," *Proceedings of the IEEE*, vol. 98, no. 12, pp. 2095–2110, Dec. 2010, doi: 10.1109/JPROC.2010.2070470.

[2] Lee, M. H., K.-T. Chen, C.-Y. Liao, S.-S. Gu, G.-Y. Siang, Y.-C. Chou, H.-Y. Chen, J. Le, R.-C. Hong, Z.-Y. Wang, S.-Y. Chen, P.-G. Chen, M. Tang, Y.-D. Lin, H.-Y. Lee, K.-S. Li, and C. W. Liu, "Extremely Steep Switch of Negative-Capacitance Nanosheet GAA-FETs and FinFET," *2018 IEEE International Electron Devices Meeting (IEDM),* San Francisco, CA, USA, 2018, pp. 31.8.1–31.8.4, doi: 10.1109/IEDM.2018.8614510.

[3] Li, K., et al., "Sub-60mV-swing Negative-capacitance FinFET without Hysteresis," *2015 IEEE International Electron Devices Meeting (IEDM)*, Washington, DC, USA, 2015, pp. 22.6.1–22.6.4, doi: 10.1109/IEDM.2015.7409760.

[4] Kwon, D., et al., "Improved Subthreshold Swing and Short Channel Effect in FDSOI n-Channel Negative Capacitance Field Effect Transistors," *IEEE Electron Device Letters*, vol. 39, no. 2, pp. 300–303, Feb. 2018, doi: 10.1109/LED.2017.2787063.

[5] Gao, W., Asif Khan, Xavi Marti, Chris Nelson, Claudy Serrao, Jayakanth Ravichandran, Ramamoorthy Rameshand, and Sayeef Salahuddin, "Room-Temperature Negative Capacitance in a Ferroelectric–Dielectric Superlattice Heterostructure," *Nano Letters,* vol. 14, no.10, pp. 5814–5819, 2014.

[6] Sharma, A., and K. Roy, "Design Space Exploration of Hysteresis-Free HfZrOx-Based Negative Capacitance FETs," *IEEE Electron Device Letters*, vol. 38, no. 8, pp. 1165–1167, Aug. 2017, doi: 10.1109/LED.2017.2714659.

[7] Navarro, C., et al., "Gate-induced vs. Implanted Body Doping Impact on Z2-FET DC Operation," *2017 IEEE SOI-3D-Subthreshold Microelectronics Technology Unified Conference (S3S)*, Burlingame, CA, 2017, pp. 1–2, doi: 10.1109/S3S.2017.8308748.

[8] Jain, Prateek, Chandan Yadav, Amit Agarwal, Yogesh Singh Chauhan, "Surface Potential Based Modeling of Charge, Current, and Capacitances in DGTFET Including Mobile Channel Charge and Ambipolar Behaviour," *Solid-State Electronics*, vol. 134, pp. 74–81, 2017, ISSN 0038-1101, doi: 10.1016/j.sse.2017.05.012.

[9] Rollo, T., and D. Esseni, "New Design Perspective for Ferroelectric NC-FETs," in *IEEE Electron Device Letters*, vol. 39, no. 4, pp. 603–606, Apr. 2018, doi: 10.1109/LED.2018.2795026.

[10] Kim, J., Z. C. Y. Chen, S. Kwon and J. Xiang, "Steep Subthreshold Slope Nanoelectromechanical Field-effect Transistors with Nanowire Channel and Back Gate Geometry," *71st Device Research Conference*, Notre Dame, IN, USA, 2013, pp. 209–210, doi: 10.1109/DRC.2013.6633867.

[11] Khatami, Y., and K. Banerjee, "Steep Subthreshold Slope n- and p-Type Tunnel-FET Devices for Low-Power and Energy-Efficient Digital Circuits," *IEEE Transactions on Electron Devices*, vol. 56, no. 11, pp. 2752–2761, Nov. 2009, doi: 10.1109/TED.2009.2030831.

[12] Kobayashi, M., K. Jang, N. Ueyama and T. Hiramoto, "Negative Capacitance for Boosting Tunnel FET Performance," *IEEE Transactions on Nanotechnology*, vol. 16, no. 2, pp. 253–258, Mar. 2017, doi: 10.1109/TNANO.2017.2658688.

[13] Jiang, C., L. Zhong and L. Xie, "Effects of Interface Trap Charges on the Electrical Characteristics of Back-gated 2D Negative Capacitance FET," *2019 IEEE 19th International Conference on Nanotechnology (IEEE-NANO),* Macao, China, 2019, pp. 163–166, doi: 10.1109/NANO46743.2019.8993936.

[14] Appleby, D. J., N. K. Ponon, K. S. Kwa, B. Zou, P. K. Petrov, T. Wang, N. M. Alford, and A. O'Neill, "Experimental Observation of Negative Capacitance in Ferroelectrics at Room Temperature," *Nano letters*, vol. 14, no. 7, pp. 3864–3868, 2014.

[15] Kobayashi, M., C. Jin, and T. Hiramoto, "Comprehensive Understanding of Negative Capacitance FET From the Perspective of Transient Ferroelectric Model," *2019 IEEE 13th International Conference on ASIC (ASICON)*, Chongqing, China, 2019, pp. 1–4, doi: 10.1109/ASICON47005.2019.8983568.

[16] Kobayashi, M., K. Jang, N. Ueyama and T. Hiramoto, "Negative Capacitance for Boosting Tunnel FET Performance," *IEEE Transactions on Nanotechnology*, vol. 16, no. 2, pp. 253–258, Mar. 2017, doi: 10.1109/TNANO.2017.2658688.

[17] Ilatikhameneh, H., T. A. Ameen, C. Chen, G. Klimeck and R. Rahman, "Sensitivity Challenge of Steep Transistors," *IEEE Transactions on Electron Devices*, vol. 65, no. 4, pp. 1633–1639, Apr. 2018, doi: 10.1109/TED.2018.2808040.

[18] You, W., and P. Su, "Design Space Exploration Considering Back-Gate Biasing Effects for 2D Negative-Capacitance Field-Effect Transistors," *IEEE Transactions on Electron Devices*, vol. 64, no. 8, pp. 3476–3481, Aug. 2017, doi: 10.1109/TED.2017.2714687.

[19] Hu, C., et al., "Prospect of Tunneling Green Transistor for 0.1V CMOS," *2010 International Electron Devices Meeting*, San Francisco, CA, USA, 2010, pp. 16.1.1–16.1.4, doi: 20.1109/IEDM.2010.5703372.

[20] Rollo, T., H. Wang, G. Han and D. Esseni, "A Simulation-based Study of NC-FETs Design: Off-state versus On-state Perspective," *2018 IEEE International Electron Devices Meeting (IEDM)*, San Francisco, CA, USA, 2018, pp. 9.5.1–9.5.4, doi: 10.1109/IEDM.2018.8614514.

[21] Alam, M. A., M. Si, and P. D. Ye, "A Critical Review of Recent Progress on Negative Capacitance Field-effect Transistors," *Applied Physics Letters,* vol. 114, p. 090401, 2019.

[22] Goyal, V., S. Tayal, S. Meena, and R. Gupta, "Impact of Interfacial Layer Thickness on Gate Stack-based DGTFET: An Analog/RF Prospective," *IOP Conference Series: Materials Science and Engineering,* 1070 012081, 2021, doi: 10.1088/1757-899x/1070/1/012081.

[23] Ionescu, A., and H. Riel, "Tunnel Field-effect Transistors as Energy-efficient Electronic Switches," *Nature,* vol. 479, pp. 329–337, 2011, doi: 10.1038/nature10679.

[24] Rahi, S. B., and B. Ghosh, "High-k Double Gate Junctionless Tunnel FET with Tunable Bandgap," *RSC Advances*, vol. 5, no. 67, pp. 54544–54550, 2015, doi: 10.1039/C5RA06954H

[25] Chung, H.-T., et al., "Effect of Crystallinity on the Electrical Characteristics of Poly-Si Tunneling FETs via Green Nanosecond Laser Crystallization," *IEEE Electron Device Letters*, vol. 42, no. 2, pp. 164–167, Feb. 2021, doi: 10.1109/LED.2021.3049329.

[26] Kim, H. W., and D. Kwon, "Low-Power Vertical Tunnel Field-Effect Transistor Ternary Inverter," *IEEE Journal of the Electron Devices Society*, doi: 10.1109/JEDS.2021.3057456.

[27] Asthana, P. K., B. Ghosh, S. B. Rahi and Y. Goswami, "Optimal Design of High Performance H-JLTFET using HfO_2 as Gate Dielectric for Ultra Low Power Applications," *RSC Advances*, vol. 4, no. 43, pp. 22803–22807, 2014, doi: 10.1039/C4RA00538D

[28] Asthana, P. K., Y. Goswami, S. Basak, S.B. Rahi and B. Ghosh, "Improved Performance of Junctionless Tunnel Field Effect Transistor with Si and SiGe Hetero-Structure for Ultra Low Power Applications," *RSC Advances*, vol.5, pp. 48779–48785, 2015, doi: 10.1039/C5RA03301B.

[29] Guenifi, N., S. B. Rahi, and T. Ghodbane, "Rigorous Study of Double Gate Tunneling Field Effect Transistor Structure Based on Silicon," *Materials Focus*, vol. 7, pp. 1–7, 2018, doi: 10.1166/mat.2018.1600.

[30] Gayduchenko, I., S. G., Xu, G. Alymov, et al., "Tunnel Field-effect Transistors for Sensitive Terahertz Detection," *Nature Communications,* vol. 12, p. 543, 2021, doi: 10.1038/s41467-020-20721-z.

[31] Sarkar, D., X. Xie, W. Liu, et al., "A Subthermionic Tunnel Field-effect Transistor with an Atomically Thin Channel," *Nature* 526, 91–95 (2015). https://doi.org/10.1038/nature15387.

[32] Rusu, A., G. A. Salvatore, D. Jiménez, and A. M. Ionescu, "Metal-Ferroelectric-Meta-Oxide-semiconductor Field Effect Transistor with sub-60mV/decade Subthreshold Swing and Internal Voltage Amplification," *2010 International Electron Devices Meeting*, San Francisco, CA, USA, 2010, pp. 16.3.1–16.3.4, doi: 10.1109/IEDM.2010.5703374.

[33] Cao, W., and K. Banerjee, "Is Negative Capacitance FET a Steep-slope Logic Switch," *National Communications,* vol. 11, p. 196, 2020, doi: 10.1038/s41467-019-13797-9.

[34] Bidenko, P., S. Lee, J. Han, J. D. Song and S. Kim, "Simulation Study on the Design of Sub-kT/q Non-Hysteretic Negative Capacitance FET Using Capacitance Matching," *IEEE Journal of the Electron Devices Society*, vol. 6, pp. 910–921, 2018, doi: 10.1109/JEDS.2018.2864593.

[35] Tu, Luqi, et al. "Ferroelectric Negative Capacitance Field Effect Transistor," *Advanced Electronic Materials,* vol. 4, no. 11, p. 1800231, 2018.

[36] Amrouch, H., G. Pahwa, A. D. Gaidhane, J. Henkel and Y. S. Chauhan, "Negative Capacitance Transistor to Address the Fundamental Limitations in Technology Scaling: Processor Performance," *IEEE Access*, vol. 6, pp. 52754–52765, 2018, doi: 10.1109/ACCESS.2018.2870916.

[37] Bacharach, J., M. S. Ullah and E. Fouad, "A Review on Negative Capacitance Based Transistors," *2019 IEEE 62nd International Midwest Symposium on Circuits and Systems (MWSCAS)*, Dallas, TX, USA, 2019, pp. 180–184, doi: 10.1109/MWSCAS.2019.8885275.

[38] Ye, P. D., "Steep-Slope Hysteresis-Free Negative-Capacitance 2D Transistors," *2018 14th IEEE International Conference on Solid-State and Integrated Circuit Technology (ICSICT)*, Qingdao, China, 2018, pp. 1–1, doi: 10.1109/ICSICT.2018.8564814.

[39] Wouters, D. J., J. -. Colinge and H. E. Maes, "Subthreshold slope in thin-film SOI MOSFETs," *IEEE Transactions on Electron Devices*, vol. 37, no. 9, pp. 2022–2033, Sept. 1990, doi: 10.1109/16.57165.

[40] Park, J., G. Jang, H. Kim, et al., "Sub-kT/q Subthreshold-Slope Using Negative Capacitance in Low-Temperature Polycrystalline-Silicon Thin-Film Transistor," *Scientific Reports,* vol. 6, p. 24734, 2016, doi: 10.1038/srep24734.

[41] Kobayashi, M., K. Jang, N. Ueyama and T. Hiramoto, "Negative Capacitance for Boosting Tunnel FET performance," *IEEE Transactions on Nanotechnology*, vol. 16, no. 2, pp. 253–258, Mar. 2017, doi: 10.1109/TNANO.2017.2658688.

[42] Saeidi, A., et al., "Negative Capacitance as Performance Booster for Tunnel FETs and MOSFETs: An Experimental Study," *IEEE Electron Device Letters*, vol. 38, no. 10, pp. 1485–1488, Oct. 2017, doi: 10.1109/LED.2017.2734943.

[43] Han, G., J. Zhou, Y. Liu, J. Li, Y. Peng and Y. Hao, "Experimental Investigation of Fundamentals of Negative Capacitance FETs," *2018 IEEE SOI-3D-Subthreshold Microelectronics Technology Unified Conference (S3S)*, Burlingame, CA, USA, 2018, pp. 1–2, doi: 10.1109/S3S.2018.8640179.

[44] Salahuddin, S., and S. Datta, "Use of Negative Capacitance to Provide Voltage Amplification for Low Power Nanoscale Devices," *Nano Letters*, vol. 8, no. 2, pp. 405–410, 2008.

[45] Karda, K., A. Jain, C. Mouli, and M. A. Alam, "An Anti-ferroelectric Gated Landau Transistor to Achieve Sub-60 mV/dec Switching at Low Voltage and High Speed," *Applied Physics Letters, 106*(16), p. 163501, 2015.

[46] Alam, Muhammad A., "A Tutorial Introduction to Negative-Capacitor Landau Transistors: Perspectives on the Road Ahead," 2015, https://nanohub.org/resources/23157.

[47] Speransky, D., and T. T. Trung, "Impactionization Effect in Deep Submicron MOSFET Features Simulation," *Proceedings of the 19th International Conference Mixed Design of Integrated Circuits and Systems—MIXDES 2012*, Warsaw, Poland, 2012, pp. 66–68.

[48] Si, M., et al., "Sub-60 mV/decferroelectric HZO MoS2 Negative Capacitance Field-effect Transistor with Internal Metal Gate: The Role of Parasitic Capacitance," *2017 IEEE International ElectronDevices Meeting (IEDM)*, San Francisco, CA, USA, 2017, pp. 23.5.1–23.5.4, doi: 10.1109/IEDM.2017.8268447.

[49] Guenifi, N., S. B. Rahiand, and M. Larbi, "Suppression of Ambipolar Current and Analysis of RF Performance in Double Gate Tunnelling Field Effect Transistors," *International Journal of Nanoparticles and Nanotechnology*, pp. 1–12, 2020, doi: 10.35840/2631-5084/5533.

[50] Tayal, Shubham, and Ashutosh Nandi, "Effect of FIBL in-conjunction with Channel Parameters on Analog and RF FOM of FinFET," *Superlattices and Microstructures*, vol. 105, 2017, pp. 152–162, ISSN 0749-6036, doi: 10.1016/j.spmi.2017.03.018.

[51] Tayal, Shubham, and Ashutosh Nandi, "Analog/RF Performance Analysis of Channel Engineered High-K gate-stack Based Junctionless Trigate-FinFET," *Superlattices and Microstructures*, vol. 112, pp. 287–295, 2017, ISSN 0749-6036, doi: 10.1016/j.spmi.2017.09.031.

[52] Jimenez, D., E. Miranda, and A. Godoy, "Analytic Model for the Surface Potential and Drain Current in Negative Capacitance Field-Effect Transistors," *IEEE Transactions on Electron Devices*, vol. 57, no. 10, pp. 2405–2409, Oct. 2010, doi: 10.1109/TED.2010.2062188.

[53] Gildenblat, G., *Compact Modeling: Principles, Techniques and Application*, Springer, Dordrecht/Heidelberg/London/New York, ISBN 978-90-481-8613-6, doi: 10.1007/978-90-481-8614-3.

[54] Pahwa, G., et al., "Analysis and Compact Modeling of Negative Capacitance Transistor with High ON-Current and Negative Output Differential Resistance—Part I: Model Description," *IEEE Transactions on Electron Devices*, vol. 63, no. 12, pp. 4981–4985, Dec. 2016, doi: 10.1109/TED.2016.2614432.

[55] Tsividis, Y., and C. McAndrew, *Operation and Modeling of the MOS Transistor*, Oxford University Press, New York, 2011.

[56] Pahwa, G., et al., "Analysis and Compact Modeling of Negative Capacitance Transistor with High ON-Current and Negative Output Differential Resistance—Part II: Model Validation," *IEEE Transactions on Electron Devices*, vol. 63, no. 12, pp. 4986–4992, Dec. 2016, doi: 10.1109/TED.2016.2614436.

[57] Khan, A. I., U. Radhakrishna, K. Chatterjee, S. Salahuddin, and D. A. Antoniadis, "Negative Capacitance Behavior in a Leaky Ferroelectric," *IEEE Transactions on Electron Devices*, vol. 63, no. 11, pp. 4416–4422, Nov. 2016.

[58] Chen, H.-P., V. C. Lee, A. Ohoka, J. Xiang, and Y. Taur, "Modeling and Design of Ferroelectric MOSFETs," *IEEE Transactions on Electron Devices*, vol. 58, no. 8, pp. 2401–2405, Aug. 2011.

[59] Pahwa, G., T. Dutta, A. Agarwal, and Y. S. Chauhan, "Compact Model for Ferroelectric Negative Capacitance Transistor With MFIS Structure," *IEEE Transactions on Electron Devices*, vol. 64, no. 3, pp. 1366–1374, Mar. 2017, doi: 10.1109/TED.2017.2654066.

[60] Devonshire, A. F., "Theory of Barium Titanate: Part I," *London Edinburgh Dublin Philosophical Magazine and Journal of Science,* vol. 40, no. 309, pp. 1040–1063, 1949.

[61] Aziz, A., S. Ghosh, S. Dutta, and S. K. Gupta, "Physics-based Circuit-compatible SPICE Model for Ferroelectric Transistors," *IEEE Electron Device Letters*, vol. 37, no. 6, pp. 805–808, June 2016.

[62] Li, Y., K. Yao, and G. S. Samudra, "Delay and Power Evaluation of Negative Capacitance Ferroelectric MOSFET Based on SPICE Model," *IEEE Transactions on Electron Devices*, vol. 64, no. 5, pp. 2403–2408, May 2017.

[63] Chenming, H., and L. Weidong, *BSIM4 and MOSFET Modeling for IC Simulation*, World Scientific, Singapore, 2011.

[64] Liang, Y., X. Li, S. K. Gupta, S. Datta, and V. Narayanan, "Analysis of DIBL Effect and Negative Resistance Performance for NCFET Based on a Compact SPICE Model," *IEEE Transactions on Electron Devices*, vol. 65, no. 12, pp. 5525–5529, Dec. 2018, doi: 10.1109/TED.2018.2875661.

[65] Dabhi, C. K., S. S. Parihar, A. Dasgupta, and Y. S. Chauhan, "Compact Modeling of Negative-Capacitance FDSOI FETs for Circuit Simulations," *IEEE Transactions on Electron Devices*, vol. 67, no. 7, pp. 2710–2716, July 2020, doi: 10.1109/TED.2020.2994018.

[66] Khandelwal, S., et al., "BSIM-IMG: A Compact Model for Ultrathin-body SOI MOSFETs with Back-gate Control," *IEEE Transactions on Electron Devices*, vol. 59, no. 8, pp. 2019–2026, Aug. 2012.

[67] Kushwaha, P., et al., "BSIM-IMG 102.9.1 Technical Manual," 2017, http://bsim.berkeley.edu/models/bsimimg/.

[68] Dutta, T., G. Pahwa, A. R. Trivedi, S. Sinha, A. Agarwal, and Y. S. Chauhan, "Performance Evaluation of 7-nm Node Negative Capacitance FinFET-based SRAM," *IEEE Electron Device Letters*, vol. 38, no. 8, pp. 1161–1164, Aug. 2017.

[69] Li, X., et al., "Enabling Energy-efficient Nonvolatile Computing with Negative Capacitance FET," *IEEE Transactions on Electron Devices*, vol. 64, no. 8, pp. 3452–3458, Aug. 2017.

[70] Kwon, D., et al., "Improved Subthreshold Swing and Short Channel Effect in FDSOI n-Channel Negative Capacitance Field Effect Transistors," *IEEE Electron Device Letters*, vol. 39, no. 2, pp. 300–303, Feb. 2018.

[71] Gaidhane, A. D., G. Pahwa, A. Dasgupta, A. Verma, and Y. S. Chauhan, "Compact Modeling of Surface Potential, Drain Current and Terminal Charges in Negative Capacitance Nanosheet FET including Quasi-Ballistic Transport," *IEEE Journal of the Electron Devices Society*, vol. 8, pp. 1168–1176, 2020, doi: 10.1109/JEDS.2020.3019927.

10 Fundamentals of 2-D Materials

Ganesan Anushya, Rasu Ramachandran, Raj Sarika, and Michael Benjamin

CONTENTS

10.1 INTRODUCTION

Nanotechnology is science, engineering, and technology conducted at the nanoscale, which is about 1 to 100 nanometers. In other words, nanotechnology is the design and fabrication of materials, devices, and systems with control at nanometer dimensions [1]. Nanoscience and nanotechnology are the study and application of extremely

DOI: 10.1201/9781003200987-10

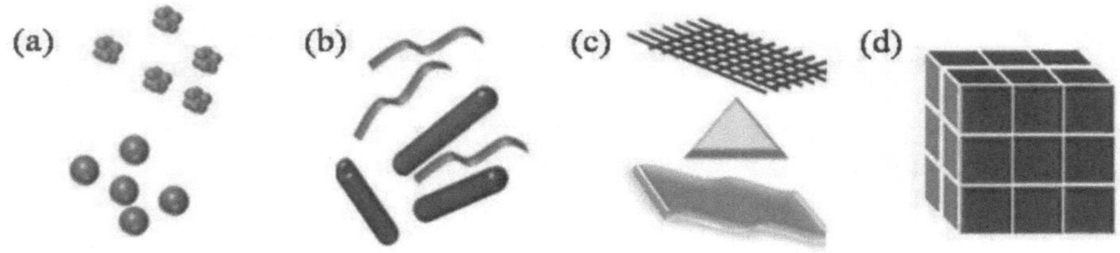

FIGURE 10.1 Classification of Nanomaterials: (a) 0-D Spheres and Clusters, (b) 1-D Nanofibers, Wires, and Rods, (c) 2-D Films, Plates, and Networks, and (d) 3-D Nanomaterials

small things and can be used across all the other science fields, such as chemistry, biology, physics, materials science, and engineering [2].

Over the past decade, nanomaterials have been the subject of enormous interest. Nanomaterials have extremely small size with at least one dimension of 100 nm or less. Nanomaterials can be metals, ceramics, polymeric materials, or composite materials. It can be nanoscale in one dimension, two dimensions, or three dimensions. They can exist in single, fused, aggregated, or agglomerated forms with spherical, tubular, and irregular shapes. Common types of nanomaterials include nanotubes, thin films, nanofibers, nanorods, nanoparticles, dendrimers, quantum dots, and fullerenes [3, 4]. Nanomaterials have applications in the field of nanotechnology and displays different physical and chemical characteristics.

In recent years, low-dimensional materials or two-dimensional (2-D) materials have more and more revolutionized increasing attention due to their extraordinary material properties compared with conventional bulk materials. The surface area of 2-D materials is increased, due to all atoms in surface getting exposed after exfoliation. These atomically thin-layered materials (2-D) have unique unexpected electrical, optical, thermal, and mechanical properties. 2-D materials can be used in wide-range applications, including flexible electronics, strain sensors, nano generators, and innovative nanoelectromechanical systems (NEMS) to catalysis and energy storage. In general, based on the structure, the 2-D materials are categorized, and few examples of 2-D materials are shown in Figure 10.2. Among these 2-D materials, the first lab-made 2-D atomic crystal graphene has shown deep interest toward fundamental research and R&D applications. This leads to several unique physical and chemical properties, and this graphene is considered as an alternative material to replace conventional materials [5, 6].

Basically, graphene is an allotrope of carbon, with a distinctive thickness of 0.34 nm. Arrangement of carbon atoms in graphene is like a honeycomb lattice (hexagonal lattice structure). Each carbon atom is sp^2 hybridized and it is covalently bonded with the other three neighboring atoms of about 1.42 Å (Figure 10.3). Graphene was first isolated at Manchester University by Russian-born scientists Andre Geim and Kostya Novoselov in 2004. They both won the Nobel Prize in 2010 for their inventive work [7].

Two-dimensional materials consist of thin layers with a thickness of at least one atomic layer. They have a layered structure with a tight in-plane bonding and a weak van der Waals within layers. In contrast to bulk materials, these nanomaterials have an elevated aspect ratio and thus have numerous atoms at their surface. The function

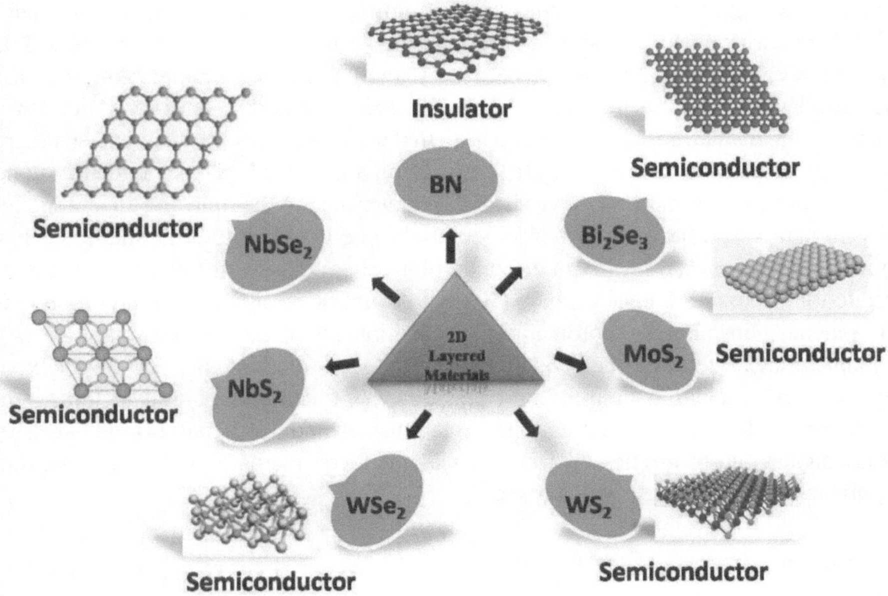

FIGURE 10.2 2-D Materials and Its Electronic Nature

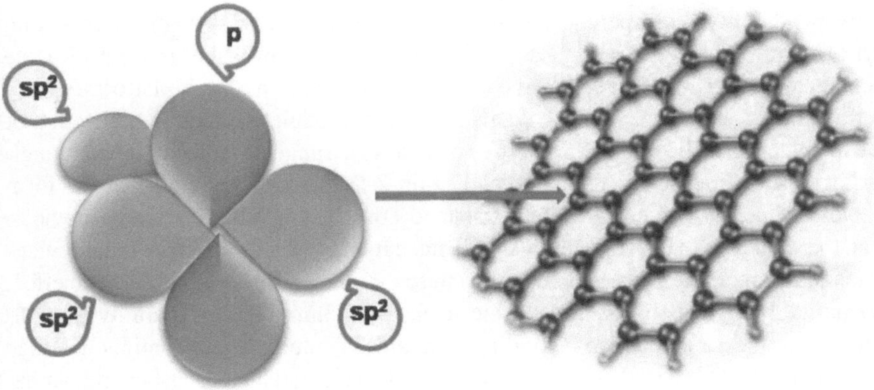

FIGURE 10.3 Hexagonal Lattice Structure of Graphene with sp^2 Hybridization

of these atoms is different from that of inner atoms, and hence, the rise in the number of surface atoms result in a difference in the behavior of 2-D nanomaterials [8]. Since they first came to light, the 2-D materials have undeniably drawn great research and professional interests due to their remarkable variety of mechanical, optical, physical, and chemical properties, which are distinct from their bulk counterparts. These unique characteristics result from the effect of their spatial confinement in one dimension. Besides, as they are atomically thin and have strong bonds in the plane, they are likely to exhibit a remarkable combination of thermal, high mechanical strength, which are ideal properties for use in different devices [9]. Since 2-D materials have an immense number of atoms on the surface layer, the properties of 2-D materials can be modified by the necessary alteration and functionalization of the surface layer. The 2-D materials are categorized as metallic, semi-metallic, semiconducting, insulating, or superconducting materials, based on its chemical structure and structural configuration. Apart from having a high surface-to-volume ratio and a high-surface activities, the rewards of 2-D materials are that they are superconductive in nature and have an immense potential to be turned into ultra small and low power transistors that are more effective than silicon-based transistors in terms of reduction of size and performance [10, 11]. A diverse array of layered compounds (graphene, phosphorene, boron nitride, transition metal dichalcogenides (TMDs), MXenes have been physically isolated or produced synthetically for a variety of applications, including field-effect transistors, photovoltaics, photodetectors, light emitting diodes, energy storage, chemical sensors, and gas separation [12–15]. In this chapter, the properties of 2-D materials, defects, methods of preparation, and applications are elaborately discussed.

10.2 Properties of 2-D Materials

The 2-D materials offer a great avenue to design a novel material, which exhibits fascinating mechanical, physicochemical, electrical, and optical properties [16, 17]. This is because they have a peculiar property like ultra low weight, high in-plane stiffness, nonlinear mechanical interactions, high Young's modulus, tunability, flexibility (bendability), transparency, atomic scale thickness, high strength, outstanding carrier mobility, as well as high anisotropy between the in-plane and out-of-plane mechanical properties. Generally, the 2-D materials exist as layer platform. Every layer is tailored with weak van der Waals forces that modulate fracture and friction of the 2-D materials. The tunable properties offer a wide range of application from engineering to scientific fields. The atomically thin 2-D materials assemble in the form of heterostructures, which is more reliable to construct flexible electronic devices. After the discovery of 2-D materials, the mechanical strength of electronic devices tremendously increases [18, 19]. This is because 2-D materials have a synergetic effect of remarkably high stiffness, impeccable surface-to-volume ratio, capability to withstand the stresses, and massive flexibility even after the multiple and continuous usage of the materials. On the other hand, 2-D materials have notable properties like amazing chemical composition, exhibiting the electronic bandgap invisible spectral region and marvelous structural arrangements, and exploiting the valley degree of freedom, which make the 2-D materials more fit for the production of highly efficient ultra-thin electronic devices. The interface between the surface and a substrate, and the presence of atoms and defects can dramatically alter the material's properties [20–23].

10.3 SIZE AND SURFACE AREA

To satisfy the needs of the society, 2-D materials help to construct the compact and efficient machines for framing the future technologies, which in turn completely rely on the size and surface area of the materials. Regarding the morphology of 2-D materials, they are ultra-thin with high surface area. Similarly, they possess a high degree of anisotropy [24, 25]. The 2-D materials are easily discriminable in terms of their optical, physicochemical, and mechanical properties in relation to their shapes and sizes. One peculiar property of the 2-D materials is that the thickness of the material can be altered for one's interest, even down to the size of a single atom. Also, the size of 2-D material is primarily responsible for the behavior of the material. Among the 2-D materials, TMDCs (transition metal dichalcogenides) are atomically thin semiconductors, which have a unique tendency to exist as monolayer assembly. A single monolayer of TMDC having less than 1 nm thickness can absorb up to 10% of incident visible light, which is equivalent to approximately 100 nm size carrying silicon [26, 27]. But still, TMDCs cannot be used for the fabrication of photovoltaic devices. Owing to the very low thickness of the TMDC monolayers, the amount of absorption of visible light is not up to the mark. Instead, we can use these materials for the fabrication of photo detectors. 2-D materials are considered the thinnest and the strongest super material. Eventually, they possess the highest specific surface areas. Hence, 2-D materials are used for various applications like catalysis, drug delivery, adsorption studies, and fabrication of energy storage devices. All these applications are needed for huge surface interactions with a small quantity of sample. In addition to that, even at very low concentrations, 2-D materials show excellent surface area to volume ratios that enhance the mechanical properties. Conversely, graphene possesses greater surface-to-mass ratio. As a result, graphene has gained much attention on engineering and scientific fields. Nevertheless, the admirable property of 2-D materials is holding less thickness, which enables them to perform quickly, accepting the external signals such as heat, light, and electrical energy. Hence, 2-D materials have a potential application in photodynamic therapy, sensors, image processing, and even conductivity studies [28–30].

10.4 ELECTRICAL CONDUCTIVITY

The 2-D materials exhibit on-local Kubo conductivity. When a bulk material is converted into 2-D thinned (atomic-monolayer) material, there is a significant change in bandgap and charge transport property. Correspondingly, its electrical property will also get changed. Typically, the conductivity of the materials depends on their own bandgap [31, 32]. The family of 2-D materials provides a full range of electrical properties, from insulating boron nitrite to semiconducting MoS_2 and to highly conducting graphene. The periodic arrangement of atoms in the crystalline structure and the mobility of electrons in 2-D material decide the nature of the bandgap. Simultaneously, the structural features of bandgap decide the electrical conductivity of 2-D materials. Hence, the newly modified bandgap is responsible for excellent conductance of the 2-D materials compared to bulk materials. Many 2-D materials have a bandgap in near infrared region. Hence, they show an excellent electrical conductivity. Especially, graphite electrodes are the major component for the fabrication of energy storage devices like solar cells, supercapacitors, and batteries. As a result of it, graphite has higher surface-to-mass ratio and can store huge number of ions. Hence, it exhibits greater conductivity. On the other hand, the electrode framed from metallic 1T phase semiconducting 2-D MoS_2 exhibits higher energy and power densities compared with the graphite electrode because it has a stacked 1T monolayer crystalline arrangement [33, 34].

2-D materials have a tendency to carry the charge faster for operations like high on/off ratio, low OFF-state conductance for low-power consumption. The 1–2 eV bandgaps of Mo and W dichalcogenides can provide high on/off ratios and with low-power dissipation. The trigonal symmetry (2H-phase) crystalline arrangement of TDMCs ($MoSe_2$, MoS_2, WS_2) possesses semiconducting character. When these 2-D materials are in bulk, they exhibit indirect bandgap. But once they exist as monolayers, they will show direct bandgap and they appear in visible spectrum. These remarkable properties attract these 2-D materials for the fabrication of optoelectronic devices. On the other hand, the charge mobility of these TDMCs is lying on the range of 100–1,000 $cm^2V^{-1}s^{-1}$. And the bandgap is 0.6–2.3 eV, which is more suitable for constructing the 2-D transistors. The monolayer puckered honeycomb black phosphorus comes under the semiconducting 2-D materials with direct bandgap. These 2-D materials have excellent stacking properties with other layers. Hence, they have tunable bandgap throughout the visible region. Simultaneously, this black phosphorus exhibits greater charge mobility around 1,000 $cm^2V^{-1}s^{-1}$. Hence, we can utilize these materials for the construction of optoelectronic devices. When we compare the conductivity of boron nitrides, they have less conductance than graphene. But they have a notable property like tunable bandgap. In spite of this, boron nitrides have Stone-Wales defects and also a large surface area. Conventionally, graphene has zero bandgap due to inversion symmetry. Hence, it has the highest charge mobility (1,000–100,000 $cm^2V^{-1}s^{-1}$). Black phosphorus has the bandgap of 2 eV (monolayer) ~0.3 eV (few layers). Furthermore, it has moderate charge mobility of about 100–1,000 $cm^2V^{-1}s^{-1}$. The increasing order of bandgap among the 2-D materials is graphene < black phosphorus < TMDs < boron nitrides. Similarly, the increasing order of charge mobility among the 2-D materials is boron nitrides < TMDs < black phosphorus < graphene. Similarly, TDMs have higher on-off ratio than graphene. Due to high power, lightweight, better racquet stability, better racquet maneuverable tendency, most of these 2-D materials are used in electrical applications [35–38].

10.5 OPTICAL PROPERTIES

Most of the 2-D materials have tunable optical properties that are directly related to doping density and type of ions, defect levels, structural response, local excitonic effect, and the charge transfer in 2-D materials. Of course, all these factors have a good impact on optical properties. Among those, electronic band structure provides a major contribution in deciding the optical properties of 2-D materials. Also, the optical properties of 2-D materials are examined by screening the electronic band structure. The optical properties of 2-D have been illustrated by studying the micron-scale flakes supplemented by theoretical calculations. Recent research has shown that the metallic and semiconducting TMDs display tunable optical property and bandgap that direct substantial attention toward the society. Especially, monolayer TMDs possess direct gap semiconductors whereas bilayer and multilayer TMDs show indirect gap semiconductors. For instance, MoS_2, $MoSe_2$, WS_2, and WSe_2 materials move from bilayer to monolayer and they undergo a change from indirect to direct bandgap. Bulk TMDs have van der Waals materials with each layer being three atoms thick. Hence, the metal layer is sandwiched between two layers of chalcogenide. Graphene is an optically transparent material that absorbs only 2% of incident visible light and also possesses high tensile strength. Another consequence of dimensional confinement is decreasing the dielectric screening between electrons and holes in semiconductors. To control the electric field, there will be a coulombic attraction and strongly bound excitons make them highly stable than excitons present in bulk materials. When the excitons are enclosed in a plane that is thinner than their Bohr radius, quantum containment can lead to an increase in their energy relative to bulk excitons, altering the light wavelength they absorb and emit. The energy of 2-D materials can be tuned by altering the number of layers, and this may influence the band structure and subsequently change the properties. MoS_2 have excellent photoluminescence properties. The monolayer MoS_2 exhibits 10^4 times stronger photoluminescence properties than that of multilayer MoS_2 [39–41].

10.6 THERMAL PROPERTIES

Researchers pay much attention toward graphene and other carbon-based nanomaterials because of their high thermal conductivity. Moreover, the other properties such as size effect, surface functionalization, vacancy, strain, isotopic doping, substrate effect, and edge configurations are explained in terms of thermal conductivity of graphene. The result obtained from thermal conductivity of graphene shows that the previously mentioned factors like boundary, structural defect, interface, and mass disorder are responsible for this property. For example, high frequency phonons are vulnerable to mass disorder and scattering of vacancies, whereas low frequency phonons exhibit little scattering from point defects. In addition, the thermal conductivity is totally repressed by boundary scattering in low-frequency phonons.

2-D materials possess exclusive crystal structure due to high thermal anisotropy. It has slower conduction of heat in cross-plane than in plane [1–3], which ranges from 50 to 300. The atoms in 2-D materials are covalently bonded in each layer and the layers are held together by van der Waals interactions. On the other hand, there are dispersions of both boundary and point defects for a graphene nanoribbon with mass disorder. So, the thermal conductivity is predicted by the associated effects of two scattering mechanisms. Also, in graphene nanoribbon, the high-frequency photon participation ratio is quelled by edge and isotope scattering, which will cause a less effect of phonon localization than that in carbon nanotubes of the same isotope density. Hence, isotopic doping can decrease the thermal conductivity of graphene nanoribbon and carbon nanotube as well as the reduction behavior is less in the former (graphene nanoribbon) than that in the latter (carbon nanotube). However, the thermal conductivity of graphene nanoribbon is decreased by 12% whereas in carbon nanotube it is 25%. Photonic engineering is an efficient tool for regulating the thermal conductivity of 2-D materials by completely controllable engineering of their phonon dispersion regimes. Li and Zhang suggested that using molecular dynamics simulations to boost the ultra-high thermal conduction of graphene nanoribbon will add a small gap at the middle, that is, create a structure of comb graphene nanoribbon. Likewise, in photonic engineering proposed to regulate the thermal conductivity of quantum wells and nanowires, this fact helps to manipulate the thermal conductivity of 2-D materials [42–45].

10.7 MECHANICAL PROPERTIES

2-D materials have numerous fascinating properties. Among those, mechanical property is the most substantial one. It directly influences the manufacturing, durability, compatibility, and lifetime of the devices. Mechanics is not only explaining about physicochemical property of materials, but also it deals with experimental and theoretical values. The interactions between the 2-D materials as well as the interaction with neighbor substrates are accountable for measuring the mechanical properties. Since the 2-D materials possess peculiar mechanical properties, moderate flexural rigidity, and extremely high stiffness and strength, the family of 2-D materials are used in a wide range of applications. Considering these concepts, herein we have explained interfacial properties (adhesion, friction), tensile deformation, stiffness, strength, and elastic behavior of 2-D materials. While comparing graphite, graphene has placed a milestone record in the total scientific society for the past few years [46, 47]. Accordingly, graphene has enormous surface area, amazing mechanical strength, massive conductivity, and remarkable flexibility. Due to its monoatomic thinness and elastic in-plane properties, graphene is extremely versatile. In the form of rippling, wrinkling, and folding, the flexural deformation of graphene is commonly observed.

Like graphene, all the 2-D materials possess nonlinearly elastic behavior. Especially, phosphorene has tunable elastic properties. This is because it has puckered atomic structural arrangement. The recent research attests that the family of 2-D materials expose negative Poisson's ratio (NPR), which implies that the 2-D materials have excellent mechanical property (sound/vibration absorption, toughness, resistance, and so on). Since graphene exists in planar crystal structure, it is more in favor of intrinsic NPR properties [48, 49]. The tensile strains of graphene exceed about 6%. Because the monolayer planar graphene sheets are involved in bond stretching process, an in-plane expansion occurs when the 2-D materials are extended with the original out-of-plane deformation to flatten the materials under stress, leading to the NPR phenomenon. The 2-D materials with monatomic or ultra-thin crystal materials exhibit similar elastic properties like those of graphene. Hence, it leads to high in-plane stiffness and low flexural rigidity. Electronically, the most needed carbon and silicon atoms have the ionization potentials that are 11:2 and 8:1 eV, respectively, but once they exist in the 2-D forms, the work function (WF) values will be less notable, graphene at 4:24 eV and silicene at 4:35 eV [50]. This indirectly explains that if the covalent character increases, tensile deformation also increases.

On the other hand, the 2-D materials, aluminum nitride (AIN) and gallium nitride (GaN) also exhibit zero strain with indirect bandgap [51]. It should be noted that if the ionization potential of AIN increases, the tensile strain also increases correspondingly. This is because more electron clouds are present closer to the nitrogen atoms. AIN possesses increase in tensile strain than that of GaN. There is a huge difference in electronegativity of the corresponding cation and anion. Simultaneously, as electron affinity increases, tensile strain also increases. Hence, the work function and tensile strain have a good relation to ionization potential and electron affinity. Moreover, ionization potential of MoS_2 decreases when the tensile strain increases [52–54].

10.8 DEFECTS

Crystal imperfection of the materials is generally pronounced as a defect. Defects profoundly affect the original properties of the materials. The size and structure of the material, especially from atomic scale 2-D materials to bigger-sized macromolecules, depend on the defects present in it. In the case of 2-D materials beyond the electrical and mechanical properties, defects are enumerated as an onset of all the properties [55]. 2-D materials have an immense impact over their mechanical, physicochemical, and electronic properties. When the 2-D materials are concerned, two different types of intrinsic defects are observed. The first one is point defects (dislocations, vacancies), and the second is line defects (grain boundaries). Among the 2-D material family, phosphorene has much less (0.40 eV) diffusion energy barrier, when compared with graphene (1.39 eV) and MoS_2 (2.27 eV). In addition to that, the hopping rate (v) of the 2-D materials can be determined by Arrhenius equations. At normal atmospheric pressure and temperature, the mobility of mono-vacancies, phosphorene is attractively hastier than graphene [56]. Contrastingly, vacancies formation energy of phosphorene (1.65 eV) has notably lesser value than graphene (7.57 eV) since the population defect has a close relation to vacancies of formation energy. Because the shortest bond distance between the two phosphorous (P-P) atoms and the C-C bond is so strong, the distance between the two-carbon atoms is so high when compared with phosphorous [57].

Defects have a higher influence on deciding the structure and size of the materials. When the dislocation defect is concerned, there are two types: edge and screw. Between the two, the screw type, which has burgers vector points in the out of-plane dimension, is the more important one. It decides the size and structure of the 2-D materials. Also, the dislocation defect is epitomized by a pentagon-heptagon pair [58, 59]. Due to the buckled atomic structure and anisotropy character of phosphorene, their atomic dislocation is massively skewed, when compared with that of graphene. In the case of MoS_2 structure, generally the dislocation defect can be observed, whereas MoS_2 also can exist in tri-atomic layer, which influences the MoS_2 structure and makes it more complicated. Because of the higher concentration of sulphur atom, the MoS_2 molecules will exist in 3-D polyhedral arrangement. Since the two-sulphur atom vacancies are created, MoS_2 favors defect energetically [60].

Grain boundaries are another type of vital defects existing in 2-D materials. It may be categorized into two different types, namely, low-angle (dislocation occurs at defined length) and high-angle (dislocation occurs as crowded, as well as overlap with each other). Graphene sheets possess high-angle grain boundaries even though graphene acts as an ultra-conducting material. It can be affirmed that graphene can accommodate strained carbon ring structure while the defects occur. For the polycrystalline graphene, it is found that numerous grain boundary junctions exist and that they exhibit higher strength. When the grain size is very small, the strength of the material is higher (pseudo Hall-Petch behavior). Phosphorene is found to have potentially stable grain boundaries and so the formation energies are much lower than graphene [61–63].

10.9 VAN DER WAALS INTERACTIONS

2-D materials have numerous physical properties to design the 2-D layered structures such as graphene, TMDs, and black phosphorus. To fabricate these materials, interlayer interactions play a vital role in property of materials. These interactions are called van der Waals (vdWs) interactions, and thus we can redefine the 2-D material properties. Layered materials are designed by enhanced crystalline planar structures; the atoms in layers (in-plane) are held together by strong covalent bonds and the layers (out-of-plane) by weak van der Waals forces [64]. Bulk 2-D materials consist of covalently bonded planes held together by weak van der Waals forces. If an external force is applied, this material can break due to very weak interactions. Conversely, the atoms in layers are held together because of strong covalent bonds. Therefore, a single monolayer has only covalent bonds. To increase the strength of this material, weak links should be removed. For example, the tensile strength of graphene is 1,000 times greater than graphite. A graphite pencil can be easily broken, whereas graphene is 100 times stronger than steel [65]. In addition, the individual layers also can be removed by breaking the van der Waals bonds with little modification in whole structure or the separated layer. For instance, graphene layers can be removed from bulk graphite with adhesive tape. Hence, layered materials can be converted to any substrate with relative ease.

Although the primary mechanism has been generally thought to be van der Waals interactions, experimental evidence indicates that other mechanisms, such as the effects of water capillary, reactive defects, and surface rigidity, might also have to be considered. New forms of quantum heterostructures with atomically sharp interfaces between 2-D layers of two different materials are possible due to the recent growth of van der Waals material systems. The rich range of 2-D electronic systems is compounded by the stacking of two or more atomic layers of various van der Waals materials and enables researchers to investigate novel collective quantum phenomena at the interfaces [66]. The correlation of electrons with spins, phonons, and other electrons can be highly increased, influencing the quantum boundary transport of charge, entropy, and energy. Quantum fluctuations are also intensified and interact with a long-range order because of the small phase-space volume. As a result of it, van der Waals heterostructures may exhibit emerging phenomena: interlayer excitons, electrons in one layer bound to holes in another, may condense into a superfluid, and exotic triplet superconducting state may result from the coupling between layers.

Considering that a large number of 2-D crystals are currently available, a large variety of heterostructures should be possible to build. Hence, 2-D materials can be framed into heterostructures, where the van der Waals forces keep the monolayers together. However, a wide variety of novel experiments and prototypes have already been conducted with van der Waals heterostructures, making it a flexible and realistic method for future experiments and applications [67, 68]. A modern concept of deterministic transfer method where the interaction of van der Waals between different 2-D materials is used to transfer flakes across the entire phase without touching

TABLE 10.1

Comparison Table for the General Properties of 2-D Materials

Properties	Graphene	Boron nitride	Transition metal dichalcogenides	Black phosphorus
Crystal structure	Hexagonal	Hexagonal	Hexagonal	Orthorhombic
Electronic structure	Semimetal	Insulator	Semiconductor	Semiconductor
Bandgap (eV)	NA	7.2	0.6 – 2.3	2 (Monolayer) 0.3 (few layer)
Charge mobility (Cm^2/Vs)	1,000 – 100,000	–	10 – 500	100 – 4,000
Type and example	Fluorographene Graphene oxide		(i) Semiconducting dichalcogenides ($MoTe_2$, ZrSe) (ii) Metallic dichalcogenides ($NbSe_2$, TiS_2) (iii) Layered semiconductors (GaSe, Bi_2Se_2)	

the 2-D materials with any polymer, eliminating impurities trapped between the layers. Hence, it provides us with a new way of tailoring the properties of layered materials by controlling interactions with van der Waals.

10.10 METHODS OF PREPARATION

Graphene (2-D popular materials) is one of the most promising SP^2-bonded atomic layered sheet-based carbon atoms, which exhibits high electrode surface area (2,630 m^2 g^{-1}) [69], reasonable thermal [70], mechanical [71], and electrochemical [72] properties. In general, there are various approaches for the preparation of large-scale, high-quality highly active functional group-based graphene oxide. Graphite oxides have been synthesized by Hummer's method using graphite powder (50 g) and Na_2NO_3 in to 2.3 L of H_2SO_4. The mixtures were vigorously stirred and maintained at 0°C; while $KMnO_4$ was added to the suspension solution, the reactions were carried out for seven days. After the reaction process, the mixture was transferred to a 1 L double distilled (DD) water, the resultant gray-colored graphite oxide materials dried in a vacuum at room temperature [73].

Chemical exfoliation method is the electrochemical reduction of pyrolytic graphite into reduced graphene oxide (rGO), the applied potential value of 0 to –1.5 V under phosphate buffer solution (pH = 5) condition [74]. The exfoliated graphene has been successfully prepared by ball milling method (grinder method) using 0.16 mM graphite solutions. The treatment was applied 100 rpm for 30 minutes; after the process, the mixture (20 mL of de-ionized water) sonicated for one minute and was placed for five days, and the resultant product is exfoliated graphene [75]. Ejigu et al. [76] have used the H-type of electrochemical cell for the electrochemical

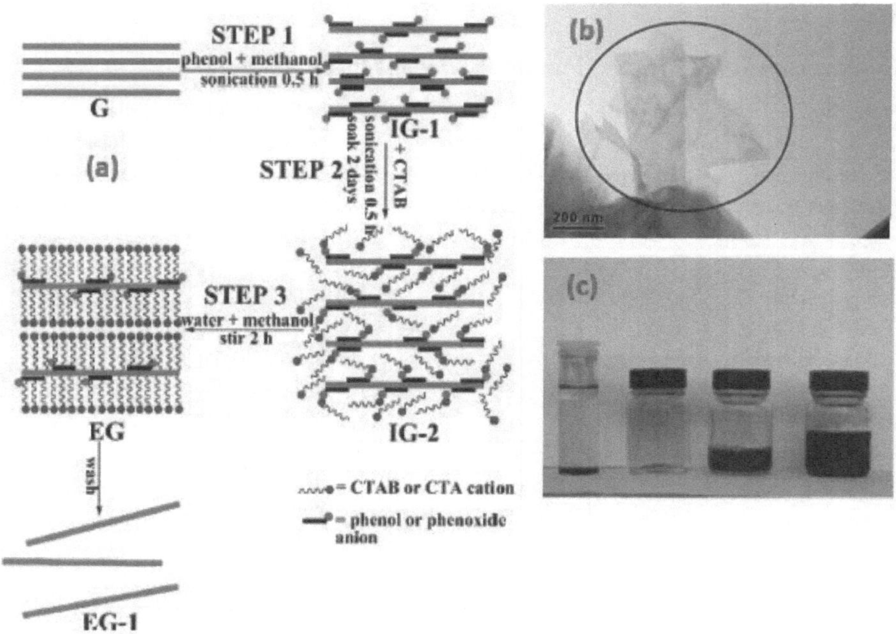

FIGURE 10.4 (a) Schematic Figure of the Exfoliation Mechanism of the UF-4 Graphite to Graphene. G Is UF-4, IG-1 Is Phenol Intercalated Graphite, IG-2 Is Phenol and CTAB Intercalated Graphite, EG Is Exfoliated Graphene, and EG-1 Is Cleaned Graphene; (b) TEM Image of EG; and (c) Digital Image of Graphene Aqueous Dispersion

(Copyright 2013 the American Chemical Society (Lin et al.) [77].)

exfoliation of graphitic electrode. During the electrolysis, the applied voltage is 10 V and the electrolyte consisted of 30 mM $CoSO_4$ and 0.2 M Na_2SO_4. The ultrathin film-based graphene has been prepared by electrophoretic deposition (EPD) method. In this method, the calculated amount of graphene was dispersed in DD water up to getting liquid suspension by using ultrasonication for one hour. After the electrophoretic treatment, the ultra-thin graphene was deposited on to the negative surface of the electrode. The high-resolution TEM images clearly indicate that sheet-like structure and the digital images are well dispersed with aqueous DD water with various concentrations (0.01, 0.1 and 1 mg mL^{-1}) (Figure 10.1) [77]. A sand-milling turbulent flow technique was used by virtue of simple preparation method. It is also considered the most cost effective, which is a facile route for the production of well-defined morphological-based activated graphene [78]. Trung's group developed the scalable synthesis of graphene by plasma chemical vapor deposition technique, indicating that grid patterns of graphene could be applied in touch screen panels [79]. The twisted bi-layer-based graphene has been successfully synthesized by a wetting transfer method using flake graphite powder. This simple and handy method offers the designable twist 2-D materials, which exhibit good electronic, mechanical, and optical properties with clean surface and interface [80].

The term of unique electronic properties of graphene was obtained from "scotch tape type of technique". It has fascinating optical, mechanical, and electrical properties, which have the impact of electrochemical applications in electrochemical sensors as well as energy storage devices. Chemical vapor deposition (CVD) is one of the well-established ones and most successful route for the growth of single-layered graphene sheets [81]. Quantum dots (Q)-based graphene has been synthesized via CVD method. The multi-layered graphene hollow nanosphere has the specific oxygen content value of 36%, and it can display folded edge with surface defects. Interestingly, Q-graphene can exhibit fast electron transfer kinetic process and play a vital potential role in the field of various electrochemical studies [82]. The novel types of screen-printed electrodes (SPEs) have attracted considerable attention because of the portable device, which is cost-effective and disposable. However, the surface topography and materials compositions of SPEs have been examined by atomic force microscopy (AFM), SEM, and XPS analysis. This type of SPEs is using electro active redox probes for the studies of biological relevant analytes, like potassium ferrocyanide, dopamine hydrochloride (DA), uric acid (UA), and L-ascorbic acid (AA) [83]. Fang et al. [84] have addressed 2-D graphene-based novel electrode materials that were employed for the detection of different types of analytes. Graphene may also cooperate with other functional species to maximize the synergistic impact and improve electroanalytical performance. Brownson and his coworkers have highlighted the inner spheres and outer spheres of graphene-based electrode catalyst, which have become a considerable research interest in the field of electrochemical applications [85]. Sahoo et al. [86] have summarized cost-effective and superior energy storage device-based 2-D electrode materials.

In general, various dimensions of electrode materials with unique structural advantages can act as highly promising electrode catalyst for charge storage devices. Multi-layered graphene supported inject printed (MLG IJP) flexible electrode has become a global concern. The fabrication required three types of materials, such as conductive silver ink, insulating dielectric ink, and multi-layered graphene ink. As a result, MLG IJP has received much attention because of good electron transfer reaction and sensitive detection nature with significant scalable fabrication of low-cost wearable sensor in the commercial biosensor practice [87]. The different layered structures (monolayer, double layer, and few layers) of graphene have been synthesized by the CVD method. Moreover, electron transfers in quasi-graphene (four layers of graphene) can be eight orders faster than in a graphene monolayer. Edge plane sites of graphene material can play a vital role in the development of superior electrochemical performance [88]. The mild and environmentally friendly natured interconnected ultra-hierarchical graphene oxide nanosheet has been synthesized by hydrothermal method. The as-synthesized interlayered reduced graphene can exhibit high specific surface area (835 F g^{-1}) with high-power conductivity (400 S m^{-1}). Nevertheless, the binder-free graphene-based electrode, which is predominantly influenced by the specific capacitance value of 169 F g^{-1} and long-term cyclic stability (93.5% after 600 cycles @ 2,000 mAh g^{-1}) [89]. Graphene nanoplate (GNP) electrode has a multi-layered structure with inner plane pores, a larger electrode surface area, and has been extensively studied for the development and electrochemical detection of endocrine-disrupting chemicals [90]. The environmentally friendly natured method has been used for the development of an efficient graphene-based *Cedrus deodara* (CD-rGO) electrode framework, which exhibits higher internal pore surface area and could be stored with hydrogen ion for the development of safe and efficient alternative power storage device applications [91]. Manchala et al. [92] have used an interesting methodology for the synthesis of high-quality soluble graphene using Jadwar roots extract and offer unique advantages in lightweight supercapacitor applications.

10.11 MORPHOLOGICAL STUDIES

The structural morphological study of 2-D graphene is greatly influenced by its physical and chemical properties. 2-D-based graphene hollow spheres have been successfully prepared by the template method, which causes significant mechanical flexibility and chemical stability and offers large electrode surface area with outstanding electrode conductivity [93]. Li et al. [94] have used the flexible fiber-based electrochemically reduced graphene oxide (ErGO) with the thickness of about 30 μm. The results clearly indicate that the thicknesses of ErGO materials deposited on Au wires increases from less than 10 mm to around 100 mm, respectively. Xie et al. [95] demonstrated a facile strategy for the preparation of high-quality folded graphene monolayer. They also modified with high-quality carbon nano scrolls (CNSs) for exploring current-voltage (I-V) studies and could explore microcircuits interconnect applications (Figure 10.5).

Low-cost hierarchical porous-based multi-layered graphene shell electrode has bent like crystal structure, which slightly expanded its lattice spacing and has excellent energy storage properties of both Li^+ and Na^+ ions [96]. Multi-layered porous 2-D-graphene nanosheet has attracted a great interest as novel electrode materials for low-cost high-power supercapacitor applications. Importantly, it has large electrode

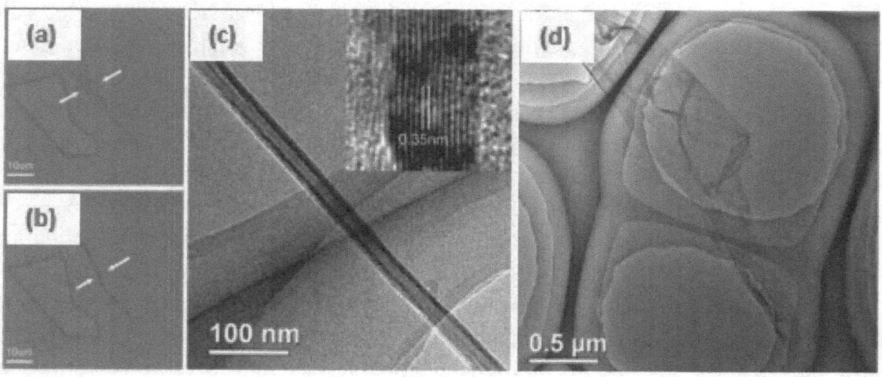

FIGURE 10.5 Optical Microscope Images of the Original (a) and Scrolled (b) Graphene Monolayer on the SiO_2 Substrate, (c) TEM Images Showing that the Distance between Adjacent Graphene Layers Is about 0.35 nm, and (d) TEM Image of a Folded Graphene Monolayer Resulting from a Harsh Fabrication Process

(Copyright 2009 the American Chemical Society (Xie et al.) [95].)

surface area with rapid ion or electron transport properties during the energy storage devices [97]. The graphene mesosponge (GMS) (nano graphene domain diameter 1–3 nm with hexagonal pattern)-based framework has unique pore channels, elastic, mechanically tough and high electrical conductivity with facilitating charge propagation, which greatly improved their capacitive behavior [98].

Li et al. [99] reported the template strategy for the preparation of hollow graphene nanospheres, while ensuring electrical conductivity to electron transport. The ultra-thin porous shell (3 nm) morphological-based graphene interconnected structures are generally prepared by a facile template route without using any surfactant. The obtained 2-D graphene ultra-thin shells can serve as an excellent candidate for lithium-ion batteries (LIBs) [100].

10.12 STUDIES OF GRAPHENE ELECTRODE SURFACES

In general, graphene is a honeycomb structure, which delocalizes the lattice geometry of SP^2-hybridized bonded carbon atom. Besides, the functionalized electrode surface platforms can also have their synergetic properties with a wide range of potential window [101]. Karaphun et al. [102] have studied the rational large specific surface area-based rGO electrode materials. The specific surface area (130.2 $m^2\ g^{-1}$) and average pore size (19.5 nm) have been calculated by using BET nitrogen adsorption-desorption isotherm studies. In this type, rGO has great attention to improve their electrical and capacitive behavior. Similarly, the active electrode surface areas have been optimized through BET surface of adsorption-desorption studies, which offer dense monolayer on the electrode flat surface. A significant progress has been made in the Nile blue-based functionalized conjugated graphene aerogel (NB-GA) electrode materials, which are the key challenges to achieve high-power capacitive nature that boosts their energy content (Figure 10.6) [103]. The mesoporous architecture-based self-organized graphene electrode has been prepared by the pulse laser deposition technique, providing high surface area with uniform surface roughness, which offers promising perspective for developing high-performance-based electrochemical sensing applications [104]. Development of 2-D nitrogen doped graphene nanosheets as promising electrode materials provides an effective way to enhance their electrode surface area and significantly improve their power efficiency [105]. Recently, the self-assembled method for the preparation of 2-D graphene hybrid hydrogel-based film electrode attained excellent gravimetric capacitance (653.7 $F\ g^{-1}$) and volumetric capacitance (~1522 $F\ cm^{-3}$) values, respectively [106].

10.13 APPLICATIONS OF 2-D MATERIALS

This section presents the various forms of graphene in diverse fields of applications, which have been carried out in the past few years.

10.13.1 Graphene

A number of graphene-related academic publications have come up day by day. Additionally, different types of graphene, such as monolayer, few-layer, multilayer,

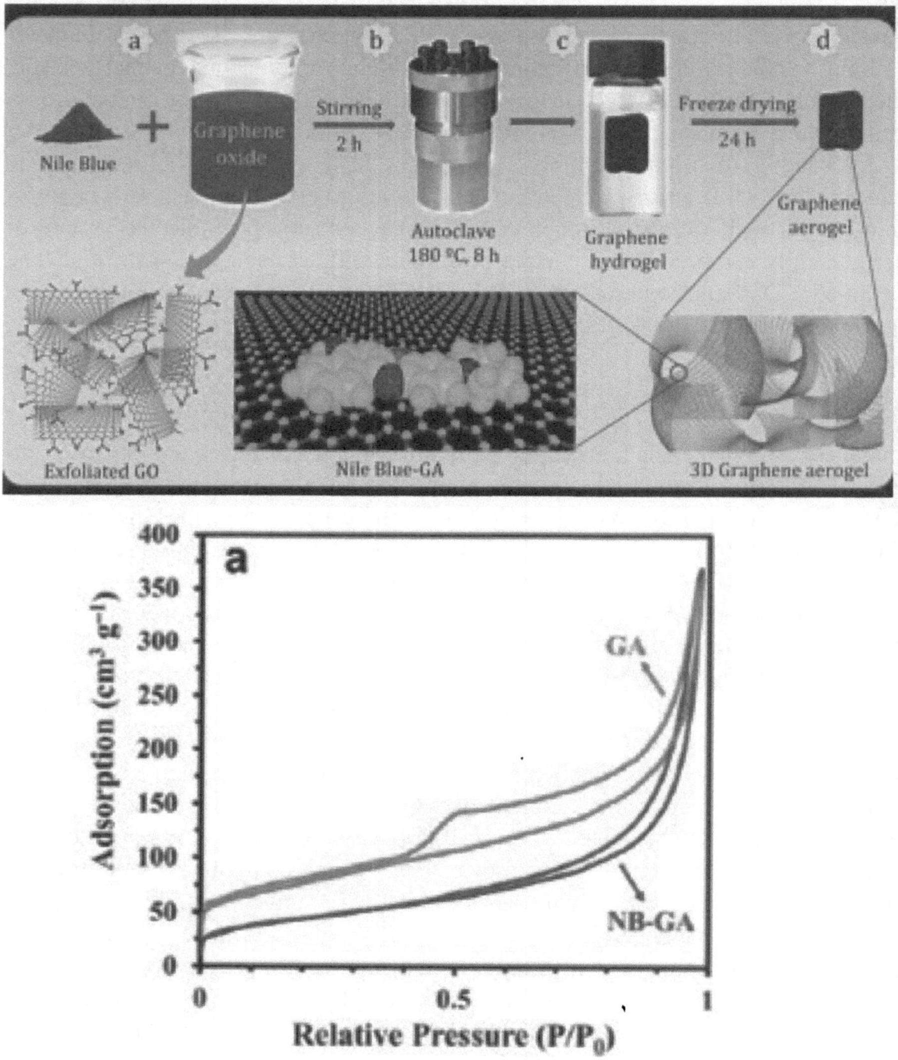

FIGURE 10.6 (a) Schematic Illustration of the Hydrothermal Synthesis Route Toward Preparation of the Nile-Blue Functionalized Graphene Aerogel (NB-GA) and (b) Nitrogen Adsorption and Desorption Isotherms Studies of the Graphene Aerogel and the NB-GA Nanocomposite

(Copyright 2019 the American Chemical Society (Shabangoli et al.) [103].)

and reduced graphene oxide, have many latent applications depending on the properties exhibited by each type of graphene. Figure 10.7 depicts the potential applications of graphene. Graphene is not used in commercial applications, but many authors have proposed under active development in the vast areas, including transistors, integrated circuits, solar cells, optics, biomedical, and energy storage.

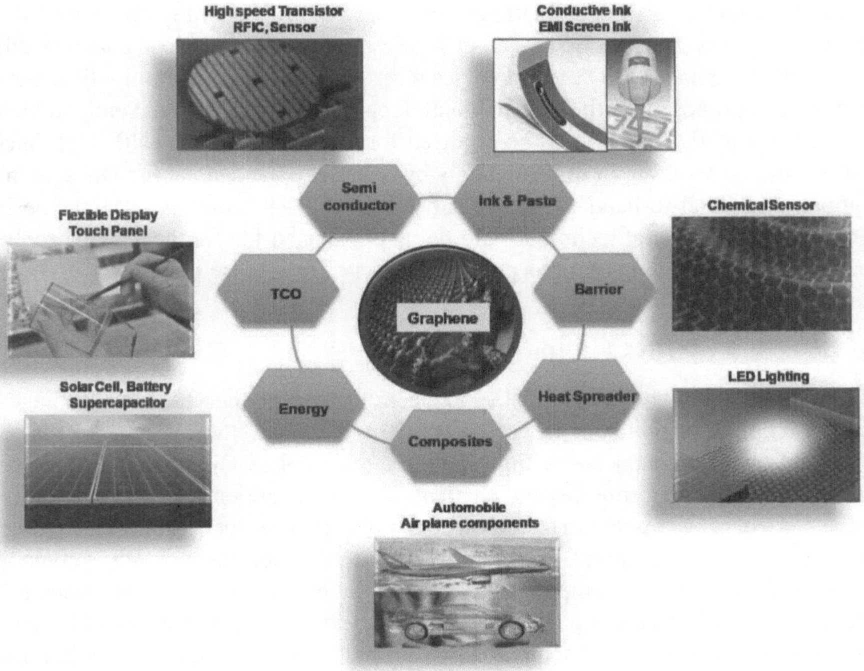

FIGURE 10.7 The Potential Applications of Graphene

10.13.2 Transistors

This section provides an overview of current development of graphene field-effect transistors. The two most common varieties of transistors are the logic transistor and the analog transistor. The former is characterized, among other things, by a high I_{on}/I_{off} ratio to ensure low energy consumption (of the OFF state) and to maintain a high logic interpretation yield. The latter is characterized primarily by its cutoff frequency and is often used in high-frequency applications as an amplifier. Both types of devices, as we shall see, could benefit from the remarkable electrical properties of graphene [107, 108].

For more than 40 years, CMOS technology has dominated the logic transistor industry with the fabrication of MOSFETs. A general trend over the years has been to reduce the length of the transistor's gate to achieve higher transistor densities and also increased performance produced by the higher electrical fields in the channel region. However, as the size of the device is reduced, more and more issues (commonly known as short-channel effects) begin to appear. These issues are not limited to such problems as hot electron effects, velocity saturation effects, and punch-through effects [109]. It has been suggested that graphene, due to its monoatomic thickness, would reduce these parasitic effects. Graphene could be included in the channel region of transistors and thus provide a high-mobility channel that would help to reduce short-channel effects. One problem is that graphene does not possess a bandgap since the conduction and valence bands touch at the Dirac point. This reduces the ON-OFF ratio of the transistor by several orders of magnitude, such that a graphene FET will typically have an ON-OFF ratio <10. This ratio is unacceptable if one wants to replace CMOS technologies where ratios >107 are commonly achieved. It has been proposed by the International Technology Roadmap of Semiconductors (ITRS) that a 10:4 ratio would be required for logic applications [110, 111]. Such a ratio could be achieved in graphene only by opening a bandgap at the Dirac point to suppress the band-to-band tunneling. A bandgap of 0.4eV could achieve the desired ON-OFF ratio. Several techniques are being envisioned for engineering a graphene bandgap. The following sections provide an overview of these methods.

10.13.3 Integrated Circuits

2-D materials could be widely applied to electronics due to their unique advantages in carrier mobility or switching performance. Other than that, the large mean-free path, excellent thermal conductance, high mobility, and gapless Dirac bands in graphene mark the unique opportunities and challenges for the adaption for electronic devices. Such electronics were used in the construction of logic gates or complex electrical elements, which are recognized as electronic circuits. An electronic circuit is composed of individual electronic components connected by conductive wires and trace [112]. Recently, researchers made graphene circuit for the first time, in which all of the circuit elements are integrated on a compact single chip. The new circuit is another important step forward for graphene-based electronics and potential applications include wireless communications and amplifiers. Despite much progress in recent years and the fact that scientists have already made some high-performance graphene-based devices, it still remains challenging to integrate graphene transistors with other components on a single chip [113]. This is mainly because graphene does not adhere very well to the metals and oxides traditionally used in semiconductor-manufacturing processes and because there are no reliable and reproducible techniques yet to make such circuits.

10.13.4 Solar Cells

Photovoltaic cells, or solar cells, are another potential application of graphene. Recent solar cell technologies contain platinum-based electrodes. The abundance of platinum on earth and its cost are considered its problems. Being an excellent conductor, graphene designed as an electrode would reduce both the cost as well as the weight with efficiency. The graphene electrode in a dye-sensitized solar cell actually exhibited an efficiency of 7.8%, which is 0.2% less than the platinum-based counter electrode, but produced at a fraction of the cost [114, 115]. Clearly, it would be better to improve efficiency, but cutting the cost is as much an issue as improving efficiency for modern-day technology. Out of all applications, graphene solar cells are furthest away from completion; in general, solar cell research has been a slow process for many years.

10.13.5 Optics

Graphene is an ideal platform for many optoelectronic devices due to its distinctive combination of high electron mobility (μ), optical transparency, and gate/

dopant-tunable carrier density. However, to truly harness the potential of this combination and make graphene-based efficient optoelectronic devices a reality, optical and electronic properties of graphene must be tuned via substitutional doping. While doping graphene with boron or nitrogen can tune the Fermi energy (FE) and lead to p- or n-type graphene, it also compromises the inherently high electron mobility in doped graphene, which is useful for electronic applications [116, 117]. Furthermore, the recombination rate of photogenerated carriers is also known to decrease with the presence of defects and dopants limiting applications of graphene in optoelectronics. Indeed, the focus of many synthesis efforts in graphene has been to achieve large-area "defect-free" graphene for electronic applications because carrier-defect scattering limits the electron mobility—an important parameter for high-speed electronics. Contrary to this conventional wisdom, we recently demonstrated that the defect configuration is more important than the defect concentration for increasing carrier concentration without compromising photogenerated carrier lifetime [118].

10.13.6 Biomedical

In the past decade, nanomaterial-based drug delivery systems have been extensively investigated for the treatment of cancer, aiming at improved therapeutic efficacy and reduced venomous side effects. Many research groups have started to explore graphene-based drug delivery systems [119]. The surface area of graphene (2,600 m^2 g^{-1}) is higher, which makes them worthy to be explored for drug delivery. Basically, a monolayer of graphene represents an extreme case in which every atom is exposed on the surface, which allows significantly higher drug loading capacity. The two prominent modifications reported in the literature for drug delivery using GBNs are chemical modification via electrostatic interaction and binding to the aromatic molecule via p–p stacking interaction [120, 121]. One more advantage of drug delivery through GBNs is the control of release rate for sustainable drug release. Single-layered GO or RGO has an ultra-high surface area available for highly efficient drug loading. Recently, GO has become quite a competitive drug delivery system with the potential to be applied for systemic targeting and local effective drug delivery. In recent years, several studies have been conducted on the delivery of anticancer drugs, genes, and peptides through graphene derivatives. Approaches such as simple physisorption via p–p stacking can be used for loading many hydrophobic drugs, such as doxorubicin and docetaxel, with antibodies for the selective killing of cancer cells. Graphene is a new promising material for drug delivery via the nanocarrier approach, due to its small size, intrinsic optical properties, large specific surface area, low cost, and useful non-covalent interactions with aromatic drug molecules [122, 123]. The large specific surface area, p–p stacking, and electrostatic or hydrophobic interactions of graphene can assist in high drug loading of less soluble drugs with high efficiency and potency.

10.13.7 Energy Storage

Graphene is also ideal for the production of energy storage devices. This is possible especially when they are treated metal oxide, hence showing limited sheet restacking.

The interconnected networks of graphene being highly conductive also contributes to the reasons behind their interest as energy storage applications. Other factors like their porous microstructure, being stable electrochemically as well as good stability mechanically are some merits of graphene when used as energy storage devices. Some of these devices include their application in a fuel cell, solar cells, batteries as well as supercapacitors. The supports in proton exchange membrane (PEM) fuel cells are often made up of graphene, while for batteries like Li batteries, they are utilized in the cell as anodic and cathodic material [124, 125]. In supercapacitors, graphene is used as the electrode material for both the double layer capacitors as well as the pseudo capacitors. They are also utilized in fuel cells as dye sensitizers.

10.14 CONCLUSION

Two-dimensional materials have been an important part of physics and materials science research in the decade and a half prior to the invention of graphene, owing to rapid scientific advancement. This newfound knowledge has ushered in new directions in the analysis of 2-D materials. The distinct physical and chemical properties of 2-D semiconductors distinguish them from other materials. They have the best attributes of both inorganic III-V semiconductors and organic materials due to their higher absorption coefficients and heavy exciton binding energy. The formation of van der Waals structures with improved properties provides a unique opportunity to develop innovative materials. 2-D materials are used in sensors and electronic and optoelectronic devices due to their high surface area, layered composition, and optical bandgap. However, despite recent advancements in 2-D materials, it is expected that more fundamental developments and exciting applications will emerge in the future.

REFERENCES

[1] Krishna, V. D., K. Wub, D. Su, M. C. J. Cheeran, J. P. Wang, and A. Perez, Nanotechnology: Review of concepts and potential application of sensing platforms in food safety, *Food Microbiology*, 75 (2018) 47–54.

[2] Jeevanandam, J., A. Barhoum, Y. S Chan, A. Dufresne, and M. K. Danquah, Review on nanoparticles and nanostructured materials: History, sources, toxicity and regulations, *Beilstein Journal of Nanotechnology*, 9 (2018) 1050–1074.

[3] Tan, C., X. Cao, X. J. Wu, Q. He, J. Yang, X. Zhang, J. Chen, W. Zhao, S. Han, G. H. Nam, M. Sindoro, and H. Zhang, Recent advances in ultrathin two-dimensional nanomaterials, *Chemical Reviews*, 117 (2017) 6225–6331.

[4] Khan, L., K. Saeed, and I. Khan, Nanoparticles: Properties, applications and toxicities, *Arabian Journal of Chemistry*, 12 (2019) 908–931.

[5] Singh, V., D. Joung, L. Zhai, S. Das, S. I. Khondaker, and S. Seal, Graphene based materials: Past, present and future, *Progress in Materials Science*, 56 (2011) 1178–1271.

[6] Allen, M. J., V. C. Tung, and R. B. Kaner, Honeycomb carbon: A review of graphene, *Chemical Reviews*, 110 (2010) 132–145.

[7] Zhu, J., D. Yang, Z. Yin, Q. Yan, and H. Zhang, Graphene and graphene-based materials for energy storage applications, *Small*, 10 (2014) 3480–3498.

[8] Rafiei-Sarmazdeh, Zahra, Seyed Morteza Zahedi-Dizaji, and Aniseh Kafi Kang, Two-dimensional nanomaterials, *Nanostructures* (2019) 1–15.

[9] Tan, C., X. Cao, X. J Wu, Q. He, J. Yang, X. Zhang, J. Chen, W. Zhao, S. Han, G. H. Nam, M. Sindoro, and H. Zhang, Recent advances in ultrathin two-dimensional nanomaterials, *Chemical Reviews*, 117 (2017) 6225–6331.

[10] Fiori, G., F. Bonaccorso, G. Iannaccone, T. Palacios, D. Neumaier, and A. Seabaugh, Electronics based on two-dimensional materials, *Nature Nanotechnology*, 9 (2014) 768–779.

[11] Late, D. J., A. Bhat, and C. S. Rout, Fundamentals and properties of 2D materials in general and sensing applications, *Fundamentals and Sensing Applications of 2D Materials* (2019) 5–24.

[12] Novoselov, K. S., A. K. Geim, S. V. Morozov, D. Jiang, Y. Zhang, S. V. Dubonos, I. V. Grigorieva, and A. A. Firsov, Electric field effect in atomically thin carbon films, *Science*, 306 (2004) 666–669.

[13] Mak, K. F., C. Lee, J. Hone, J. Shan, and T. F. Heinz, Atomically thin MoS_2: A new direct-gap semiconductor, *Physics Review Letters*, 105 (2010) 136805.

[14] Allen, M. J., V. C. Tung, and R. B. Kaner, Honeycomb carbon: A review of graphene, *Chemical Reviews,* 110 (2010) 132–145.

[15] Matte, H. S. S. R., A. Gomathi, A. K. Manna, D. J. Late, R. Datta, S. K. Pati, and C. N. R. Rao, MoS_2 and WS_2 analogues of graphene, *Angewandte Chemie International Edition*, 49 (2010) 4059–4062.

[16] Khan, K., A. K. Tareen, M. Aslam, R. Wang, Y. Zhang, A. Mahmood, Z. Ouyang, H. Zhang, and Z. Guo, Recent developments in emerging two-dimensional materials and their applications, *Journal of Materials Chemistry C*, 8 (2020) 387–440.

[17] Choi, W., N. Choudhary, G. H. Han, J. Park, D. Akinwande, and Y. H. Lee, Recent development of two-dimensional transition metal dichalcogenides and their applications, *Materials Today*, 20 (2017) 116–130.

[18] Zhan, H., D. Guo, and G. Xie, Two-dimensional layered materials: From mechanical and coupling properties towards applications in electronics, *Nanoscale*, 11 (2019) 13181–13212.

[19] Lemme, M. C., L. J. Li, T. Palacios, and F. Schwierz, Two-dimensional materials for electronic applications, *MRS Bulletin*, 39 (2014) 711–718.

[20] Briggs, N., S. Subramanian, Z. Lin, X. Li, X. Zhang, K. Zhang, K. Xiao, D. Geohegan, R. Wallace, L. Q. Chen, and M. Terrones, A roadmap for electronic grade 2D materials, *2D Materials*, 6 (2019) 022001.

[21] Zhang, R., and R. Cheung, Mechanical properties and applications of two-dimensional materials-synthesis, characterization and potential applications, *Rijeka Crotia: InTech*, (2016) 219–246.

[22] Kolobov, A. V., and Tominaga, J. Two-dimensional transition-metal dichalcogenides, *Springer*, 239 (2016).

[23] Broholm, C., I. Fisher, J. Moore, M. Murnane, A. Moreo, J. Tranquada, D. Basov, J. Freericks, M. Aronson, A. MacDonald, and E. Fradkin, *Basic Research Needs Workshop on Quantum Materials for Energy Relevant Technology*, USDOE Office of Science (SC) (United States), 2016.

[24] Zavabeti, A., A. Jannat, L. Zhong, A. A. Haidry, Z. Yao, and J. Z. Ou, Two-dimensional materials in large-areas: Synthesis, properties and applications, *Nano-Micro Letters*, 12 (2020) 1–34.

[25] Zhang, H., Ultrathin two-dimensional nanomaterials, *ACS Nano*, 9 (2015) 9451–9469.

[26] Duan, X., C. Wang, A. Pan, R. Yu, and X. Duan, Two-dimensional transition metal dichalcogenides as atomically thin semiconductors: Opportunities and challenges, *Chemical Society Reviews*, 44 (2015) 8859–8876.

[27] Zhang, X., S. Y. Teng, A. C. M. Loy, B. S. How, W. D. Leong, and X. Tao, Transition metal dichalcogenides for the application of pollution reduction: A review, *Nanomaterials*, 10 (2020) 1012.

[28] Manzeli, S., D. Ovchinnikov, D. Pasquier, O. V. Yazyev, and A. Kis, 2D transition metal dichalcogenides, *Nature Reviews Materials*, 2 (2017) 1–15.

[29] Wang, Q. H., K. Kalantar-Zadeh, A. Kis, J. N. Coleman, and M. S. Strano, Electronics and optoelectronics of two-dimensional transition metal dichalcogenides, *Nature Nanotechnology*, 7 (2012) 699–712.

[30] Khan, K., A. K. Tareen, M. Aslam, R. Wang, Y. Zhang, A. Mahmood, Z. Ouyang, H. Zhang, and Z. Guo, Recent developments in emerging two-dimensional materials and their applications, *Journal of Materials Chemistry C*, 8 (2020) 387–440.

[31] Velický, M., and P. S. Toth, From two-dimensional materials to their heterostructures: An electrochemist's perspective, *Applied Materials Today*, 8 (2017) 68–103.

[32] Postorino, S., D. Grassano, M. D'Alessandro, A. Pianetti, O. Pulci and M. Palummo, Strain-induced effects on the electronic properties of 2D materials, *Nanomaterials and Nanotechnology*, 10 (2020) 1847980420902569.

[33] Mujib, S. B., Z. Ren, S. Mukherjee, D. M. Soares and G. Singh, Design, characterization, and application of elemental 2D materials for electrochemical energy storage, sensing, and catalysis, *Materials Advances*, 1 (2020) 2562–2591.

[34] Sangwan, V. K., and M. C. Hersam, Electronic transport in two-dimensional materials, *Annual Review of Physical Chemistry*, 69 (2018) 299–325.

[35] Huang, H. H., X. Fan, D. J. Singh, and W. T. Zheng, Recent progress of TMD nanomaterials: Phase transitions and applications, *Nanoscale*, 12 (2020) 1247–1268.

[36] Choi, W., N. Choudhary, G. H. Han, J. Park, D. Akinwande, and Y. H. Lee, Recent development of two-dimensional transition metal dichalcogenides and their applications, *Materials Today*, 20 (2017) 116–130.

[37] Chaves, A., J. G. Azadani, H. Alsalman, D. R. da Costa, R. Frisenda, A. J. Chaves, S. H. Song, Y. D. Kim, D. He, J. Zhou, and A. Castellanos-Gomez, Bandgap engineering of two-dimensional semiconductor materials, *NPJ 2D Materials and Applications*, 4 (2020) 1–21.

[38] Zhu, J., G. Xiao, and X. Zuo, Two-dimensional black phosphorus: An emerging anode material for lithium-ion batteries, *Nano-Micro Letters*, 12 (2020) 1–25.

[39] Wang, J., F. Ma, and M. Sun, Graphene, hexagonal boron nitride, and their heterostructures: Properties and applications, *RSC Advances*, 7 (2017) 16801–16822.

[40] Ma, Q., G. Ren, K. Xu, and J. Z. Ou, Tunable optical properties of 2D materials and their applications, *Advanced Optical Materials*, 9 (2021) 2001313.

[41] Yao, F., X. Zhou, and A. Xiong, Tunable electronic and optical properties of two-dimensional ZnSe/AlAs van der Waals heterostructure, *Applied Physics A*, 126 (2020) 1–10.

[42] Ma, Q., G. Ren, K. Xu, and J. Z. Ou, Tunable optical properties of 2D materials and their applications, *Advanced Optical Materials*, 9 (2021) 2001313.

[43] Zhang, G., and Y. W. Zhang, Thermal properties of two-dimensional materials, *Chinese Physics B*, 26 (2017) 034401.

[44] Song, H., J. Liu, B. Liu, J. Wu, H. M. Cheng, and F. Kang, Two-dimensional materials for thermal management applications, *Joule*, 2 (2018) 442–463.

[45] Wang, Y., N. Xu, D. Li, and J. Zhu, Thermal properties of two dimensional layered materials, *Advanced Functional Materials*, 27 (2017) 1604134.

[46] Kasirga, T. S., *Thermal Conductivity Measurements in Atomically Thin Materials and Devices*, Springer Nature, Singapore, 2020.

[47] Ibrahim, Y., A. Mohamed, A. M. Abdelgawad, K. Eid, A. M. Abdullah, and A. Elzatahry, The recent advances in the mechanical properties of self-standing two-dimensional MXene-based nanostructures: Deep insights into the supercapacitor, *Nanomaterials*, 10 (2020) 1916.

[48] Akinwande, D., C. J. Brennan, J. S. Bunch, P. Egberts, J. R. Felts, H. Gao, R. Huang, J. S. Kim, T. Li, Y. Li, and K. M. Liechti, A review on mechanics and mechanical properties of 2D materials—Graphene and beyond, *Extreme Mechanics Letters*, 13 (2017) 42–77.

[49] Fang, R., X. Cui, C. Stampfl, S. P. Ringer, and R. Zheng, Negative Poisson's ratio in 2D life-boat structured crystals, *Nanoscale Advances*, 1 (2019) 1117–1123.

[50] Yuan, J. H., G. Q. Mao, K. H. Xue, J. Wang and X. S. Miao, A new family of two-dimensional ferroelastic semiconductors with negative Poisson's ratios, *Nanoscale*, 12 (2020) 14150–14159.
[51] Yang, N., D. Yang, L. Chen, D. Liu, M. Cai, and X. Fan, Design and adjustment of the graphene work function via size, modification, defects, and doping: A first-principle theory study, *Nanoscale Research Letters*, 12 (2017) 1–7.
[52] Pinhal, G. B., N. L. Marana, G. S. Fabris, and J. R. Sambrano, Structural, electronic and mechanical properties of single-walled AlN and GaN nanotubes via DFT/B3LYP, *Theoretical Chemistry Accounts*, 138 (2019) 31.
[53] Wang, W., C. Yang, L. Bai, M. Li, and W. Li, First-principles study on the structural and electronic properties of monolayer MoS_2 with S-vacancy under uniaxial tensile strain, *Nanomaterials*, 8 (2018) 74.
[54] Siskins, M., M. Lee, F. Alijani, M. R. Van Blankenstein, D. Davidovikj, H. S. van der Zant, and P. G. Steeneken, Highly anisotropic mechanical and optical properties of 2D layered As_2S_3 membranes, *ACS Nano*, 13 (2019) 10845–10851.
[55] Li, X., M. Sun, C. Shan, Q. Chen, and X. Wei, Mechanical properties of 2D materials studied by in situ microscopy techniques, *Advanced Materials Interfaces*, 5 (2018) 1701246.
[56] Turiansky, M. E., A. Alkauskas, and C. G. Van de Walle, Spinning up quantum defects in 2D materials, *Nature Materials*, 19 (2020) 487–489.
[57] Akinwande, D., C. J. Brennan, J. S. Bunch, P. Egberts, J. R. Felts, H. Gao, R. Huang, J. S. Kim, T. Li, Y. Li, and K. M. Liechti, A review on mechanics and mechanical properties of 2D materials—Graphene and beyond, *Extreme Mechanics Letters*, 13 (2017) 42–77.
[58] Cai, Y., Q. Ke, G. Zhang, B. I. Yakobson, and Y. W. Zhang, Highly itinerant atomic vacancies in phosphorene, *Journal of the American Chemical Society*, 138 (2016) 10199–10206.
[59] Lee, G. D., C. Z. Wang, E. Yoon, N. M. Hwang and K. M. Ho, The role of pentagon—heptagon pair defect in carbon nanotube: The center of vacancy reconstruction, *Applied Physics Letters*, 97 (2010) 093106.
[60] Lee, G. D., C. Z. Wang, E. Yoon, N. M. Hwang, and K. M. Ho, The formation of pentagon-heptagon pair defect by the reconstruction of vacancy defects in carbon nanotube, *Applied Physics Letters*, 92 (2008) 043104.
[61] Han, S. W., G. B. Cha, K. Kim, and S. C. Hong, Hydrogen interaction with a sulfur-vacancy-induced occupied defect state in the electronic band structure of MoS_2, *Physical Chemistry Chemical Physics*, 21 (2019) 15302–15309.
[62] Zhao, X., Y. Ji, J. Chen, W. Fu, J. Dan, Y. Liu, S. J. Pennycook, W. Zhou, and K. P. Loh, Healing of planar defects in 2D materials via grain boundary sliding, *Advanced Materials*, 31 (2019) 1900237.
[63] Zou, X., and B. I. Yakobson, An open canvas 2D materials with defects, disorder, and functionality, *Accounts of Chemical Research*, 48 (2015) 73–80.
[64] Zhang, L., W. Chu, Q. Zheng, A. V. Benderskii, O. V. Prezhdo and J. Zhao, Suppression of electron–hole recombination by intrinsic defects in 2D monoelemental material, *The Journal of Physical Chemistry Letters*, 10 (2019) 6151–6158.
[65] Dai, M., Z. Wang, F. Wang, Y. Qiu, J. Zhang, C. Y. Xu, T. Zhai, W. Cao, Y. Fu, D. Jia, and Y. Zhou, Two-dimensional van der Waals materials with aligned in-plane polarization and large piezoelectric effect for self-powered piezoelectric sensors, *Nano Letters*, 19 (2019) 5410–5416.
[66] Yu, W., L. Sisi, Y. Haiyan, and L. Jie, Progress in the functional modification of graphene/graphene oxide: A review, *RSC Advances*, 10 (2020) 15328–15345.
[67] Velický, M., and P. S. Toth, From two-dimensional materials to their heterostructures: An electrochemist's perspective, *Applied Materials Today*, 8 (2017) 68–103.
[68] Liand, X., and H. Zhu, Two-dimensional MoS_2: Properties, preparation, and applications, *Journal of Materiomics*, 1 (2015) 33–44.

[69] Stoller, M. D., S. Park, Y. Zhu, J. An, and R. S. Ruoff, Graphene-based ultracapacitors, *Nano Letters*, 8 (2008) 3498–3502.

[70] Balandin, A. A., S. Gosh, W. Z. Bao, I. Calizo, D. Teweldebrhan, F. Miao, and C. N. Lau, Superior thermal conductivity of single-layer graphene, *Nano Letters*, 8 (2008) 902–907.

[71] Lee, C., X. Wei, J. W. Kysar, and J. Hone, Measurement of the elastic properties and intrinsic strength of monolayer graphene, *Science*, 321 (2008) 385–388.

[72] Bolotin, K. I., K. J. Sikes, Z. Jiang, M. Klima, G. Fudenberg, J. Hone, P. Kim, and H. L. Stormer, Ultrahigh electron mobility in suspended graphene, *Solid State Communications*, 146 (2008) 351–355.

[73] Hummers, W. S., and R. E. Offeman, Preparation of graphitic oxide, *Journal of The American Chemical Society*, 80 (1958) 1339–1339.

[74] Liu, J., M. Notarianni, G. D. Will, V. T. Tiong, H. Wang, and N. Motta, Electrochemically exfoliated graphene for electrode films: Effect of graphene flake thickness on the sheet resistance and capacitive properties, *Langmuir*, 29 (2013) 13307–13314.

[75] Leon, V., A. M. Rodriquez, P. Prieto, M. Prato, and E. Vazquez, Exfoliation of graphite with triazine derivatives under ball-milling conditions: Preparation of few-layer graphene via selective noncovalent interactions, *ACS Nano*, 8 (2014) 563–571.

[76] Ejigu, A., L. W. L. Fevre, K. Fujisawa, M. Terrones, A. J. Forsyth, R. A. W. Dryfe, Electrochemically exfoliated graphene electrode for high-performance rechargeable chloroaluminate and dual-ion batteries, *ACS Applied Mater Interfaces*, 11 (2019) 23261–23270.

[77] Lin, Y., J. Jin, O. Kusmartsevab, and M. Song, Preparation of pristine graphene sheets and large-area/ultrathin graphene films for high conducting and transparent applications, *Journal of Physical Chemistry C*, 117 (2013) 17237–17244.

[78] Chen, H., S. Chen, Y. Zhang, H. Ren, X. Hu, and Y. Bai, Sand-milling fabrication of screen-printable graphene composite inks for high-performance planar micro-supercapacitors, *ACS Applied Mater Interfaces*, 12 (2020) 56319–56329.

[79] Trung, T. N., D. O. Kim, J. H. Lee, V. D. Dao, H. S. Choi, and E. T. Kim, Simple and reliable lift-off patterning approach for graphene and graphene–Ag nanowire hybrid films, *ACS Applied Mater Interfaces*, 9 (2017) 21406–21412.

[80] Hou, Y., X. Ren, J. Fan, G. Wang, Z. Dai, C. Jin, W. Wang, Y. Zhu, S. Zhang, L. Liu, and Z. Zhang, Preparation of twisted bilayer graphene via the wetting transfer method, *ACS Applied Mater Interfaces*, 12 (2020) 40958–40967.

[81] Brownson, D. A. C., and C. E. Banks, The electrochemistry of CVD graphene: Progress and prospects, *Physical Chemistry Chemical Physics*, 14 (2012) 8264–8281.

[82] Randviir, E. P., D. A. C. Brownson, M. G. Mingot, D. K. Kampouris, J. Iniesta, and C. E. Banks, Electrochemistry of Q-graphene, *Nanoscale*, 4 (2012) 6470–6480.

[83] Randviir, E. P., D. A. C. Brown, J. P. Metters, R. O. Kadara, and C. E. Banks, The fabrication, characterisation and electrochemical investigation of screen-printed graphene electrodes, *Physical Chemistry Chemical Physics*, 16 (2014) 4598–4611.

[84] Fang, Y., and E. Wang, Electrochemical biosensors on platforms of graphene, *Chemical Communications*, 49 (2013) 9526–9539.

[85] Brownson, D. A. C., L. J. Munro, D. K. Kampouris, and C. E. Banks, Electrochemistry of graphene: Not such a beneficial electrode material, *RSC Advances*, 1 (2011) 978–988.

[86] Saho, R., A. Pal, and T. Pal, 2D materials for renewable energy storage devices: Outlook and challenges. *Chemical Communications*, 52 (2016) 13528–13542.

[87] Pandhi, T., C. Cornwell, K. Fujimoto, P. Barnes, J. Cox, H. Xiong, P. H. Davis, H. Subraman, J. E. Koehne, and D. Estrada, Fully inkjet-printed multilayered graphene-based flexible electrodes for repeatable electrochemical response, *RSC Advances*, 10 (2020) 38205–38219.

[88] Brownson, D. A. C., S. A. Varey, F. Hussain, S. J. Haigh, and C. E. Banks, Electrochemical properties of CVD grown pristine graphene: Monolayer-vs. quasi-graphene, *Nanoscale*, 6 (2014) 1607–1621.

[89] Zuo, Z., T. Y. Kim, I. Kholmanov, H. Li, H. Chou, and Y. Li, Ultra-light hierarchical graphene electrode for binder-free supercapacitors and lithium-ion battery anodes, *Small*, 37 (2015) 4922–4930.

[90] Wan, Q., H. Cai, Y. Liu, H. Song, H. Liao, S. Liu, and N. Yang, Graphene nanoplatelets: Electrochemical properties and applications for oxidation of endocrine-disrupting chemicals, *Chemistry-A European Journal,* 19 (2013) 3483–3489.

[91] Jindal, H., A. S. Oberoi, I. S. Sandhu, M. Chitkara, and B. Singh, Graphene for hydrogen energy storage – A comparative study on GO and rGO employed in a modified reversible PEM fuel cell, *International Journal of Energy Research,* (2020) 1–12.

[92] Manchala, S., V. Tandava, D. Jampaiah, S. K. Bhargava, and V. Shanker, A novel strategy for sustainable synthesis of soluble-graphene by a herb delphinium denudatum root extract for use as light-weight supercapacitors, *Chemistry Select,* 5 (2020) 2701–2709.

[93] Shao, Q., J. Tang, Y. Lin, F. Zhang, J. Yuan, H. Zhang, N. Shinya, and L. C. Qin, Synthesis and characterization of graphene hollow spheres for application in supercapacitors, *Journal of Materials Chemistry A*, 1 (2013) 15423.

[94] Li, Y., K. Sheng, W. Yuan, and G. Shi, A high-performance flexible fibre-shaped electrochemical capacitor based on electrochemically reduced graphene oxide, *Chemical Communications,* 49 (2013) 291.

[95] Xie, X., L. Ju, X. Feng, Y. Sun, R. Zhou, K. Liu, S. Fan, Q. Li, and K. Jiang, Controlled fabrication of high-quality carbon nanoscrolls from monolayer graphene, *Nano Letters,* 9 (2009) 2565–2570.

[96] Wang, Y., X. Liu, C. Yagn, N. Li, K. Yan, H. Chi, F. Sun, J. Zhao, and Y. Li, Low-cost fabrication of three-dimensional hierarchical porous graphene anode material for sodium ion batteries application, *Surface & Coating Technology,* 360 (2019) 110–115.

[97] Sun, L., C. Tian, M. Li, X. Meng, L. Wang, R. Wang, J. Yin, and H. Fu, From coconut shell to porous graphene-like nanosheets for high-power supercapacitors, *Journal of Materials Chemistry A*, 1 (2013) 6462.

[98] Nishihara, H., T. Simura, S. Kobayashi, K. Nomura, R. Berenguer, M. Ito, M. Uchimura, H. Iden, K. Arihara, A. Ohma, Y. Hayasaka, and T. Kyotani, Oxidation-resistant and elastic mesoporous carbon with single-layer graphene walls, *Advanced Functional Materials*, 26 (2016) 6418–6427.

[99] Li, X., S. He, L. Sang, F. Zhang, Y. Song, B. Zhai, and X. Wang, Facile synthesis of three-dimensional porous graphene nanostructures from coordination complexes for supercapacitor electrode, *Advanced Power Technology*, 31 (2020) 4157–4165.

[100] Cai, D., L. Ding, S. Wang, Z. Li, M. Zhu, and H. Wang, Facile synthesis of ultrathin-shell graphene hollow spheres for high-performance lithium-ion batteries, *Electrochimica Acta*, 139 (2014) 96–103.

[101] Li, B., G. Zhang, I. B. Tahirbegi, M. J. Morten, and H. Tan, Monitoring amyloid-β 42 conformational changes using a spray-printed graphene electrode, *Electrochemical Communications*, 123 (2021) 106927.

[102] Karaphun, A., C. Phrompet, W. Tuichai, N. Chanlek, C. Sriwong, C. Ruttanapun, The influence of annealing on a large specific surface area and enhancing electrochemical properties of reduced graphene oxide to improve the performance of the active electrode of supercapacitor devices, *Materials Science & Engineering B*, 264 (2021) 114941.

[103] Shabangoli, Y., M. S. Rahmanifar, A. Noori, M. F. Kady, R. B. Kaner, M. F. Mousavi, Nile blue functionalized graphene aerogel as a pseudocapacitive negative electrode material across the full pH range, *ACS Nano*, 13 (2019) 12567–12576.

[104] Fortgang, P., T. Tite, V. Barnier, N. Zehani, C. Maddi, F. Lagarde, A. S. Loir, N. J. Renault, C. Donnet, F. Gaorrelie, and C. Chaix, Robust electrografting on self-organized 3D graphene electrodes, *ACS Applied Mater Interfaces*, 8 (2016) 1424–1433.

[105] Li, Q., A. Bai, Z. Xue, Y. Zheng, and H. Sun, Nitrogen and sulfur co-doped graphene composite electrode with high electrocatalytic activity for vanadium redox flow battery application, *Electrochimica Acta*, 362 (2020) 137223.

[106] Dutta, P., A. Sikdar, A. Majumdar, M. Borah, N. Padma, S. Gosh, and U. N. Maiti, Graphene aided gelation of MXene with oxidation protected surface for supercapacitor electrodes with excellent gravimetric performance, *Carbon*, 169 (2020) 225–234.

[107] Schwierz, F., Graphene transistors, *Nature Nanotechnology*, 5 (2010) 487–496.

[108] Wang, X. R., Y. J. Ouyang, X. L. Li, H. L. Wang, J. Guo, and H. J. Dai, Room-temperature all-semiconducting sub-10-nm graphene nanoribbon field-effect transistors, *Physical Review Letters*, 100 (2008) 206803.

[109] Sarma, S. D., S. Adam, E. Hwang, and E. Rossi, Electronic transport in two dimensional graphene, *Reviews of Modern Physics*, 83 (2011) 407.

[110] Stauber, T., N. Peres, and F. Guinea, Electronic transport in graphene: A semiclassical approach including midgap states, *Physical Review B.*, 76 (2007) 205423.

[111] Bolotin, K. I., K. Sikes, Z. Jiang, M. Klima, G. Fudenberg, and J. Hone, Ultrahigh electron mobility in suspended graphene, *Solid State Communications*, 146 (2008) 351–355.

[112] Huang, X., Z. Zeng, Z. Fan, J. Liu, and H. Zhang, Graphene-based electrodes, *Advanced Materials*, 24 (2012) 5979–6004.

[113] Lin, Y. M., A. V. Garcia, S. J. Han, D. B. Farmer, I. Meric, Y. Sun, Y. Wu, C. Dimitrakopoulos, A. Grill, P. Avouris, and K. A. Jenkins, Wafer-scale graphene integrated circuit, *Science*, 332 (2011) 1294–1297.

[114] Ahmed, F., R. K. Brajpuriya, and Y. Handa, A review on graphene based solar cells, *International Journal of Recent Scientific Research*, 8 (2017) 16893–16896.

[115] Mahmoudi, T., Y. Wang, and Y. B. Hahn, Graphene and its derivatives for solar cells application, *Nano Energy*, 47 (2018) 51–65.

[116] Neto, A. H. C., F. Guinea, N. M. R. Peres, K. S. Novoselov, A. K. Geim, The electronic properties of graphene, *Reviews of Modern Physics*, 81 (2009) 109–162.

[117] Li, X. L., X. R. Wang, L. Zhang, S. W. Lee, H. J. Dai, Chemically derived, ultrasmooth graphene nanoribbon semiconductors, *Science*, 319 (2008) 1229–1232.

[118] Banerjee, S. K., L. F. Register, E. Tutuc, D. Reddy, and A. H. MacDonald, Bilayer PseudoSpin field-effect transistor (BiSFET): A proposed new logic device, *IEEE Electron Device Letters*, 30 (2009) 158–160.

[119] Tonelli, F. M. P., V. A. M. Goulart, K. N. Gomes, M. S. Ladeira, A. K. Santos, E. Lorençon, L. O. Ladeira, and R. R. Resende, Graphene-based nanomaterials: Biological and medical applications and toxicity, *Nanomedicine*, 10 (2015) 2423–2450.

[120] Singh, Z. S., Applications and toxicity of graphene family nanomaterials and their composites, *Nanotechnology Science and Applications*, (2016) 15–28.

[121] Shim, G., M.-G. Kim, J. Y. Park, and Y.-K. Oh, Graphene-based nanosheets for delivery of chemotherapeutics and biological drugs, *Advanced Drug Delivery Reviews*, 105 (2016) 205–227.

[122] Kuila, T., S. Bose, A. K. Mishra, P. Khanra, N. H. Kim, and J. H. Lee, Chemical functionalization of graphene and its applications, *Progress in Materials Science*, 57 (2012) 1061–1105.

[123] Yang, K., L. Feng, X. Shi, and Z. Liu, Nano-graphene in biomedicine: Theranostic applications, *Chemical Society Reviews*, 42 (2013) 530–547.

[124] Chowdhury, S. M., C. Surhland, Z. Sanchez, P. Chaudhary, M. A. Suresh Kumar, et al., Graphene nanoribbons as a drug delivery agent for lucanthone mediated therapy of glioblastoma multiforme, *Nanomedicine: Nanotechnology, Biology and Medicine*, 11 (2015) 109–118.

[125] Ivanovskii, A. L., Graphene-based and graphene-like materials, *Russian Chemical Reviews,* 81 (2012) 571–605.

11 Two-Dimensional Transition Metal Dichalcogenide (TMD) Materials in Field-Effect Transistor (FET) Devices for Low Power Applications

R. Sridevi and J. Charles Pravin

CONTENTS

DOI: 10.1201/9781003200987-11

11.1 INTRODUCTION

The size of a semiconductor device has drastically reduced toward nanoscale range to increase the density of the integrated circuits. Field-effect transistor (FET) is preferred for IC applications due to the virtue of its construction and biasing, large input impedance, and low noise level compared to a bipolar transistor. The metal oxide semiconductor field-effect transistor (MOSFET) has some advantages over the junction FET, which includes high input impedance, enhancement of channel width, and low cost of production. MOSFET is used for low-energy functional devices. MOSFET uses silicon as an active region in the semiconductor industry for the past few decades [1]. FinFET is a promising device for memory applications in 22 nm technology due to its good short-channel effect (SCE) controllability [2]. Tunnel FET (TFET) is an upcoming device with the structure almost similar to MOSFET with the different switching mechanism. TFET has more advantages such as ultra low power, ultra low voltage, high ON/OFF ratio, and low leakage current [3]. Using double gate TFET with high k gate dielectric improves the characteristics like high ON current, less OFF current, and improved average subthreshold swing. The double-gate transistor is mostly used for the reduction of SCEs, better electrical characteristics, and easy fabrication process. Tunnel FET technology is more suitable for low standby power (LSTP) applications; this technology is a better replacement for MOSFET technology [4]. There has been considerable interest in developing TMD-based FET devices like SOI MOSFET [5], SOI FinFET [6], TFET [6], and junction less nanowire transistors [7] using various techniques such as mechanical exfoliation [8] and chemical vapor deposition (CVD) [9]. Initially, the graphene materials were used as an active region in the transistor device, but they produced high OFF current and were of toxic quality. Thus, the interest toward beyond-graphene 2-D material increases.

The device size has shrunk successfully with the help of 2-D materials. In recent years, 14 nm high k dielectric material/TMD material n-type MOSFET devices have been developed [10]. These materials are preferred for micro and nano electronic applications with high integration in the future [11–20]. The 2-D material has promising characteristics such as easy accessibility, very high carrier mobility, and high thermal conductivity. Moreover, this material has not been affected by mobility degradation, unlike silicon and bulk materials.

From 2010, researchers have been developing the transition metal dichalcogenide (TMD) material. TMDs combined by M is a transition metal element from Group IV (Ti, Zr, Hf, and so on), Group V (V, Nb, Ta), or Group VI (Mo, W, and so on), and X is a chalcogen [21]. TMD materials are grouped based on their properties of the bandgap, carrier mobility, and ON/OFF current ratio. Recently, FET with 2-D materials have been used due to a great deal of interest in the electrical behavior. The FET devices have superior performance in ON/OFF current ratio, contact resistance, and carrier mobility [22–23]. For recent advancement in the low power nanowire FET with low operating voltage, a new class of 2-D materials is being used. Here, intensively reviewed 2-D TMD materials such as MoS_2, $MoSe_2$, $MoTe_2$, WS_2, WSe_2, and WTe_2 can be used in the semiconductor devices. Additionally, few techniques and structures have been reviewed.

The remaining part of this chapter is divided into the following sections. Section 11.2 describes the evolution of the transistor based on the other existing techniques. Section 11.3 shows the analysis of the 2-D materials. Section 11.4 elaborately explains the existing method of FET devices with simulation of experimental results and their implications. Finally, Section 11.5 concludes with an overview of the role of 2-D materials in FET devices for low power applications.

11.2 EVOLUTION OF TRANSISTORS

This section describes the evolution of transistors. Initially, the bipolar device was introduced and this device is still used in analog circuit applications. However, it does not have significant scaling property because of its complicated physical structure [24]. After that, the unipolar device of FET was presented, which has been differentiated from a junction transistor [25–27]. This FET structure is shown in Figure 11.1. Figure 11.2 shows the channel enhancement mode of silicon insulated gate FET. An insulated type gate holds a continuous transfer characteristic as well as low gate leakage current [28–30]. A thin gate dielectric and high doping concentration are required to control the OFF current in the conventional MOSFET.

The SOI device is used for low source/drain capacitance and better body factor as shown in Figure 11.3, where the structure of lightly doped drain (LDD) region has minimized the drain electric field and short-channel effects [6]. GAA MOSFET is a promising device because it provides the best electrostatic control and reduced short-channel effect (SCE) [31]. Here, the calculation of the subthreshold potential has been solved using Poisson equation. Then, FinFET is the most preferred device for memory applications because of its better short-channel controllability. This structure is also called a 3-D transistor or bulk FinFET, which obtains the devices at the 22-nm technology node [6] that is shown in Figure 11.4. The FinFET device will not be suited in the future because of its difficult patterning and doping structures [6]. A tunnel transistor is similar to a MOSFET, including the fabrication process except for the switching mechanism. TFETs present improved analog performance like voltage gain (Av) and early voltage (VEA) compared to other transistors [6]. Figure 11.5 shows the tunnel FET structure that displays the result smaller than 60 mV/dec of the subthreshold swing, which permits faster-switching circuits than in MOSFET

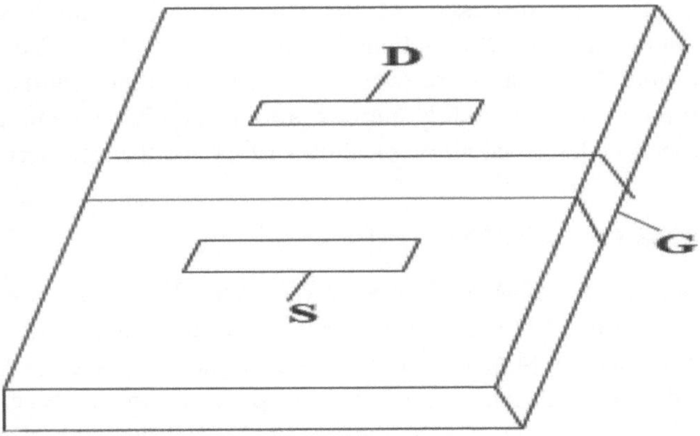

FIGURE 11.1 Field-Effect Transistor

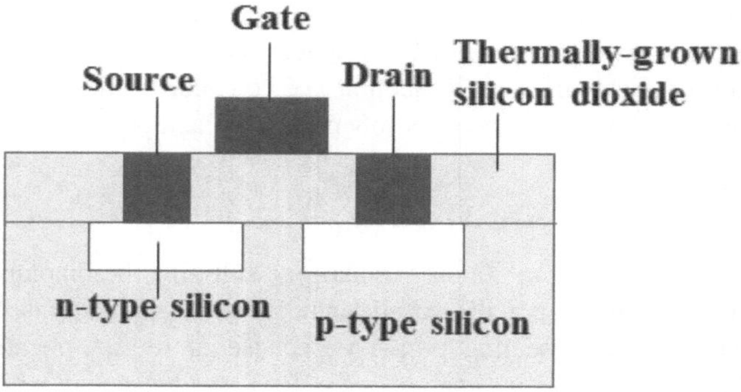

FIGURE 11.2 Enhancement Mode of Silicon IGFET

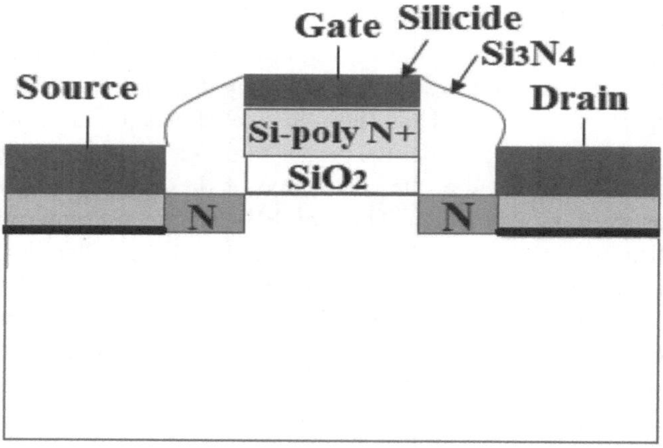

FIGURE 11.3 SOI MOSFET

technology [6]. Then the discussion of junctionless transistor has less fabrication complexity compared to other junctions-based transistors [32]. The junctionless accumulation mode GAA nanowire transistor is used for their simulation process, and it has more current and slightly better drain-induced barrier lowering (DIBL) [33–39]. This section described all types of transistors and their properties.

11.3 ANALYSIS OF 2-D MATERIALS

2-D TMD materials such as MoS_2, $MoSe_2$, $MoTe_2$, WS_2, WSe_2, and WTe_2 are studied in detail in this section. 2-D material will be used in future applications because it is flexible and transparent [40]. The libraries of 2-D material are graphene, h-Boron nitride, and TMD materials. Here, graphene is a conductor, the h-Boron nitride is an insulator, and TMD materials are semiconductors. The developing scalability of 2-D materials have been synthesized using chemical vapor deposition (CVD) technology

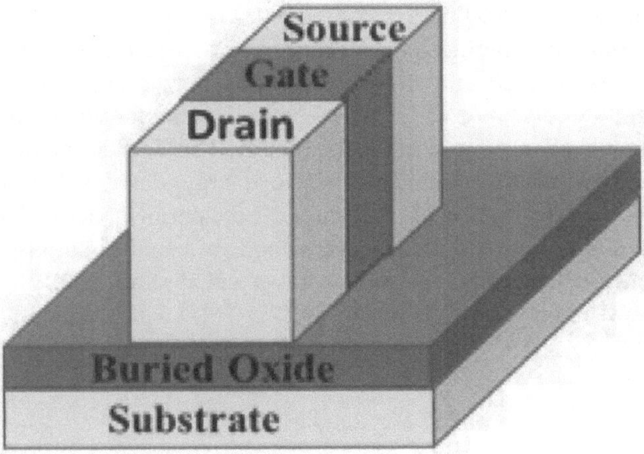

FIGURE 11.4 Tri-Gate SOI FinFET

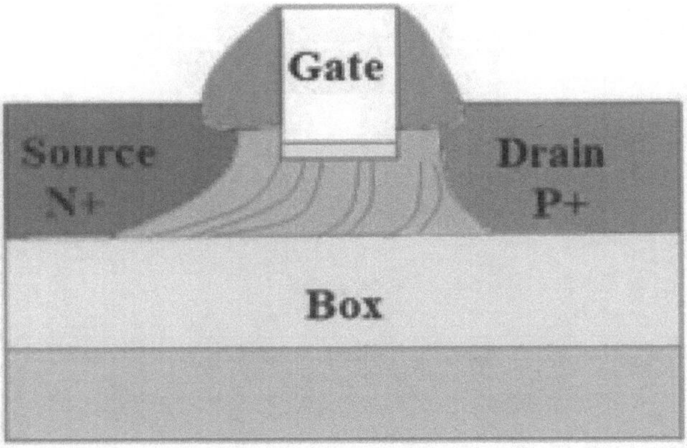

FIGURE 11.5 Tunnel FET

instead of mechanical exfoliation and liquid-phase exfoliation for large-scale and low-cost factors. 2-D material van der Waals heterostructure are meant to assemble different 2-D materials into vertical stacks; this heterostructure should be cautious in controlling sheet order [41]. The simulations of dissipative quantum transport are used to evaluate the scalability and performance of monolayer/multilayer 2-D semiconductor for 10 nm. Researchers have succeeded in using MoS_2 as the channel material since they could meet 6.6 nm gate length and WSe_2 have been focused on high mobility and suitability of the low effective mass of their material. Here 2-D materials are utilized in an energy storage device for improving their performance at high temperature. Future research of 2-D materials will be used for more efficient energy generation and storage device as well as in nano medicine [42] and aircraft application [43].

Figure 11.6 illustrates the characteristics of 2-D materials and also differentiates the cross-sectional view of 2-D and bulk material. The 2-D material structure consists of atomically smooth, dangling bond free, uniform, and fixed thickness [44–45]. This aid suppresses trap generation intrinsically, scattering of carrier, and thickness variation. High ON current of TFET device is well suited for upcoming application, but this is a great challenge for researchers. Finally, TFET device achieves 60mV/dec of subthreshold swing and smaller intrinsic energy delay values compared to conventional CMOS devices [46–47]. To enhance the performance of TMD materials with TFET, it is not only required to have a good gate control but also an optimum bandgap of channel materials, effective mass, and doping level of source/drain. Figure 11.7 shows the types of 2-D materials. 2-D TMD with atomically thin-layered

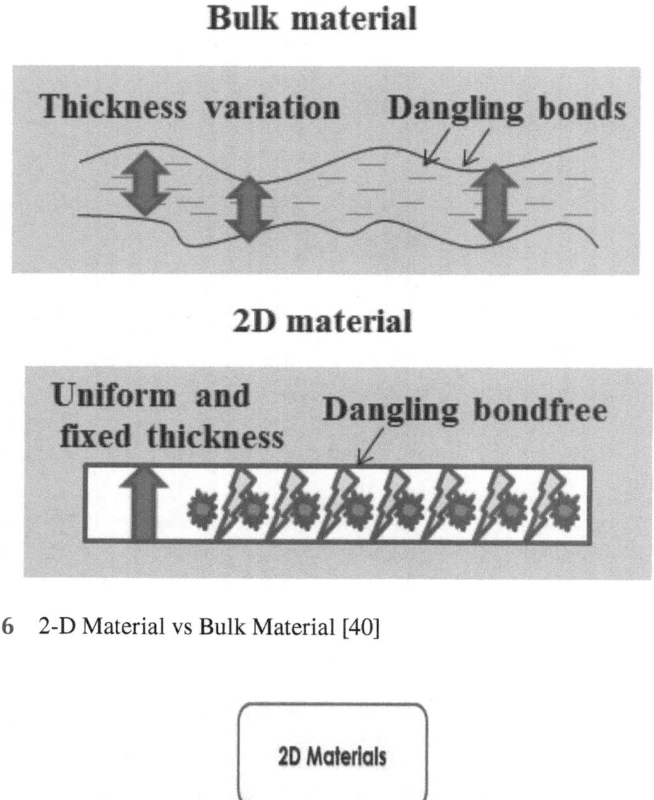

FIGURE 11.6 2-D Material vs Bulk Material [40]

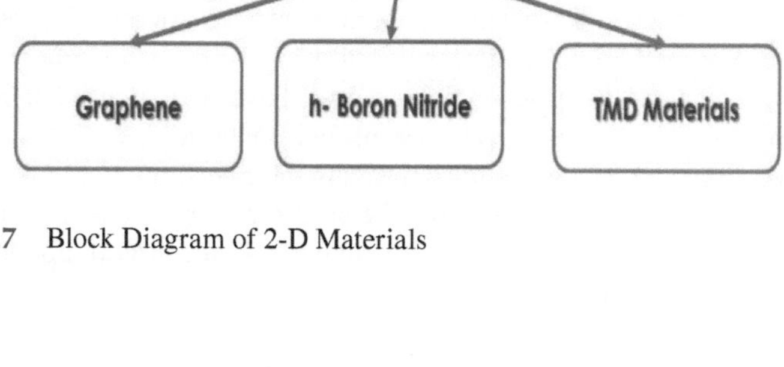

FIGURE 11.7 Block Diagram of 2-D Materials

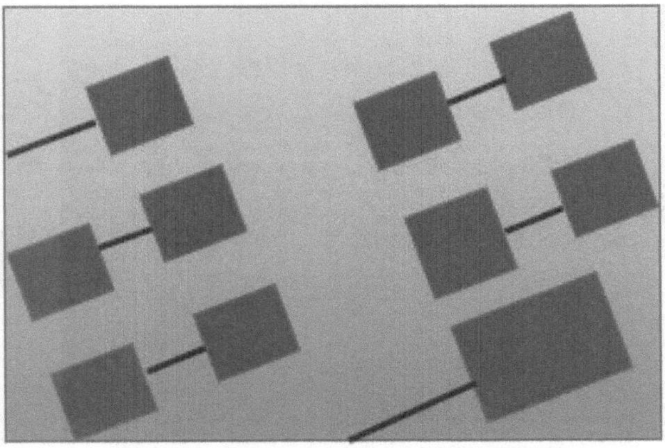

FIGURE 11.8 Structure of Graphene

films are widely used in photovoltaic [48], phototransistor [49], optoelectronic device [50], photodetector [51], and water purification applications [21].

11.3.1 Graphene

The 2-D material graphene was discovered by two scientists at the University of Manchester. Figure 11.8 displays graphene, a crystalline allotrope of carbon with 2-D properties. Other allotropes, such as graphite, carbon nanotubes, charcoal, and fullerenes, have it as their essential structural feature. It is also possible to think of it as an infinitely large aromatic molecule, the limiting case of the family of flat polycyclic aromatic hydrocarbons. Graphene is pure carbon in the form of a one-atom-thick layer that is nearly transparent. It is incredibly heavy for its light weight (100 times stronger than steel) and efficiently conducts heat and electricity. With its unusual combination of bonded carbon atom structures and myriad and nuanced physical properties, graphene is poised to have a significant effect on the future of material sciences, electronics, and nanotechnology. Since its first isolation in 2004, research in graphene materials has expanded rapidly.

The study was aided by theoretical explanations of graphene's composition, structure, and properties, all of which had been determined decades before. High-quality graphene was also remarkably simple to isolate, allowing for further analysis. The properties and applications of this 2-D type of carbon structure have opened up new possibilities for future devices as well as systems, according to the researchers.

11.3.2 Transition Metal Dichalcogenide (TMD) Materials

TMDs combined by M is a transition metal element from Group IV (Ti, Zr, Hf, and so on), Group V (V, Nb, Ta), or Group VI (Mo, W and so on), and X is a chalcogen. The chemical formula is MX_2, where M is a transition metal atom (groups 4–12 in the periodic table) and X is a chalcogen (group 16). Figure 11.9 shows the types of 2-D TMD materials. The periodic table of TMD materials is shown in Figure 11.10.

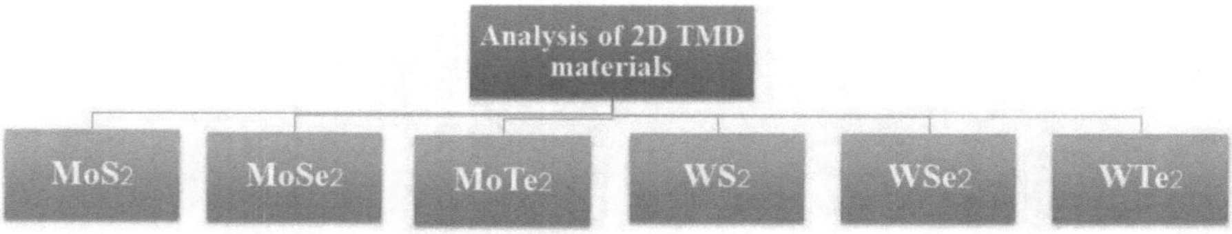

FIGURE 11.9 Types of 2-D TMD Materials

MX_2	
M = Transition Metal	**X = Chalcogen**
Mo	Se
W	Te

FIGURE 11.10 Periodic Table for TMD Materials

11.3.2.1 Molybdenum Disulfide (MoS_2) Material

This section provides an overview of the MoS_2 material. This material has unique properties of ideal 2-D structure, sizable energy bandgaps [52], chemical and optical properties, low power consumption [53], light emission, valleytronics, high efficiency catalysts [54], and atomically thin geometry. For this reason, MoS_2 is used for future electronics [55]. The 2-D TMD materials can be developed from monolayer to multilayer structures.

Monolayer Structure of MoS_2

Monolayer MoS_2 has the thickness of 0.65 nm, the larger bandgap of 1.8 eV. The electron effective mass is 0.48 m_o, the dielectric permittivity is 8.29 ε_0, and the electron affinity is 4 eV. Figure 11.11 shows the layer structure of monolayer MoS_2, which shows the layer thickness of 0.65 nm. Figure 11.12 shows the transistor of the MoS_2 structure.

Properties of MoS_2

Bulk Properties

MoS_2 is derived naturally from the mineral "molybdenite". Molybdenite appears in its raw form as a dark, glittery material. Sheets easily slide over one another due to poor interlayer interactions.

Optical and Electrical Properties

N-type behavior is common in MoS_2 monolayer transistors, with carrier mobilities of around 350 $cm^2V^{-1}s^{-1}$. They can have large ON/OFF ratios of 108 when fabricated into field-effect transistors, making them desirable for high-efficiency switching and logic circuits.

Mechanical Properties

Thin-film FETs have been shown to maintain their electronic properties when bent to a radius of curvature of 0.75 mm using MoS_2 monolayers, which are flexible. They

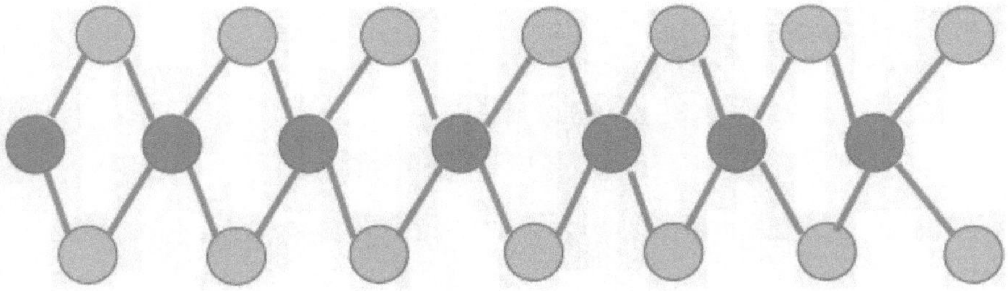

FIGURE 11.11 Layer of Molybdenum Atoms (Blue) and Sulphur Atom (Yellow)

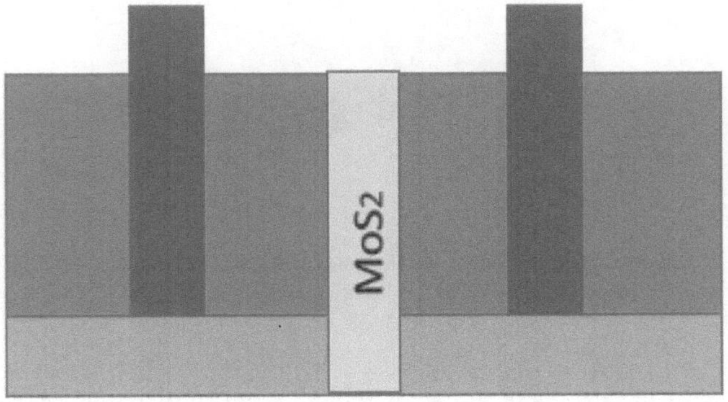

FIGURE 11.12 Molybdenum Disulfide Transistor

have a stiffness coefficient comparable to steel, and a higher breaking strength than flexible plastics, making them ideal for flexible electronics.

Valleytronics

MoS_2 and other 2-D TMDCs could pave the way for technologies other than electronics, where degrees of freedom (other than charge) can be used for data storage and processing. At both the K and K' (often called -K) points of the Brillouin field, the electronic band structure of MoS_2 shows energy maxima in the valence band and minima in the conduction band. There are two distinct "valleys" in this region.

Electrical Performance

Single-layer MoS_2 has a large direct bandgap of 1.8 eV, being suitable acting as switching nano devices. A study proposed a single-layer MoS_2 transistor adopting a hafnium oxide as the gate dielectric material, in which the mobility of MoS_2 could be up to 200 $cm^2/(Vs)$ at room temperature with the current ON/OFF ratio to be 1×10^8. Figure 11.13 shows the electrical performance of MoS_2.

Van der Waals Force

Mo and S are arranged in a sandwich structure by covalent bonds in a series of S-Mo-S in single-layer MoS_2 films, and the sandwich layers interact through a

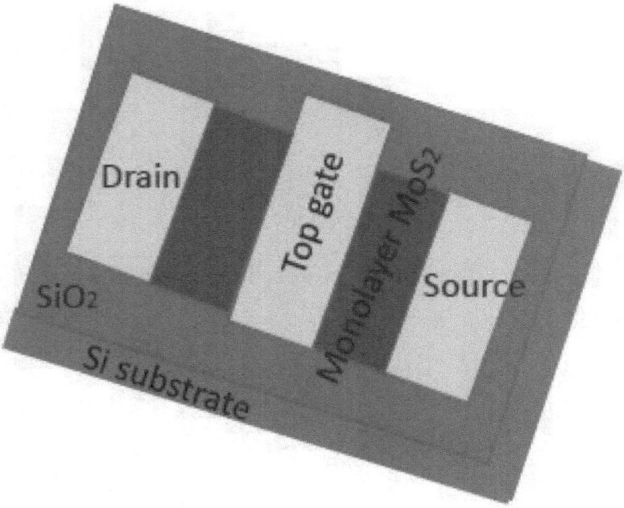

FIGURE 11.13 Electrical Performance of Monolayer MoS_2 Transistor with High-k Dielectric Material

relatively weak van der Waals force. It includes intermolecular forces as well as attraction and repulsion among atoms, molecules, and surfaces. An atom or molecule-to-atom or molecule interaction is distance-dependent. This is a weaker force than the covalent and ionic bonds. The van der Waals interaction between chalcogen layers is 1,000 times weaker than the layer-to-layer interaction between chalcogen layers. One layer of molybdenum atoms and two layers of sulphur atoms make up the MoS_2 monolayer. Because of the van der Waals interaction between the layers, it can be stacked. The interface between the monolayer MoS_2 and substrate contributes to this interaction. The 2-D material has the slightly stronger binding energy interlayer distance of 3–4 A°.

2-D Van der Waals Heterostructures

The binding energy of per C atom binding to MoS_2 is 23 meV, and the forming interlayer spacing between graphene and MoS_2 is 3.32, according to Yandong Ma's party. The band structure of graphene was mostly retained in this hybrid structure due to the variance in on-site energy caused by MoS_2, thus adding a small bandgap of 2 meV that was almost negligible. Further study revealed that the bandgap could be modified by adjusting the interlayer spacing, suggesting the possibility of developing devices with tunable bandgaps and high electron mobility at the same time. The van der Waals heterostructure is shown in Figure 11.14.

3-D MoS_2-Based Structures

The range of applications for 2-D heterostructures was primarily based on electronic and optoelectronic devices. In general, 3-D self-assembled structures were needed

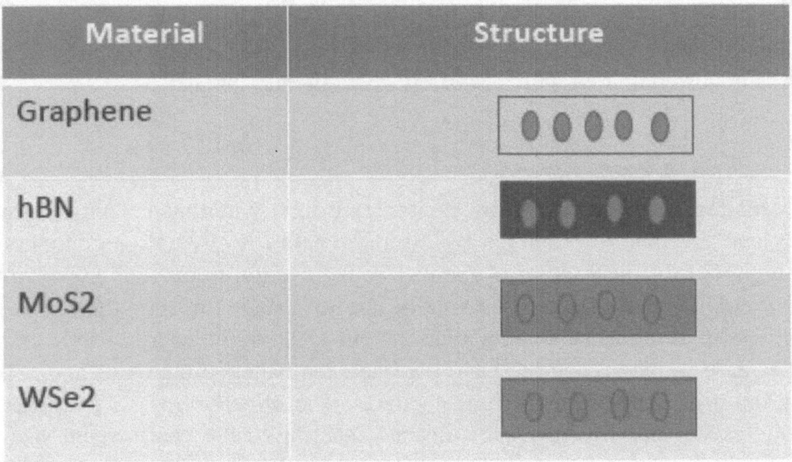

FIGURE 11.14 2-D van der Waals Heterostructure

for applications such as super capacitors, lithium ion batteries, and HER. Because of MoS_2's tunable bandgap, they could achieve photo responsivity over a broad spectrum of wavelengths, from ultraviolet to infrared, with high stability.

Furthermore, MoS_2 with sufficient interlayer spacing provides a convenient structure for ion storage. However, electrical conductivity and cycling stability of MoS_2 electrodes remain difficult to achieve. Fortunately, carbonaceous materials like carbon nanotubes (CNT), graphene, and even other organic conducting polymers can thoroughly demonstrate these two properties.

Quantum Mechanical Effects

Quantum effects are known to occur in the channel of MOSFET, where the confinement is in the perpendicular direction of the channel. Solving Schrodinger and Poisson equation is needed to find the actual position of the charge and quantum potential, where the Schrodinger equation stands for quantum mechanics. When the carrier is confined to a region where one or more of the dimensions reach the range of less than 100 nm, the quantum energy level begins to spread out and the quantum nature becomes detectable. This carrier exhibits the quantum effects imposed by the nanostructure. Quantum effect is dominant when the devices size reaches a nanometer range. This effect is the inherent level of uncertainty that exists within any physical system due to the impossibility of simultaneously knowing both of its most fundamental properties: position and momentum. Transistors are scaled down to a nanometer regime, and quantum effects in both transverse and transport directions start playing a major role in determining device characteristics.

The effects of quantum confinement are modeled, where the quantum allows the self-consistent solution to the Schrodinger and Poisson equation.

- The double gate transistor is to control the MoS_2 channel very efficiently by choosing the MoS_2 channel width to be very small.
- This concept helps to suppress short-channel effects and leads to higher current as compared with a MOSFET having only one gate.

Quantum Confinement Effects

In semiconductor nanocrystals, the excitons are tightly confined within the nanocrystals and are therefore defined by discrete energy levels unlike continuous in micro regime. In nanostructure, the ability of the exciton to move is impeded. The quantum confinement effect is essentially due to changes in the atomic structure because of direct influence of ultra small length scale on the energy band structure. With decrease in the size of the nano particles, the bandgap increases and exhibits more prominent quantum confinement effects. The effective mass of the electron and hole, as well as the dielectric constant, determines the confinement size limits. When the particle's size is too small to be equal to the electron's wavelength, it depends on the (i) material property and (ii) Bohr radius. Quantum effects are a function of channel length. When the size of nanomaterial is less than exciton Bohr radius, the quantum confinement effect is noticed in terms of discrete energy level.

The exciton in a bulk semiconductor will travel freely in all directions. When a semiconductor's length is reduced to the same order as the radius of an exciton, electron motion is confined in a closed space. The Schrodinger equation can be solved using the Ritz Galerkin Finite element technique.

Wave Function

Electronic wave functions for the conducting electrons are displaced over the entire particle in semiconductors. The particle size is equal to its wave function. Inside the confining area, the wave function appears continuous, but it is zero at the quantum well's walls. Owing to the confinement of the electron wave function in nanoscale structures, electron states are discrete rather than continuous. Quantum confinement considers the electronic structure of nanoparticles, which is highly dependent on the size of the particles. The confining dimension reduces until it reaches a limit.

Quantum Potential

The curvature of the amplitude of the wave function determines quantum potential. It explains the wave nature by describing the behavior of atomic particles. The probability of finding the particle at a given location is given by the wave function. The Schrodinger equation directs quantum particle movement. Quantum energy levels have been calculated using the Schrodinger equation. This wave equation considers the quantum mechanical effects.

Poisson Equation

The surface potential could be obtained in the channel region by Poisson equation in the presence of significant doping concentration in Equation 11.1,

$$\frac{d^2\phi}{dy^2} = \frac{q}{\in_{mos2}}\left[p(y) - n(y) + N_D - N_A\right] \tag{11.1}$$

Here:
Φ—surface potential
q—electron charge
N_A—channel doping concentration
$p(x)$—hole concentration
$n(x)$—electron concentration

Exciton Bohr Radius

Since the particle dimension of a MoS_2 is smaller than the bulk semiconductor Bohr exciton radius, quantum effects become more prevalent. The exciton Bohr radius of the monolayer MoS_2 is given in Equation 11.2.

$$\alpha_B^* = \frac{4\pi\varepsilon\hbar^2}{\mu e^2} \tag{11.2}$$

Here:
$\hbar$ is the reduced Planck's constant
μ is the electron-hole reduced effective mass
ε is the permittivity of MoS_2
e is the electron charge.

The layer number of the film is reduced to five, and the exciton in MoS_2 films is required to experience confinement, which may result in a reduction in exciton size. The quantum confinement effect becomes more pronounced as the particle size exceeds the Bohr exciton radius. As a result, the exciton's ability to travel in the confinement direction is hampered. Quantum entanglement is deteriorating.

Bandgap

A semiconductor's size exceeds the exciton Bohr radius. The size dependence of the bandgap in Equation 11.3,

$$E^* E_g + \frac{\hbar^2}{8\mu r^2} - \frac{1.8e^2}{4\varepsilon_o \varepsilon_r} \tag{11.3}$$

E_g is the bandgap
$\hbar$ is the reduced Planck's constant
μ is the electron-hole reduced effective mass
ε is the permittivity of MoS_2
e is the electron charge.

Electron Energy State

Since the excitons in the MoS_2 nanostructure we studied are closely confined within the nano regime, the energy state is discrete. Equation 11.4 gives the analytical equation for the electron energy state.

$$E = \frac{\hbar^2 \pi^2}{2mL^2} n^2 \tag{11.4}$$

where L is the channel dimension
$\hbar$ is the reduced Planck's constant
m is the effective mass
n is the number of states.

The energy state of one of the quantum numbers has been changed from continuous to discrete in the nanoscale devices. The discrete energy state, unlike the classical explanation, depicts the quantum confinement effect caused by the channel thickness. As the number of layers are decreased, the valence band maximum (VBM) of MoS_2 has a lower energy state, whereas the conduction band minimum (CBM) has a higher energy state; both of these things can be deduced from the quantum confinement effect. The permitted energy state is a wave vector in k space in the y direction. The monolayer MoS_2 has a confinement effect due to its higher bandgap and higher electron state.

Density of States (DOS)

In Equation 11.5, the number of permitted electron states in the conduction band of MoS_2 can be calculated,

$$D(E) = g_s g_v \frac{m^*}{2\pi\hbar^2}(E - E_c) \tag{11.5}$$

where, g_s and g_v are the spin and valley degeneracy of the single-layer MoS_2, respectively.

Schrodinger Equation

1-D Schrodinger wave equation is given in Equation 11.6,

$$\frac{-\hbar^2}{2m^*}\frac{d^2\Psi}{dy^2} + (-q)U_{(y)}\Psi = E.\Psi \tag{11.6}$$

Here:
$\hbar$ (6.626×10^{-34}) is the reduced Planck's constant
$m^* = 0.48\ m_0$ is the effective mass of electrons
m_0 (9.1×10^{-31}) is the free mass of an electron
Ψ is the wave function
U(x) is the value of the surface potential, which is influenced by quantum effects.

Electron Concentration and Wave Function

The electron concentration distribution, n(y), can be calculated with the help of the wave function. $n(y)$ is given in Equation 11.7,

$$n(y) = D(E)F(E)|\psi(y)|^2 \tag{11.7}$$

Rui Cheng et al. [56] have shown the achievement of the self-aligned device with performance, including gain over 30, cutoff frequency up to 42 GHz, and maximum oscillation frequency f_{max} up to 50 GHz. The MoS_2 material-based transistor device has shown great potential for high-speed flexible electronics [56–57]. The MoS_2 material can be used to conquer the following problem of ON/OFF ratio, high OFF current, and an intrinsic voltage gain of the graphene.

An optical microscope, atomic force microscopy (AFM), and Raman spectroscopy were used to examine the surface characteristics and crystalline qualities of MoS_2. At three different temperatures, the characteristics of mechanically exfoliated monolayer MoS_2 and CVD grown MoS_2 were investigated using AFM [10]. In MoS_2 devices, the mobility significantly degrades due to phonon and roughness scattering [58]. Furthermore, enhancement of mobility and scattering of carrier have been done by using high-quality high-k dielectric instead of silicon dioxide, because high-k dielectric improves the gate capacitance without leakage effects [59].

An ultra-short-channel monolayer MoS_2 FET device defines the feature size of gate/channel length, and the self-aligned mask has been successfully served for source/drain metallization [60–62]. MoS_2 as channel material is focused on the temperature variation and its effects, while also calculating the scattering rates using deformation potential approximation method [63–65]. The use of h-BN cap layers affects the results, which can improve sensor stability and prevent degradation [66]. This device produces 30–50 cm^2/Vs of mobility and ON/OFF ratio on the order of 10^5–10^6. Figure 11.15 demonstrates the characteristics of the uncapped device and capped devices, showing that the uncapped MoS_2 was degraded 60% after two days of aging while the capped device did not degrade much within a week [67–68]. The Al_2O_3 top-gated CVD grown monolayer MoS_2 FET was first focused on the thermal and hysteresis effects [69], including thermoelectric power factor values, which are determined by the structure of this material [70]. Single-layer MoS_2 with high-k dielectric fabrication has been a complicated process, and it also has its compatibility limits [71]. Getting the mobility from MOSFET using MoS_2 [63] is obtained in Equation 11.8,

$$\mu = g_c q \frac{\Sigma_{kn} \vartheta_{kn}(E)^2 \, X \, D(E) X \tau_{kn}(E) X \dfrac{df^{o}kn}{dE}}{\Sigma_{kn} D(E) X f^{o}_{kn}(E)} \tag{11.8}$$

where μ is the electron mobility, q is the electron charge, g_c s catering the spin degeneracy of the $MoS_{2,}$ D(E) is the density of state of $MoS_{2,}$ and f^{o}_{kn} is the distribution function.

In MoS_2, the metal-semiconductor metal (MSM) device shows improved results at low temperature [72]. Identical I-V characteristics are also present that prove the mobility model [72–76]. MoS_2 FETs with Al_2O_3 top-surface passivation have been conducted with the help of the electrical characterization and low frequency noise (LFN) measurements [77–78]. Molecular beam epitaxy (MBE) technique is used to synthesize MoS_2 materials on SiO_2/Si substrate [79–83]. Quantum transport will be used to calculate the charge effect, I-V characteristics of single-layer zigzag nanoribbon MoS_2 [84–91].

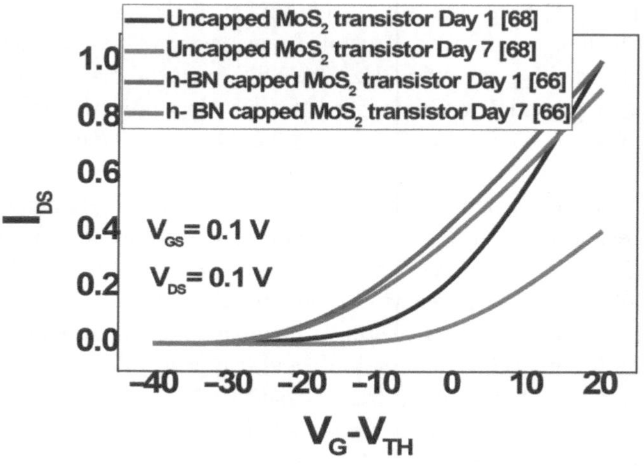

FIGURE 11.15 The Ambient Aging Comparison of Uncapped [68] and Capped Device [66]

TABLE 11.1
2H, 3R, Monolayer and Dual Layer Bandgap Details [92–95]

	2H	3R	Monolayer	Dual Layer
Bandgap	0.9 [94]	1.1 [93]	1.6 [95]	1.26 [92]
Bandgap characteristic	Indirect [94]	Indirect [93]	Direct [95]	Indirect [92]

Upcoming MoS_2 materials of FET devices are used in solar cell, lithium ion battery, memory device, photo detector, and electro catalysts due to the enhanced absorption observed in single-layer MoS_2. The density functional theory is used to learn the characteristics of MoS_2 electronic structure [92–95]. The bandgap properties of MoS_2 materials are given in Table 11.1.

11.3.2.2 Molybdenum Diselenide ($MoSe_2$) Material

The $MoSe_2$ material is grown and synthesized by chemical vapor deposition (CVD) method. In $MoSe_2$ device, sulfur vacancy and other defects somewhat affect the electrical and optoelectronic properties [96–98]. From $MoSe_2$ film, ribbon channel is defined by SU-8 patterning method. Remarkably, the narrowing of the conduction channel area with SU-8 patterning method is critical to suppress the leakage current. Perylene tetracarboxylic acid tetrapotassium (PTAS) salt has been used to encourage the atomically thin layer material growth in CVD. The attained results are ON/OFF ratio of 10^2 at 37.5 V gate overdrive, low resistance, and the mobility of 0.02 cm^2/Vs at 300 K of the $MoSe_2$ film [99–102]. "Liquid state" glass is an extremely capable substrate for minimum rate development of high-quality bulky size 2-D TMDs [103–107].

In relation to graphene, TMD material has some advantages of properties such as intrinsic bandgap and high performance in terms of carrier mobility, high ON current, ON/OFF ratio, and thermal conductivity [108–111].

11.3.2.3 Molybdenum Ditelluride ($MoTe_2$) Material

Among TMD materials, the Te-based nanometer thickness still has not been explored, particularly molybdenum ditelluride ($MoTe_2$) crystal. The $MoTe_2$ materials were grown by chemical vapor transport method that yields 2H-$MoTe_2$ crystal with lateral dimensions. Lezama et al. have reported the above 4 nm thickness of $MoTe_2$ material and concentrated on surface transport as well as gap structure of quantitative calculation [112]. In optoelectronic applications, $MoTe_2$ can play a powerful role in improving the potential of materials. The results of subthreshold swing of electron and hole, respectively, S = 140 mV/dec and S = 125 mV/dec were examined [113]. Moreover, the measurement of the optical transmission of $MoTe_2$ having different thickness of 20 μm, 120 nm, and 32 nm were determined by AFM. $MoTe_2$ and MoS_2 materials have been used in van der Waals heterostructure for the band-to-band tunneling (BTBT) [113]. MoS_2 and $MoTe_2$ heterojunction have exhibited the improvement of BTBT correlated to homojunction applying density functional theory (DFT) simulations. To calculate the field-effect mobility, Equation 11.9 is used,

$$\mu = \frac{L}{W} X \frac{\partial I_{ds}}{\partial \left(C_{tg} V_{tg} \right)} \tag{11.9}$$

where L represents length of the channel, W represents the width of the channel, I_{ds} is the drain current, V_{tg} is the gate voltage, and C_{tg} is the gate capacitance.

11.3.2.4 Tungsten Ditelluride (WTe_2) Material

The transmission valley is determined in the conduction band owing to the distinct electronic structure of tungsten ditelluride (WTe_2). This structure produces a huge negative differential resistance at topmost valley ratio of 10^3 for the proper selection of doping, gate as well as supply voltages [114–115]. Figure 11.6 shows the schematic structure of DG WTe2 TFET. In this structure, HfO_2 has been widely used as the gate insulator. Here, gate oxide of HfO_2 that has 2 nm thickness with a dielectric constant of 25 is presented. WTe_2 can be used as channel material because of its properties of smaller bandgap and carrier effective mass. These properties of WTe_2 are superior to the other TMD materials such as $MoSe_2$, $MoTe_2$, WS_2, and WSe_2 [116–121]. Drain current is calculated in monolayer WTe_2 TFET at different gate voltages. The drain current [115] is defined in Equation 11.10,

$$I_D = \frac{2e}{hL} \int dE \sum\nolimits_k T(E,k)\left[f_s(E) - f_D(E) \right] \tag{11.10}$$

where the channel length is L, the transmission coefficient is $T\,(E,\,k)$, f_s is the source Fermi distribution function, and f_d is the drain Fermi distribution function.

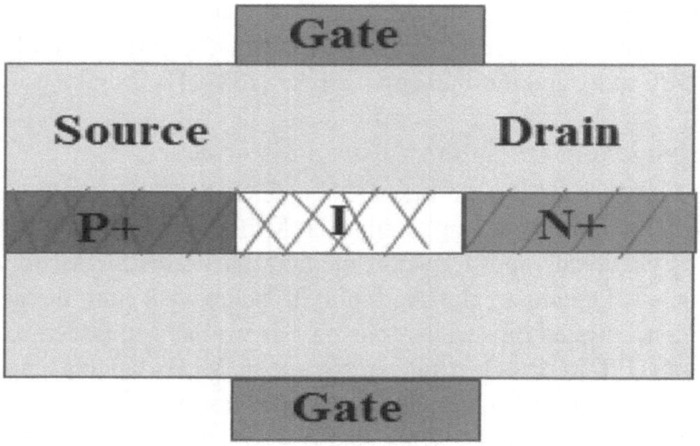

FIGURE 11.16 Structure of Double-Gate WTe2 TFET [115]

11.3.2.5 Tungsten Disulfide (WS_2) Material

Tungsten disulfide belongs to the transition metal dichalcogenide, which are 2-D semiconductor nanomaterials. Bulk WS_2 forms dark gray hexagonal crystals with a layered structure like MoS_2. It exhibits properties of a dry lubricant. The nonequilibrium Green's function formalism technique that has been used for the investigation of tungsten disulfide (WS_2) performance and enhancement of drive current through the heterojunction is reviewed. A heterojunction is described as an interface that arises between the two layers of different semiconductors. An order for both n-type and p-type TFET is used to improve the ON current and ON/OFF current ratio [122]. Carbon materials and TMD (MX_2) materials have been used for the development of electrostatic gate control. In p-type MoS_2–WS_2 FET structure, MoS_2 was used as source region material and WS_2 was used as the channel and drain region material. This structure has eight orders of ON/OFF current ratio along with 12.1 μA/μm ON current. Computational efficiency is calculated through Green's function method. Moreover, the potential profile is attained with 2-D Poisson's equation [123–124].

The advanced nano transistors have some major challenges like supply voltage and power consumption. As a result of the scaling of transistor size, the voltages cannot be reduced. When the supply voltage is minimized, accordingly, rapid transition occurs between ON/OFF state [125]. Gate tunable thermionic TFET is working under low supply voltage as 0.5V. Different approaches of concurrent tunneling of thermionic and tunneling components can be used to overcome the important issues of this TFET device [126–129].

11.3.2.6 Tungsten Diselenide (WSe_2) Material

Single-layer WSe_2 has an attractive property like direct bandgap in 2-D semiconductor [130–132]. In this paper [133], the density functional theory (DFT) calculation is used for the support of device measurements. Dielectric of Al_2O_3 is deposited on top of WSe_2 structure, which can improve the performance [134–136].

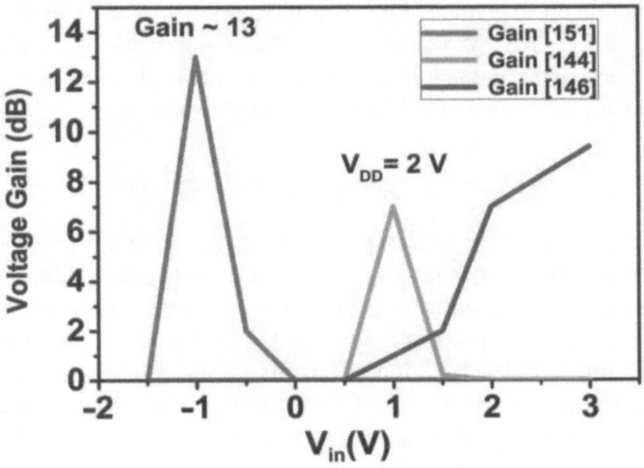

FIGURE 11.17 Voltage Gain of the WSe_2 Inverter [144, 151] and Graphene-on-Silicon FET [146]

Performance of WSe_2 device is less affected by phonon scattering that displays a performance parameter of ballisticity of 83% for 10 nm channel length [137–139]. MoS_2/WSe_2 heterostructures have been used in dual gate architecture, resulting in 80% of gate coupling efficiency. MoS_2/WSe_2 heterostructures depict the importance of current transport in van der Waals semiconductor heterostructures through forward bias operation mode van der Waals heterostructures in tunneling transistor can be used for the low power electronics in future application [140–143]. WSe_2 complementary FET paper has attained an ON/OFF current ratio higher than 10^4 for both n and p-type FETs [144, 145]. Different electrodes in ambipolar device characteristics of WSe_2 FET are used to improve the efficiency of both electron and hole injections [146–153]. We compared the voltage gain of the WSe2 and graphene material as shown in Figure 11.17. WSe_2-based inverter device exhibits the voltage gain 13 dB, defined as the negative of dV_{out}/dV_{in} [151], the peak dc voltage gain of 7 dB for V_{DD} is 2 V [144], and graphene-on-silicon FET exhibits a maximum voltage gain of 9 dB at a load resistance [146]. Highly doped graphene can be used in FET contact, which is a versatile, flexible, and transparent low-resistance ohmic contact [154–157].

WSe_2 p-FET has produced a high value of hole mobility and perfect subthreshold swing at room temperature [158–160]. Even though van der Waals heterostructure has reported the new approach of 2-D/2-D low-resistance ohmic contacts of the new device, which overcomes an important bottleneck in the performance [161–163], still, fabrication of air-stable transistors with CMOS technology is a challenging an issue that needs to be overcome [164–166].

In this chapter, the past of 2-D materials-based FET device has been reviewed with existing works. The characteristics of TMD materials have been illustrated, their types have been discussed, and the structures of 2-D and bulk materials have been differentiated. Here, 2-D material is fabricated using chemical vapor deposition method that has been explained. First, we have discussed 2-D TMD materials of MoS_2, $MoSe_2$, $MoTe_2$, WS_2, WSe_2, and WTe_2. Additionally, the material's properties, I-V characteristics, and bandgap details as well as material-based devices like FET, MOSFET, FinFET, and tunnel FET have also been discussed.

11.4 RESEARCH OUTCOMES OF CURRENT STUDY

This section describes the analysis of previously obtained results. Moreover, the I-V characteristics and comparison tables have also been considered. Matin Amani et al. have reported that the chemical vapor deposition was used to grow the molybdenum disulfide (MoS_2) [167] and graphene. CVD is better than the remaining techniques of mechanical and liquid exfoliation due to its large-scale, high quality, and low cost. The gate voltage vs drain current for both depletion and enhancement mode devices and field-effect mobility as a function of gate voltage for three different devices are shown in Figure 11.18. This figure clearly indicates the perceive current saturation at higher values of V_{DS}, which is anticipated with long channel FETs [167–170].

Fan Chen et al. have reported that the WTe_2-MoS_2 grouping was selected because the creation of broken gap heterojunction was looked for in TFET. Quantum transport simulation is used for this device structure [171]. This structure has two geometry parameters of an overlap region and an extension region such as $L_{overlap}$ and L_{ext}, respectively. The values were optimized at 30 nm and 15 nm, correspondingly [171]. When L_{ext} has zero value, high leakage current will occur. Figure 11.19 shows -0.3 V supply voltage and high ON current of 1,000 uA/um [172–174].

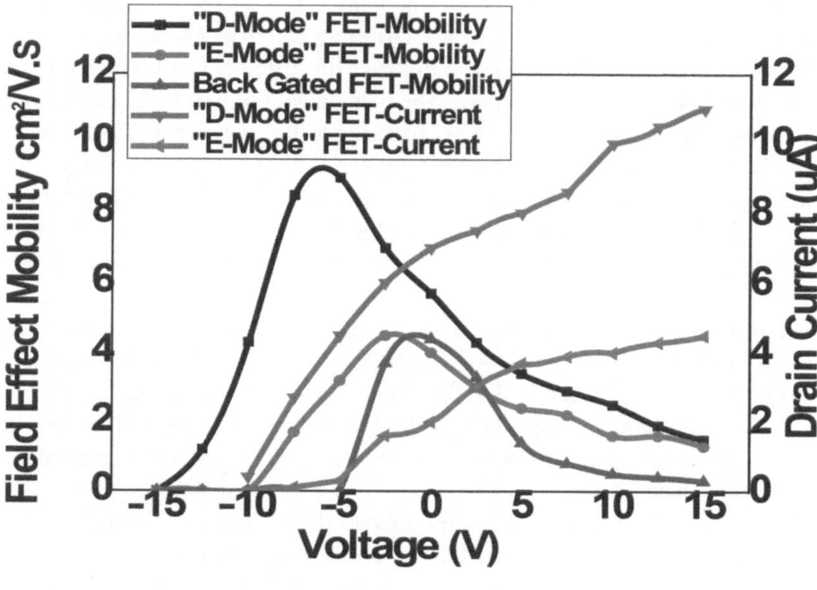

FIGURE 11.18 Field-Effect Mobility and Drain Current vs Voltage [167]

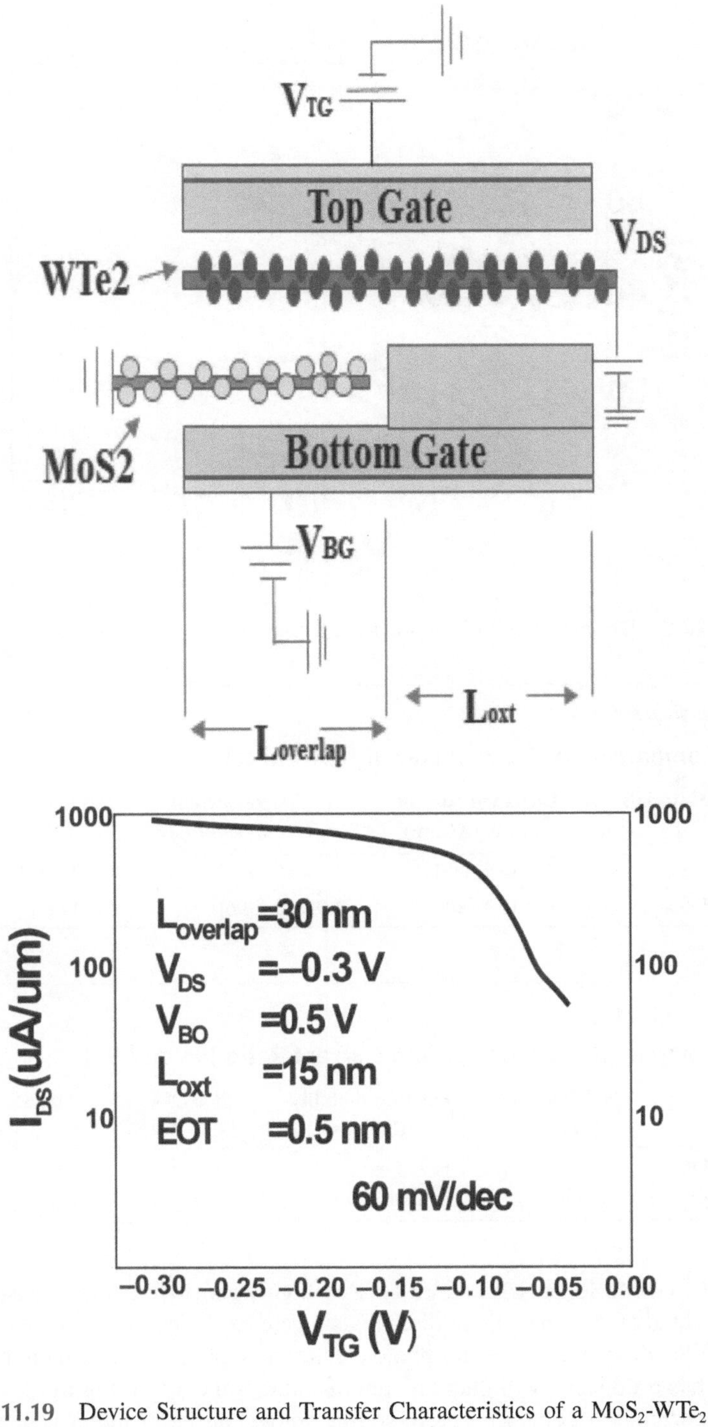

FIGURE 11.19 Device Structure and Transfer Characteristics of a MoS_2-WTe_2 Interlayer TFET [171]

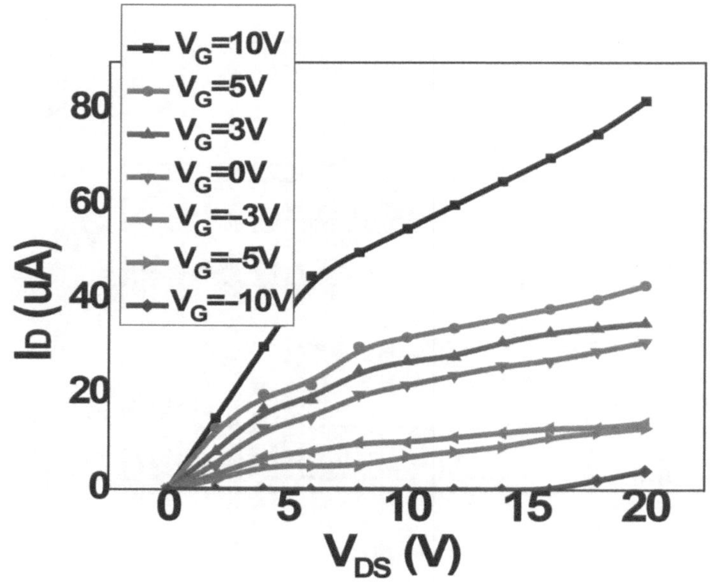

FIGURE 11.20 Drain-Source Voltage vs Drain Current [175]

TABLE 11.2
Comparison of Current Density [96 and 113]

Materials	Exchange current density (A/Cm²)	Exchange current per site (A/Site)	TOF (s⁻¹)
MoS_2	2.2×10^{-6} [113]	4.1×10^{-21}[113]	0.013 [113]
$MoSe_2$	2.0×10^{-6} [96]	4.5×10^{-21}[96]	0.014 [96]

TABLE 11.3
Comparison of Mobility and Current Ratio [96 and 97]

	Hole mobility (Cm^2/ Vs)	Electron mobility (Cm^2/ Vs)	ON/OFF	Type
MoS_2		~17	10^4–10^5	n-type
$MoSe_2$	~15 [96]	~23 [96]	10^4–10^5[96]	Ambipolar

Few-layered $MoSe_2$ back-gated FET-based highly sensitive phototransistors have been fabricated [175]. The silicon dioxide as a gate oxide is a limiting factor in this device. $MoSe_2$ devices are used for photo electronic applications in which the phototransistor has produced the higher amount of saturation current. Figure 11.20 shows the drain current variation with drain-source voltage at the different values of gate voltage. In low power application, the high threshold voltage is an unsuitable one.

TABLE 11.4
Bandgap Energy of Different Materials [92, 112, 115, and 120]

	MoS_2	WS_2	$MoSe_2$	WSe_2	$MoTe_2$	WTe_2
E_g (eV)	1.66 [92]	1.81	1.43	1.54	1.07 [112]	1.07 [120]
E_p (eV)	0.56	0.77	0.49	0.61	0.35	0.53
E_w (eV)	0.13	0.05	0.05 [115]	0.11	0.20	0.26

TABLE 11.5
ON Current, Bandgap, Reduced Effective Mass, Scaling Length, Effective Scaling Length, and η Values [46, 92, and 120]

Material	ION	Eg	mr*	Λ	Λ	H
WTe_2	127	0.75 [120]	0.17	0.45	2.45	3.15 [46]
WSe_2	4.6	1.56	0.21	0.41	2.5	5.2
WSe_2	2.3	1.08	0.32	0.5	2.7	5.8
MoS_2	0.3	1.68 [92]	0.29	0.38	2.5	6.3

Therefore, using high k dielectric material instead of SiO2 as a gate oxide will reduce the threshold voltage and improve the ON/OFF current ratio [176–178].

This study shows that the TMD material has been reviewed with experimental results. The comparison Table 11.2 and Table 11.3 show the MoS_2 and $MoSe_2$ parameter values such as mobility, ON/OFF ratio, current density, and turn of frequency [179–193]. The comparison of various kinds of bandgap energy for different 2-D materials is shown in Table 11.4. The performance results of WTe_2, WSe_2, WSe_2, and MoS_2 materials are shown in the parameters of ON current, bandgap, reduced effective mass, scaling length, effective scaling length, and η value in Table 11.5 [194–200].

Here, specific surface area-based normalized kinetic current is presented as the current. Fitting the linear portion of Tafel is used for the measurement of exchange current density. In addition, the transmission electron microscopy (TEM) examines the edge site density, every site of edge, yielding turn over frequency (TOF) for MoS_2 and $MoSe_2$. Here, MoS_2 current density is better than the $MoSe_2$ material. TOF and exchange current per site of MoS_2 values are minimum.

High ON/OFF current ratio for both MoS_2 and $MoSe_2$ has been obtained. $MoSe_2$ presents the similar electron mobility and ON/OFF ratio like MoS_2. For this reason, $MoSe_2$ is used for the FET application. Both materials have the same ON/OFF ratios, but holes and electron mobilities are different.

E_g is bandgap energy, E_p is the energy difference between conductance valley energy minimum and conduction band minimum, and Ew is conductance valley width. This result exhibits transmission valley that is responsible for the presence of negative differential resistance (NDR) in TMD material in the form of MX_2. Monolayer TMD TFET is quite common and not limited to the exact band structure of WTe_2 due to the NDR effect. The comparison table shows that the $MoTe_2$ and WTe_2 materials have lower bandgap energy although high bandgap material is better.

WTe_2 TFET provides higher performance of ON current and subthreshold swing, where WTe_2 has smallest bandgap and effective mass compared to other materials. The MoS_2 material is used for tunneling applications and ultra-scale MOSFET applications because it has a high effective mass. Furthermore, the MoS_2 material to suppress the source-drain tunneling due to high effective mass. Effective masses of electron and hole, bandgap, body thickness, source-to-channel potential difference, and source and drain doping level parameters are considered for ON current state of a TFET device [46] in Equation 11.11,

$$I \,\alpha \exp\left(\frac{-4\sqrt{2m_r^{*}\,E_g^{\,3/2}}}{3q\hbar E}\right) \tag{11.11}$$

where E_g is the bandgap energy, q is the charge of electron, E is the electric field, and $\hbar$ is the reduced Planck's constant.

From the existing results, the parameters of 2-D materials like current density, mobility, current ratio, energy bandgap, scaling length, reduced effective mass effective scaling length, and η values have been analyzed and compared. To conclude, the van der Waals heterostructure materials are better than the homostructure material. In future, gate engineering technique will be used for research work.

11.5 CONCLUSION AND FUTURE SCOPE

In this chapter, the importance of 2-D TMD material has been proposed. The 2-D material of molybdenum disulphide (MoS_2) will be enhanced because of its high carrier mobility, high ON/OFF current ratio, low leakage current, low short-channel effect, high thermal conductivity, and high performance of low power operating voltage. The existing result shows the I-V characteristics of perceived saturation current in MoS_2, high ON current in heterojunction of MoS_2-WTe_2, and high saturation current in $MoSe_2$. In FET technology, the tunnel FET is an upcoming device because it has an ultra low power voltage. TMD materials are most suitable for tunnel FET due to their low dielectric constant and atomically sharp and self-passivated interfaces. Chemical vapor deposition (CVD) technology is preferred over other technologies for high yield and low cost of 2-D material preparation. Applying high-k dielectric material as the gate oxide instead of SiO_2 because of high k-dielectric can reduce the threshold voltage and enhance the ON/OFF current ratio. The combination of MoS_2 and $MoTe_2$ heterojunction has been used in tunnel FET device for low standby power (LSTP) applications, which replace the MOSFET technology. No wonder MoS_2 channel material is an attractive candidate for future ultra low power applications.

The new material of molybdenum disulfide material device makes a suitable prospect for low power applications due to its enhanced carrier injection, reduced threshold voltage, and decreased contact resistance. Molybdenum disulfide material-based device will be suitable for biomedical applications, because it has direct bandgap, easy-to-fabricate, and sensing layer property. MoS_2 device performs at gigahertz frequencies, with high current drives. This operating frequency proves the practicability of fabricating the device, as it will be suitable for high-frequency applications.

ACKNOWLEDGMENT

The authors are grateful to the Centre for VLSI Design, Department of Electronics and Communication Engineering, Kalasalingam Academy of Research and Education for supporting this research work.

REFERENCES

[1] www.talkingelectronics.com/Download%20eBooks/Principles%20of%20electronics/CH19.pdf.

[2] Parvais, Bertrand, Thomas Chiarella, Nadine Collaert, R. Rooyackers, "*FinFETs and their futures*" (Springer, Berlin) pp. 141–153, 2011.

[3] Turkane, Satish M. and A. K. Kureshi, "Review of tunnel field effect transistor (TFET)" *International Journal of Applied Engineering Research,* Vol. 11, No. 7 pp. 4922–4929, 2016, ISSN 0973-4562

[4] Boucart, Kathy and Adrian Mihai Ionescu, "Double-gate tunnel FET with high-κ gate dielectric" *IEEE Transactions on Electron Devices*, Vol. 54, No. 7, July 2007.

[5] Charles Pravin, J., D. Nirmal, P. Prajoon, and J. Ajayan, "Implementation of nanoscale circuits using dual metal gate engineered Nanowire MOSFET with high-k dielectrics for low power applications" *Physica E: Low-Dimensional Systems and Nanostructures,* Vol. 83, pp. 95–100, 2016.

[6] Arns, Robert G. "The other transistor: Early history of the metal-oxide-semiconductor field-effect transistor" *Engineering Science And Education Journal*, Vol. 7, No. 5, pp. 233–240, Oct. 1998.

[7] Charles Pravin, J., D. Nirmal, P. Prajoon and M. Anuja Menokey, "A new drain current model for a dual metal junctionless transistor for enhanced digital circuit performance" *IEEE Transactions on Electron Devices*, Vol. 63, No. 9, Sept. 2016.

[8] Li, Xiao and Hongwei Zhu, "Two-dimensional MoS2: Properties, preparation, and applications" *Journal of Materiomics,* Vol. 1, No. 1, pp. 33–44, 2015.

[9] Wang, Shanshan, Xiaochen Wang, and Jamie H. Warner, "All chemical vapor deposition growth of MoS_2:h-BN vertical van der Waals heterostructures" *ACS Nano*, Vol. 9, No. 5, pp. 5246–5254, 2015.

[10] Noor Faizah, Z. A., I. Ahmad, P. J. Ker, P. S. AkmaaRoslan, and A. H. Afifah Maheran, "Modeling of 14 nm gate length n-Type MOSFET" *IEEE Regional Symposium on Micro and Nanoelectronics*, pp. 1–4, 2015.

[11] Nichols, B. M., A. L. Mazzoni, M. L. Chin, P. B. Shah, S. Najmaei, R. A. Burke, M. Dubey, "Advances in 2D materials for electronic devices" *Semiconductor and Semimetals*, Vol. 95, pp. 221–227, 2016.

[12] Aguirre-Morales, J.-D., S. Fregonese, C. Mukherjee, C. Maneux, and T. Zimmer, "An accurate physics-based compact model for dual-gate bilayer graphene FETs" *IEEE Transaction Electron Devices*, Vol. 62, pp. 4333–4339, 2015.

[13] Akinwande, D., N. Petrone, and J. Hone. "Two-dimensional flexible nanoelectronics" *Nature Communication,* Vol. 5, p. 5678, 2014.

[14] Alam, A. U., K. D. Holland, M. Wong, S. Ahmed, D. Kienle, and M. Vaidyanathan, "RF linearity performance potential of short-channel graphene field-effect transistors" *IEEE Transaction Microwave Theory Techniques*, Vol. 63, pp. 3874–3887, 2015.

[15] Gupta, Ankur, Tamilselvan Sakthivel, and Sudipta Seal, "Recent development in 2D materials beyond graphene" *Progress in Material Science*, Vol. 73, pp. 44–126, 2015.

[16] Perera, M. M., M-W. Lin, H-J. Chuang, B. P. Chamlagain, C. Wang, X. Tan, et al. "Improved carrier mobility in few-layer MoS_2 field-effect transistors with ionic-liquid gating" *ACS Nano*, Vol. 7, pp. 4449–4458, 2013.

[17] Li, H., Z. Yin, Q. He, H. Li, X. Huang, G. Lu, et al. "Fabrication of single- and multilayer MoS2 film-based field-effect transistors for sensing NO at room temperature" *Small*, Vol. 8, pp. 63–67, 2012.

[18] Chen, Chang-Hsia, Chun-Lan Wu, Jiang Pu, Ming-Hui Chiu, Pushpendra Kumar, Taishi Takenobu, and Lain-Jong Li, "Hole mobility enhancement and p-doping in monolayer WSe2 by gold decoration" *2D Materials*, Vol. 1, No. 1, 2014.

[19] Novoselov, K. S., V. I. Falko, L. Colombo, P. R. Gellert, M. G. Schwab, and K. Kim, "A roadmap for graphene" *Nature*, Vol. 490, pp. 192–200, 2012.

[20] Chen, C. H., C. T. Lin, Y. H. Lee, K. K. Liu, C. Y. Su, W. Zhang, and L. J. Li, "Electrical probing of submicroliter liquid using graphene strip transistors built on a nanopipette" *Small*, Vol. 8, pp. 43–46, 2012.

[21] McDonnell, Stephen J. and Robert M. Wallace, "Atomically-thin layered films for device applications based upon 2D TMDC materials" *Thin Solid Films*, Vol. 616, pp. 482–501, 2016.

[22] Novoselov, K., D. Jiang, F. Schedin, T. Booth, V. Khotkevich, S. Morozov, and A. Geim, "Two-dimensional atomic crystals" *Proceedings of the National Academy of Sciences of the United States of America,* Vol. 102, pp. 10451–10453, 2005.

[23] Geim, A. K. and I. V. Grigorieva, "Van der Waals heterostructures" *Nature*, Vol. 499, pp. 419–425, 2013.

[24] https://electronics.stackexchange.com/questions/102378/why-is-bjt-scaling-not-as-significant-as-mosfet-scaling.

[25] Huang, C., M. Marshall, and B. H. White, "Field effect transistor applications" *Transactions of the American Institute of Electrical Engineers, Part I: Communication and Electronics*, Vol. 75, No. 3, pp. 323–329, Aug. 1956.

[26] Shockley, W. "A unipolar field effect transistor" *Proceedings, Institute of Radio Engineers, New York, N. Y.*, Vol. 40, pp. 1365–76, Nov. 1952.

[27] Dacey, G. C. and I. M. Ross. "Unipolar 'field effect' transistor" Vol. 41, pp. 970–79, Aug. 1953.

[28] "News of radio" column, 1st July 1948, *The New York Times*, p. 46.

[29] Garner, L. E., Jr. "Transistors and their applications in television-radio electronics" Coyne Electrical School, Chicago, p. 36, 1953.

[30] Martino, Joao A., Paula G. D. Agopian, Eddy Simoen and Cor Claeys, "Field effect transistors: From MOSFET to tunnel-FET analog performance perspective" *12th IEEE International Conference on Solid-State and Integrated Circuit Technology (ICSICT),* pp. 1–4, 2014.

[31] Zhang, Lining, Chenyue Ma, Jin He, Xinnan Lin, Mansun Chan, "Analytical solution of subthreshold channel potential of gate underlap cylindrical gate-all-around MOSFET" *Solid-State Electronics*, Vol. 54, No. 8, pp. 806–880, 2010.

[32] Charles Pravin, J., D. Nirmal, P. Prajoon, N. Mohan Kumar, and J. Ajayan, "Investigation of 6T SRAM memory circuit using high-k dielectrics based nano scale junctionless transistor" *Superlattices and Microstructures*, Vol. 104, pp. 470–476, 2017.

[33] Ajayan, J., D. Nirmal, P. Prajoon, and J. Charles Pravin, "Analysis of nanometer-scale InGaAs/InAs/InGaAs composite channel MOSFETs using high-K dielectrics for high speed applications" *International Journal of Electronics and Communication,* Vol. 79, pp. 151–157, 2017.

[34] Kahng, D., "Historical perspective on the development of MOS transistors and related devices" *IEEE Transactions on Electron Devices*, Vol. 123, No. 7, pp. 655–657, 1976.

[35] Faggin, F., T. Klein, and L. Vadasz, "Silicon gate technology developed for ICs" *IEEE Transactions on Electron Devices*, Vol. 16, No. 2, p. 236, 1969.

[36] Colinge, J. P. "*Silicon-on-insulator technology: Materials to VLSI*" (Springer Science & Business Media, Berlin), 2004.

[37] Charles Pravin, J., P. Prajoon, Flavia Princess Nesamania, P. Senthil Kumar, D. Nirmal, and G. Srikesh, "Nanoscale high-k dielectrics for junctionless nanowire transistor for drain current analysis" *Journal of Electronic Materials,* Vol. 47, No. 5, pp. 2679–2686, 2018.

[38] Choi, W. Y., B.-G. Park, I. D. Lee, and T.-1. K. Liu, "Tunneling field-effect transistors (TFETs) with subthreshold swing (SS) less than 60 mV/dec" *IEEE Electron Device Letters*, Vol. 28, No. 8, pp. 743–745, 2007.

[39] Razavi, P., G. Fagas, I. Ferain, N. Dehdashti Akhavan, R. Yu, and J. P. Colinge, "Performance investigation of short-channel junctionless multigate transistors" *Ulis 2011 Ultimate Integration on Silicon,* pp. 1–3, 2011.

[40] Cao, Wei, Jiahao Kang, Deblina Sarkar, Wei Liu, and Kaustav Banerjee, "2D semiconductor FETs—projections and design for sub-10 nm VLSI" *IEEE Transactions on Electron Devices*, Vol. 62, No. 11, pp. 3459–3469, 2015.

[41] Qi, Haimei, Lina Wang, Jie Sun, Yi Long, Peng Hu, Fucai Liu, and Xuexia He, "Production methods of van der Waals heterostructures based on transition metal dichalcogenides" *Crystals,* Vol. 8, No. 1, pp. 1–17, 2018.

[42] www.manchester.ac.uk/discover/news/major-award-for-graphene-and-2d-materials-in-biomedicine/.

[43] Yanga, Guohai, Chengzhou Zhub, Dan Dub, Junjie Zhua, and Yuehe Lin, "Graphene-like two-dimensional layered nanomaterials: Applications in biosensors and nanomedicine" *Nanoscale,* Vol. 7, No. 34, pp. 14217–14231, 2015.

[44] Radisavljevic, B., A. Radenovic, J. Brivio, V. Giacometti, and A. Kis, "Single-layer MoS2 transistors" *Nature Nanotechnology*, Vol. 6, pp. 147–150, Jan. 2011.

[45] Liu, L., Y. Lu, and J. Guo, "On monolayer MoS2 field-effect transistors at the scaling limit" *IEEE Transaction Electron Devices*, Vol. 60, No. 12, pp. 4133–4139, Dec. 2013.

[46] Ilatikhameneh, Hesameddin, Yaohua Tan, Bozidar Novakovic, Gerhard Klimeck, Rajib Rahman, and Joerg Appenzeller, "Tunnel field-effect transistors in 2D transition metal dichalcogenide materials" *IEEE Journal on Exploratory Solid-State Computational Devices and Circuits,* Vol. 1, pp. 12–18, 2015.

[47] Appenzeller, J., Y.-M. Lin, J. Knoch, and Ph. Avouris, "Band-to-band tunneling in carbon nanotube field-effect transistors" *Physical Review Letters*, Vol. 93, p. 196805, 2004.

[48] Lin, Y., D. Xie, Y. Chen, T. Feng, Q. Shao, H. Tian, T. Ren, X. Li, X. Li, L. Fan, K. Wang, D. Wu, and H. Zhu, "Optimization of graphene/silicon heterojunction solar cells" *Proceedings of 38th IEEE Photovoltaic Specialists Conference (PVSC),* 2012, pp. 2566–2570.

[49] Zhang, W., J. K. Huang, C. H. Chen, Y. H. Chang, Y. J. Cheng, and L. J. Li, "High-gain phototransistors based on a CVD MoS_2 monolayer" *Advanced Material (Weinheim, Ger.)*, Vol. 25, pp. 3456–3461, 2013.

[50] Desai, S. B., S. R. Madhvapathy, M. Amani, D. Kiriya, M. Hettick, M. Tosun, Y. Zhou, M. Dubey, J. W. Ager, D. Chrzan, and A. Javey, "Gold-mediated exfoliation of ultra large optoelectronically-perfect monolayers" *Advanced Material (Weinheim, Ger.),* Vol. 28, pp. 4053–4058, 2016.

[51] Xia, J., X. Huang, L.-Z. Liu, M. Wang, L. Wang, B. Huang, D.-D. Zhu, J.-J. Li, C.-Z. Gu, and X.- M. Meng, "CVD synthesis of large-area, highly crystalline $MoSe_2$ atomic layers on diverse substrates and application to photodetectors" *Nanoscale,* Vol. 6, pp. 8949–8955, 2014.

[52] Sridevi, R. and J. Charles Pravin, "Investigation of quantum confinement effects on molybdenum disulfide (MoS2) based transistor using Ritz Galerkin finite element technique" *Silicon*, pp. 1–7, 2021.

[53] Sridevi, R. and J. Charles Pravin, “Lowering the Schottky barrier height by titanium contact for high-drain current in mono-layer MoS2 transistor” *Journal of Electronic Materials*, Vol. 50, No. 6, pp. 3295–3301, 2021.

[54] Yang, Lei, Ping Liu, Jing Li, and Bin Xiang, “Two-dimensional material molybdenum disulfides as electrocatalysts for hydrogen evolution” *Catalysts*, Vol. 7, No. 10, pp. 1–18, 2017.

[55] Sridevi, R. and J. Charles Pravin, “High performance double gated molybdenum disulfide (MoS2) transistor for low power applications” *IEEE International Conference on Clean Energy and Energy Efficient Electronics Circuit for Sustainable Development (INCCES)*, pp. 1–3, 2019.

[56] Cheng, Rui, Shan Jiang, Yu Chen, Yuan Liu, Nathan Weiss, Hung-Chieh Cheng, Hao Wu, Yu Huang and Xiangfeng Duan, “Few-layer molybdenum disulfide transistors and circuits for high-speed flexible electronics” *Nature Communications*, Vol. 5, No. 1, pp. 1–9, 2014.

[57] Wang, Q. H., K. Kalantar-Zadeh, A. Kis, J. N. Coleman, and M. S. Strano, “Electronics and optoelectronics of two-dimensional transition metal dichalcogenides” *Nature Nanotechnology*, Vol. 7, pp. 699–712, 2012.

[58] Ma, Nan and Debdeep Jena, “Charge scattering and mobility in atomically thin semiconductors” *Physical Review X*, Vol. 4, No. 1, p. 011043, 2014.

[59] Ky, Dinh Le Cao, Bien-Cuong Tran Khac, Chinh Tam Le, Yong Soo Kim, Koo-Hyun Chung, “Friction characteristics of mechanically exfoliated and CVD grown single-layer MoS_2” *Friction*, Vol. 6, No. 4, pp. 395–406, 2017.

[60] Cao, Wei, Wei Liu, Jiahao Kang, and Kaustav Banerjee, “An ultra-short channel monolayer MoS_2 FET defined by the curvature of a thin nanowire” *IEEE Electron Device Letter*, Vol. 37, No. 11, pp. 1497–1500, 2016.

[61] Kang, J., W. Liu, D. Sarkar, D. Jena, and K. Banerjee, “Computational study of metal contacts to monolayer transition-metal dichalcogenide semiconductors” *Physical Review X,* Vol. 4, No. 3, p. 031005, July 2014.

[62] Allain, A., J. Kang, K. Banerjee, and A. Kis, “Electrical contacts to two-dimensional semiconductors” *Nature Materials*, Vol. 14, pp. 1195–1205, Nov. 2015, DOI: 10.1038/nmat4452.

[63] Tiwaria, Sabyasachi, Subhashish Dolaia, H. Rahamana, and P. S. Guptaa, “Effect of Temperature & phonon scattering on the Drain current of a MOSFET using SL-MoS_2 as its channel material” *Superlattices and Microstructures*, Vol. 111, pp. 912–921, 2017.

[64] Sarkar, D., W. Liu, X. Xie, A. C. Anselmo, S. Mitragotri, and K. Banerjee, “MoS_2 field effect transistor for next-generation label-free biosensors” *ACS Nano*, Vol. 8, No. 4, pp. 3992–4003, 2014.

[65] Yoon, Y., K. Ganapathi, and S. Salahuddin, “How good can monolayer MoS_2 transistors be?” *Nano Letters*, Vol. 11, No. 9, pp. 3768–3773, 2011.

[66] Liu, G., S. L. Rumyantsev, C. Jiang, M. S. Shur, and A. A. Balandin, “Selective gas sensing with h-BN capped MoS2 heterostructure thin film transistors” *IEEE Electron Device Letters*, Vol. 36, No. 11, pp. 1202–1204, 2015.

[67] Schedin, F., A. K. Geim, S. V. Morozov, E. W. Hill, P. Blake, M. I. Katsnelson, and K. S. Novoselov, “Detection of individual gas molecules adsorbed on graphene” *Nature Materials*, Vol. 6, No. 9, pp. 652–655, July 2007.

[68] Samnakay, R., C. Jiang, S. L. Rumyantsev, M. S. Shur, and A. A. Balandin, “Selective chemical vapor sensing with few-layer MoS_2thinfilm transistors: Comparison with graphene devices” *Applied Physics Letters,* Vol. 106, No. 2, pp. 023115–1–5, Jan. 2015.

[69] Shah, B., M. Amani, M. L. Chin, T. P. O’Regan, F. J. Crowne, and M. Dubey, “Analysis of temperature dependent hysteresis in MoS_2 field effect transistors for high frequency applications” *Solid-State Electronics*, Vol. 91, pp. 87–90, 2013.

[70] Hippalgaonkar, Kedar, Ying Wang, Yu Ye, Diana Y. Qiu, Hanyu Zhu, Yuan Wang, Joel Moore, Steven G. Louie, and Xiang Zhang, "High thermoelectric power factor in two-dimensional crystals of MoS2" *Physical Review B*, Vol. 95, pp. 115407–115409, 2017.

[71] Kim, Sunkook, Aniruddha Konar, Wan-sik Hwang, Jong Hak Lee, Jiyoul Lee, Jaehyun Yang, Changhoon Jung, Hyoungsub Kim, Ji-BeomYoo, Jae-Young Choi, Yong Wan Jin, sang Yoon Lee, Debdeep Jena, Woong Choi, and Kinam Kim, "High-mobility and low-power thin-film transistors based on multilayer MoS_2 crystals" *Nature Communications*, Vol. 3, No. 1, pp. 1–7, 2012.

[72] Zhang, Shan, Hongbin Pu, and Yong Yang, "Physical models for MoS_2 and their application to simulations of MoS2 MSM device" *IOP Conference Series: Materials Science and Engineering*, Vol. 284, No. 1, 2017.

[73] Lombardi, C., Manzini, S., Saporito, A., et al. "A physically based mobility model for numerical simulation of nonplanar devices" *IEEE Transactions on Computer-Aided Design of Integrated Circuits and Systems*, Vol. 7, No. 11, pp. 1164–1171, 1988.

[74] Reggiani, S., Valdinoci, M., Colalongo, L., et al. "A unified analytical model for bulk and surface mobility in Si n-and p-Channel MOSFET's" *IEEE Solid-State Device Research Conference, Proceeding of the 29th European*, Vol. 1, pp. 240–243, 1999.

[75] Schwarz, S. and Russek, S. E. "Semi-empirical equations for electron velocity in silicon: Part II—MOS inversion layer" *IEEE Transactions on Electron Devices*, Vol. 30, No. 12, pp. 1629–1633, 1983.

[76] Klaassen, D. B. M. "A unified mobility model for device simulation" *IEEE Electron Devices Meeting. IEDM'90. Technical Digest. International*, Vol. 1991, pp. 357–360, 1990.

[77] Sharma, Deepak, Matin Amani, Abhishek Motayed, Pankaj B. Shah, A. Glen Birdwell, Sina Najmaei, Pulickel M Ajayan, Jun Lou Madan Dubey, Qiliang Li, and Albert V. Davydov, "Electrical transport and low-frequency noise in chemical vapor deposited single-layer MoS2 devices" *Nanotechnology*, Vol. 25, No. 15, p. 155702, 2014.

[78] Wu, Y. Q., Y. M. Lin, A. A. Bol, K. A. Jenkins, F. N. Xia, D. B. Farmer, Y. Zhu, and P. Avouris, "High-frequency, scaled graphene transistors on diamond-like carbon" *Nature*, Vol. 472, pp. 74–78, 2011.

[79] Zhan, Linjie, Wen Wan, Zhenwei Zhu, Tien-Mo Shih, and Weiwei Cai, "MoS_2 materials synthesized on SiO_2/Si substrates via MBE" *Journal of Physics: Conference Series*, Vol. 864, No. 1, 2017.

[80] Mak, K. F., C. Lee, J. Hone, J. Shan, and T. F. Heinz, "Atomically thin MoS 2: A new direct-gap semiconductor" *Physics Review Letters*, Vol. 105, p. 136805, 2010.

[81] del Corro, E., et al. "Atypical exciton–phonon interactions in WS2 and WSe2 monolayers revealed by resonance Raman spectroscopy" *Nano Letters*, Vol. 16, pp. 2363–2368, 2016.

[82] Zhang, Y. W., et al. " Negative to positive crossover of the magnetoresistance in layered WS2" *Applied Physics Letters*, Vol. 108, p. 153114, 2016.

[83] Liu, H., et al. "Line and point defects in MoSe2 bilayer studied by scanning tunneling microscopy and spectroscopy" *ACS Nano*, Vol. 9, No. 6, pp. 6619–6625, 2015.

[84] Bao, W. Z., X. H. Cai, D. Kim, K. Sridhara, and M. S. Fuhrer, "High mobility ambipolar MoS2 field-effect transistors: Substrate and dielectric effects" *Applied Physics Letters*, Vol. 102, No. 4, p. 042104, 2013.

[85] Dianat, B., N. Taghizade, and M. Afshar, "Charge effect on current-voltage characteristics of MoS_2 nanoribbon" *Journal of Physics: Conference Series*, Vol. 869, p. 012019, 2017.

[86] Chandra, Y., R. Chowdhury, S. Adhikari, and F. Scarpa, "Elastic instability of bilayer graphene using atomistic finite element" *Physica E: Low-Dimensional Systems and Nanostructures*, Vol. 44, No. 1, pp. 12–16, 2011.

[87] Bejan, Doina, and Ecaterina Cornelia Niculescu, "Intense laser effects on the optical properties of asymmetric GaAs double quantum dots under applied electric field" *The European Physical Journal B*, Vol. 89, No. 6, pp. 1–10, 2016.

[88] Jun Kan, Zhenyu Li, Jinlong Yang, and J. G. Hou, "Will zigzag graphene nanoribbon turn to half metal under electric filed" *Applied Physics Letters*, Vol. 91, No. 24, p. 243116, 2007.
[89] Yandong Ma, Ying Dai, Meng Guo, Chengwang Niu, Yingtao Zhu, and Baibiao Huang, "Evidence of the existence of magnetism in pristine VX2 monolayer and their strain induced tunable magnetic properties" *ACS Nano*, Vol. 6, No. 2, pp. 1695–1701, 2012.
[90] Xu, Yuehua, Jun Dai, and Xiao Cheng Zeng, "Electron-transport properties of few-layer black phosphorus" *The Journal of Physical Chemistry Letters*, Vol. 6, No. 11, pp. 1996–2002, 2015.
[91] Peng, Li, Kailun Yao, Sicong Zhu, Yun Ni, Fengxia Zu, Shuling Wang, Bin Guo, and Yong Tian, "Spin transport properties of partially edge-hydrogenated MoS2 nanoribbon heterostructure" *Journal of Applied Physics*, Vol. 115, No. 22, p. 223705, 2014.
[92] Rahman, Imam Abdul and Acep Purqon, "First principles study of molybdenum disulfide electronic structure" *Journal of Physics: Conference Series*, Vol. 877, No. 1, p. 012026, 2016.
[93] Ahmad, S. and S. Mukherjee, "A comparative study of electronic properties of bulk mos2 and its monolayer using DFT technique: Application of mechanical strain on mos2 monolayer" *Gaphene*, Vol. 3, pp. 52–59, 2014.
[94] Tong, Xin, Eric Ashalley, Feng Lin, Handong Li, and Zhiming M. Wang, "Advances in MoS_2-based field effect transistors" *Nano-Micro Letters*, Vol. 7, No. 3, pp. 203–218, 2015.
[95] Tsai, Meng-Lin, Sheng-Han Su, Jan-Kai Chang, Dung-Sheng Tsai, Chang-Hsiao Chen, Chih-I Wu, Lain-Jong Li, Lih-Juann Chen, and Jr-Hau He, "Monolayer Mos_2 heterojunction solar cells" *ACS Nano*, Vol. 8, No. 8, pp. 8317–8322, 2014.
[96] Chang, Yung-Huang, Wenjing Zhang, Yihan Zhu, Yu Han, Jiang Pu, Jan-Kai Chang, Wei-Ting Hsu, Jing-Kai Huang, Chang-Lung Hsu, Ming-Hui Chiu, Taishi Takenobu, Henan Li, Chih-I Wu, Wen-Hao Chang, Andrew Thye Shen Wee, and Lain-Jong Li, "Monolayer $MoSe_2$ grown by chemical vapor deposition for fast photodetection" *ACS Nano,* Vol. 8, No. 8, pp. 8582–8590, 2014.
[97] Chhowalla, M., H. S. Shin, G. Eda, L.-J. Li, K. P. Loh, and H. Zhang, "The chemistry of two-dimensional layered transition metal dichalcogenide nanosheets" *Nature Chemistry,* Vol. 5, pp. 263–275, 2013.
[98] Wang, Q. H., K. Kalantar-Zadeh, A. Kis, J. N. Coleman, and M. S. Strano, "Electronics and optoelectronics of two-dimensional transition metal dichalcogenides" *Nature Nanotechnology,* Vol. 7, pp. 699–712, 2012.
[99] Iqbal Bakti Utama, M., Xin Lu, Da Zhan, Son Tung Ha, Yanwen Yuan, Zexiang Shena and Qihua Xiong, "Etching-free patterning method for electrical characterizations of atomically thin CVD-grown MoSe2 film" *Nanoscale,* Vol. 6, No. 21, pp. 12376–12382, 2014.
[100] Wang, Q. H., K. Kalantar-Zadeh, A. Kis, J. N. Coleman, and M. S. Strano, "Electronics and Optoelectronics of two dimensional TMD" *Nature Nanotechnology*, Vol. 7, pp. 699–712, 2012.
[101] Butler, S. Z., S. M. Hollen, L. Cao, Y. Cui, J. A. Gupta, H. R. 90 Gutiérrez, T. F. Heinz, S. S. Hong, J. Huang, A. F. Ismach, E. Johnston-Halperin, M. Kuno, V. V. Plashnitsa, R. D. Robinson, R. S. Ruoff, S. Salahuddin, J. Shan, L. Shi, M. G. Spencer, M. Terrones, W. Windl, and J. E. Goldberger, "Progress, challenges, and opportunities in 2D materilas beyond graphene" *ACS Nano*, Vol. 7, pp. 2898–2926, 2013.
[102] Geim, A. K. and I. V. Grigorieva, "Van der Waals heterostructures" *Nature*, Vol. 499, pp. 419–425, 2013.
[103] Iqbal Bakti Utama, M., Xin Lu, Yanwen Yuan, and Qihua Xiong, "Detrimental influence of catalyst seeding on the device properties of CVD-grown 2D layered materials: A case study on $MoSe_2$" *Applied Physics Letters*, Vol. 105, p. 253102, 2014.

[104] Mak, K. F., C. Lee, J. Hone, J. Shan, and T. F. Heinz, "Atomically thin MoS 2: A new direct-gap semiconductor" *Physics Review Letters*, Vol. 105, p. 136805, 2010.

[105] Splendiani, A., L. Sun, Y. Zhang, T. Li, J. Kim, C. Y. Chim, G. Galli, and F. Wang, "Emerging photoluminescence in monolayer MoS2 "*Nano Letters,* Vol. 10, p. 1271, 2010.

[106] Zeng, H., J. Dai, W. Yao, D. Xiao, and X. Cui, "Valley polarization in MoS2 monolayers by optical pumping" *Nature Nanotechnology*, Vol. 7, p. 490, 2012.

[107] Mak, K. F., K. He, J. Shan, and T. F. Heinz, "Control of valley polarization in monolayer MoS 2 by optical helicity" *Nature Nanotechnology*, Vol. 7, p. 494, 2012.

[108] Chen, Jianyi, Xiaoxu Zhao, Sherman J. R. Tan, Hai Xu, Bo Wu, Bo Liu, Deyi Fu, Wei Fu, Dechao Geng, Yanpeng Liu, Wei Liu, Wei Tang, Linjun Li, Wu Zhou, Tze Chien Sum, and Kian Ping Loh, "Chemical vapor deposition of large-size monolayer MoSe2 crystals on molten glass" *Journal of the American Chemical Society,* Vol. 139, No. 3, pp. 1073–1076, 2017.

[109] Cheng, Ruiqing, Feng Wang, Lei Yin, Kai Xu, Tofik Ahmed Shifa, Yao Wen, Xueying Zhan, Jie Li, Chao Jiang, Zhenxing Wang, and Jun He, "Multifunctional tunneling devices based on graphene/h-BN/MoSe2 van der Waals heterostructures" *Applied Physics Letters*, Vol. 110, No. 17, p. 173507, 2017.

[110] Wang, Q. H, K. Kalantar-Zadeh, A. Kis, J. N. Coleman, and M. S. Strano, "Measurement of mobility in dual-gated MoS 2 transistors" *Nature Nanotechnology*, Vol. 7, p. 699, 2012.

[111] Zhu, Z. Y, Y. C. Cheng, and U. Schwingenschlögl, "Giant spin-orbit-induced spin splitting in two-dimensional transition-metal dichalcogenide semiconductors" *Physics Review B*, Vol. 84, p. 153402, 2011.

[112] Lezama, Ignacio Gutiérrez, Alberto Ubaldini, Maria Longobardi, Enrico Giannini, Christoph Renner, Alexey B. Kuzmenko, and Alberto F Morpurgo, "Surface transport and bandgap structure of exfoliated 2H-$MoTe_2$ crystals" *2D Materials,* Vol. 1, No. 2, p. 021002, 2014.

[113] Balaji, Yashwanth, Quentin Smets, Cesar J. Lockhart de la Rosa, Anh Khoa Augustin Lu, Daniele Chiappe, Tarun Agarwal, Dennis Lin, Cedric Huyghebaert, Iuliana Radu, Dan Mocuta, and Guido Groeseneken, "Tunneling transistors based on MoS_2/$MoTe_2$ Van der Waals heterostructures" *IEEE Journal of the Electron Devices Society,* Vol. 6, pp. 1048–1055, 2018.

[114] Chin, S. K., D. W. Seah, K. Lam, G. S. Samudra, and G. Liang, "Device physics and characteristics of graphene nanoribbon tunneling FETs" *IEEE Transactions on Electron Devices*, Vol. 57, pp. 3144–3152, 2010.

[115] Liu, Fei, Jian Wang, and Hong Guo, "Negative differential resistance in monolayer WTe_2 tunneling transistors" *Nanotechnology*, Vol. 26, No. 17, p. 175201, 2015.

[116] Wang, Kangpeng, Jun Wang, Jintai Fan, Mustafa Lotya, Arlene O'Neill, Daniel Fox, Yanyan Feng, "Ultrafast saturable absorption of two-dimensional MoS2 nanosheets" *ACS Nano,* Vol. 7, No. 10, pp. 9260–9267, 2013.

[117] Jariwala, D., V. K. Sangwan, L. J. Lauhon, T. J. Marks, and M. C. Hersam, "Emerging device applications for semiconducting two-dimensional transition metal dichalcogenides" *ACS Nano*, Vol. 8, pp. 1102, 2014.

[118] Mak, K. F., C. Lee, J. Hone, J. Shan, and T. F. Heinz, "Atomically thin MoS 2: A new direct-gap semiconductor" *Physics Review Lett*ers, Vol. 105, p. 136805, 2010.

[119] Radisavljevic, B., A. Radenovic, J. Brivio, V. Giacometti, and A. Kis, "Single layer MoS2 transistors" *Nature Nanotechnology*, Vol. 6, pp. 147–150, 2011.

[120] Yoon, Y., K. Ganapathi, and S. Salahuddin, "How good can monolayer MoS2 transistors be?" *Nano Letters,* Vol. 11, p. 3768, 2011.

[121] Radisavljevic, B., M. B. Whitwick, and A. Kis, "Integrated circuits and logic operations based on single-layer MoS2" *ACS Nano*, Vol. 5, p. 9934, 2011.

[122] Lam, Kai-Tak, Xi Cao, and Jing Guo, "Device performance of heterojunction tunneling field-effect transistors based on transition metal dichalcogenide monolayer" *IEEE Electron Device Letters*, Vol. 34, No. 10, Oct. 2013.

[123] Gopalakrishnan, K., P. B. Griffin, and J. D. Plummer, "I-MOS: A novel semiconductor device with a subthreshold slope lower than kT/q" *Proceeding of IEEE IEDM*, pp. 289–292, Dec. 2002.

[124] Mohata, D. K., R. Bijesh, Y. Zhu, et al. "Demonstration of improved heteroepitaxy, scaled gate stack and reduced interface states enabling heterojunction tunnel FETs with high drive current and high on-off ratio" *Proceedings pf the Symposium on VLSI Technology*, pp. 53–54, June 2012.

[125] Bhattacharjee, Shubhadeep, Kolla Lakshmi Ganapathi, Sangeneni Mohan, and Navakanta Bhat, "A sub-thermionic MoS2 FET with tunable transport" *Applied Physics Letters,* Vol. 111, No. 11, p. 163501, 2017.

[126] www.iisc.ac.in/breaking-the-boltzmann-limit-for-low-power-nano-transistor/.

[127] https://phys.org/news/2017-12 scientists-device-capable-dual-transistor.html.

[128] https://researchmatters.in/news/union-transistors-novel-upgrade-combines-two-transistors-one.

[129] https://thewire.in/219192/new-transistor-design-combines-best-high-low-performance-versions/.

[130] Allain, Adrien and Andras Kis, "Electron and hole mobilities in single-layer WSe2" *ACS Nano*, Vol. 8, No. 7, pp. 7180–7185, 2014.

[131] Lopez-Sanchez, O., D. Lembke, M. Kayci, A. Radenovic, and A. Kis, "Ultrasensitive photodetectors based on monolayer MoS_2" *Nature Nanotechnology*, Vol. 8, pp. 497–501, 2013.

[132] Lee, P. A. and T. V. Ramakrishnan, "Disordered electronic systems" *Reviews of Modern Physics*, Vol. 57, pp. 287–337, 1985.

[133] Liu, We, Jiahao Kang, Deblina Sarkar, Yasin Khatami, Debdeep Jena and Kaustav Banerjee, "Role of metal contacts in designing high-performance monolayer n-Type WSe2 field effect transistors" *Nano Letters*, Vol. 13, No. 5, pp. 1983–1990, 2013.

[134] Novoselov, K. S., A. K. Geim, S. V. Morozov, D. Jiang, Y. Zhang, S. V. Dubonos, I. V. Grigorieva, and A. A. Firsov, "Electric field effect in atomically thin carbon films" *Science*, Vol. 306, pp. 666–669, 2004.

[135] Schwierz, F., "Graphene transistors" *Nature Nanotechnology*, Vol. 5, pp. 487–496, 2010.

[136] Luisier, M., M. Lundstrom, D. A. Antoniadis, and J. Bokor, " Ultimate device scaling: Intrinsic performance comparisons of carbon-based, InGaAs, and Si field-effect transistors for 5 nm gate length" *International Electron Devices Meeting*, pp. 251–254, 2011.

[137] Sengupta, Amretashis, Anuja Chanana, and Santanu Mahapatra, "Phonon scattering limited performance of monolayer MoS_2 and WSe_2 n-MOSFET" *AIP Advances*, Vol. 5, No. 2, p. 027101, 2015.

[138] Liu, L., S. B. Kumar, Y. Ouyang, and J. Guo, "Performance limits of monolayer transition metal dichalcogenide transistors" *IEEE Transaction Electron Devices,* Vol. 58, No. 9, pp. 3042–3047, Sept. 2011.

[139] Nikonov, D., G. Bourianoff, P. Gargini, and H. Pal, "Scattering in NEGF: Made simple" [Online]. Available: https://nanohub.org/resources/7772.

[140] Roy, Tania, Mahmut Tosun, Xi Cao, Hui Fang, Der-Hsien Lien, Peida Zhao, Yu-Ze Chen, Yu-Lun Chueh, Jing Guo, and Ali Javey, "Dual-Gated MoS2/WSe2 van der Waals tunnel diodes and transistors" *ACS Nano,* Vol. 9, No. 2, pp. 2071–2079, 2015.

[141] Ionescu, A. M. and H. Riel, "Tunnel field-effect transistors as energy-efficient electronic switches" *Nature*, Vol. 479, pp. 329–337, 2011.

[142] Choi, W. Y., B.-G. Park, J. D. Lee, and T.-J. K. Liu, "Tunneling field-effect transistors (TFETs) with subthreshold swing (SS) less than 60 mV/dec" *IEEE Electron Device Letters*, Vol. 28, pp. 743–745, 2007.

[143] Bhuwalka, K. K., J. Schulze, and T. Eisele, "Performance enhancement of vertical tunnel field-effect transistor with SiGe in the delta P(þ) layer" *Japanese Journal of Applied Physics*, Vol. 43, pp. 4073–4078, 2004.

[144] Tosun, Mahmut, Steven Chuang, Hui Fang, Angada B. Sachid, Mark Hettick, Yongjing Lin, Yuping Zeng and Ali Javey, "High-gain inverters based on WSe_2 complementary field-effect transistors" *ACS Nano,* Vol. 8, No. 5, pp. 4948–4953, 2014.

[145] Yang-Kyu, C., K. Asano, N. Lindert, V. Subramanian, K. Tsu-Jae, J. Bokor, and H. Chenming, "Ultra-thin body SOI MOSFET for deep-sub-tenth micron era" *Technical Digest — International Electron Devices Meeting*, p. 919921, 1999.

[146] Song, Seung Min, Jae Hoon Bong, Wansik Hwang and Byung Jin Cho, "Improved drain current saturation and voltage gain in graphene-on-silicon field effect transistors" *Scientific Reports*, Vol. 6, p. 25392, 2016.

[147] Das, Saptarshi and Joerg Appenzeller, "WSe_2 field effect transistors with enhanced ambipolar characteristics" *Applied Physics Letters*, Vol. 103, No. 10, p. 103501, 2013.

[148] Sze, S. M., *Semiconductor Devices*, 2nd ed. (John Wiley & Sons, Hoboken, NJ), 2003.

[149] Chen, Z., J. Appenzeller, J. Knoch, Y. Lin, and P. Avouris, "The role of metal–nanotube contact in the performance of carbon nanotube field-effect transistors" *Nano Letters,* Vol. 5, pp. 1497–1502, 2005.

[150] Kedzierski, J., P. Xuan, V. Subramanian, J. Bokor, T.-J. King, C. Hu, and E. Anderson, "A 20 nm gate-length ultra-thin body p-MOSFET with silicide source/drain" *Superlattices Micro-structures*, Vol. 28, pp. 445–452, 2000.

[151] Huang, Jing-Kai, Jiang Pu, Chang-Lung Hsu, Ming-Hui Chiu, Zhen-Yu Juang, Yung-Huang Chang, Wen-Hao Chang, Yoshihiro Iwasa, Taishi Takenobu, and L. Lain-Jong, "Large-area synthesis of highly crystalline WSe2 monolayers and device applications" *ACS Nano*, Vol. 8, No. 1, pp. 923–930, 2014.

[152] Chhowalla, M., H. S. Shin, G. Eda, L.-J. Li, K. P. Loh, and H. LohZhang, "The chemistry of two-dimensional layered transition metal dichalcogenide nanosheets" *Nature Chemistry*, Vol. 5, pp. 263–275, 2013.

[153] Ataca, C., H. Sahin, and S. Ciraci, "Stable, single-LayerMX_2 transition-metal oxides and dichalcogenides in a honeycomb-like structure" *Journal of Physical Chemistry C*, Vol. 116, pp. 8983–8999, 2012.

[154] Chuang, Hsun-Jen, Xuebin Tan, Nirmal Jeevi Ghimire, Meeghage Madusanka Perera, Bhim Chamlagain, Mark Ming-Cheng Cheng, Jiaqiang Yan, David Mandrus, David Tomanek, and Zhixian Zhou, "High mobility WSe2 p- and n-type field-effect transistors contacted by highly doped graphene for low-resistance contacts" *Nano Letters*, Vol. 14, No. 6, pp. 3594–3601, 2014.

[155] Fuhrer, M. and J. Hone, "Measurement of mobility in dual gated MoS2 transistor" *Nature Nanotechnology*, Vol. 8, pp. 146–147, 2013.

[156] Liu, H., A. T. Neal, and P. D. Ye, "Channel length scaling of MoS2 MOSFET" *ACS Nano,* Vol. 6, No. 10, pp. 8563–8569, 2012.

[157] Zhang, Lu, Dawei He, Jiaqi He, Yang Fu, Ang Bian, Xiuxiu Han, Shuangyan Liu, Yongsheng Wang, and Hui Zhao, "Ultrafast charge transfer in a type II MoS2-ReSe2 van der waals heterostructure" *Optics Express*, Vol. 27, No. 13, pp. 17851–17858, 2019.

[158] Fang, Hui, Steven Chuang, Ting Chia Chang, Kuniharu Takei, Toshitake Takahashi, and Ali Javey, "High performance single layered WSe2 *p*-FETs with chemically doped contacts" *NANO Letters*, Vol. 12, No. 7, pp. 3788–3792, 2012.

[159] Taur, Y. "CMOS design near the limit of scaling" *IBM Journal of Review & Development*, Vol. 46, pp. 213–222, 2002.

[160] Liu, Keng-Ming, and Yung-Yu Hsieh. "Investigation of the dimension effects of 30-nm below multiple-gate SOI MOSFETs by TCAD simulation" *The 4th IEEE International NanoElectronics Conference*, pp. 1–2, 2011.

[161] Hsun-Jen Chuang, Bhim Chamlagain, Michael Koehler, Meeghage Madusanka Perera, Jiaqiang Yan, David Mandrus, David Tomanek, and Zhixian Zhou, "Low-resistance 2D/2D ohmic contacts: A universal approach to high-performance WSe_2, MoS_2, and $MoSe_2$ transistors" *NANO Letters,* Vol. 16, No. 3, pp. 1896–1902, 2016.

[162] Lee, G.-H., Y.-J. Yu, X. Cui, N. Petrone, C.-H. Lee, M. S. Choi, D.-Y. Lee, C. Lee, W. J. Yoo, K. Watanabe, T. Taniguchi, C. Nuckolls, P. Kim, and J. Hone, "Flexible and transport MoS2 FET on hexagonal BN-graphene heterostructures" *ACS Nano*, Vol. 7, No. 9, pp. 7931–7936, 2013.

[163] Yoon, J., W. Park, G.-Y. Bae, Y. Kim, H. S. Jang, Y. Hyun, S. K. Lim, Y. H. Kahng, W.-K. Hong, B. H. Lee and H. C. Ko, "MoS2 field-effect transistors with graphene/metal heterocontacts" *Small*, Vol. 9, No. 19, pp. 3295–3300, 2013.

[164] Yu, Lili, Ahmad Zubair, Elton J. G. Santos, Xu Zhang, Yuxuan Lin, Yuhao Zhang and Tomaś Palacios, "High-performance WSe_2 complementary metal oxide semiconductor technology and integrated circuits" *Nano Letters*, Vol. 15, No. 8, pp. 4928–4934, 2015.

[165] Fiori, G., et al. "Electronics based on two-dimensional materials" *Nature Nanotechnology*, Vol. 9, pp. 768–779, 2014.

[166] Hsu, A., et al. "Large-area 2-D electronics: Materials, technology and devices" *Proceedings of the IEEE*, Vol. 101, pp. 1638–1652, 2013.

[167] Amani, Matin, Robert A. Burke, Robert M. Proie, and Madan Dubey, "Flexible integrated circuits and multifunctional electronics based on single atomic layers of MoS_2 and graphene" *Nanotechnology,* Vol. 26, No. 11, p. 115202, 2015.

[168] Cao, Q., H. S. Kim, N. Pimparkar, J. P. Kulkarni, C. Wang, M. Shim, K. Roy, M. A. Alam, and J. A. Rogers, "Medium-scale carbon nanotube thin-film integrated circuits on flexible plastic substrates" *Nature,* Vol. 454, pp. 495–500, 2008.

[169] Zhai, Y., M. Leo, R. Rao, D. Xu, and S. K. Banerjee, "High-performance thin-film transistors exfoliated from bulk wafer" *Nano Letters,* Vol. 12, pp. 5609–5615, 2012.

[170] Radisavljevic, B. and A. Kis, "Mobility engineering and a metal-insulator transition in monolayer MoS2" *Nature Material,* Vol. 12, pp. 815–820, 2013.

[171] Chen, Fan, Hesameddin Ilatikhameneh, Yaohua Tan, Daniel Valencia, Gerhard Klimeck and Rajib Rahman, "Transport in vertically stacked hetero-structures from 2D materials" *Journal of Physics: Conference Series,* Vol. 864, No. 1, p. 012053, 2017.

[172] Geim, A. and I. Grigorieva, "2D van der Waals materials" *Nature,* Vol. 499, pp. 419–425, 2013, ISSN 0028-0836.

[173] Szabo, A., S. J. Koester, and M. Luisier, "Transport in vertically stacked heterostructures from 2D materials" *Device Research Conference (DRC)*, 2014 72nd Annual (IEEE), pp. 19–20, ISBN 1479954055.

[174] Eunseon Yu, Seongjae Cho, Kaushik Roy, and Byung Good Park, "A quantum-well charge-trap synaptic transistor with highly linear weight tunability" *Journal of the Electron Device Society*, Vol. 8, pp. 834–840, 2020.

[175] Abderrahmane, A., P. J. Ko, T. V. Thu, S. Ishizawa, T. Takamura, and A. Sandhu, "High photosensitivity few-layered $MoSe_2$ back-gated field-effect phototransistors" *Nanotechnology*, Vol. 26, No. 36, p. 365202, 2014.

[176] Lopez-Sanchez, O., D. Lembke, M. Kayci, A. Radenovic, and A. Kis, "Ultrasensitive photodetectors based on monolayer MoS2" *Nature Nanotechnology,* Vol. 8, p. 497, 2013.

[177] Yang, S., Y. Li, X. Wang, N. Huo, J. B. Xia, S. S. Li, and J. Li, "High performance few-layer GaS photodetector and its unique photo-response in different gas environments" *Nanoscale*, Vol. 6, pp. 2582–2587, 2014.

[178] Hu, P., Z. Wen, L. Wang, P. Tan, and K. Xiao, "Synthesis of few-layer GaSe nanosheets for high performance photodetector" *ACS Nano,* Vol. 6, pp. 5988–5994, 2012.

[179] Novoselov, K. S., D. Jiang, F. Schedin, T. J. Booth, V. V. Khotkevich, S. V. Morozov, A. K. Geim, "2D atomic crystals" *Proceedings of the National Academy of Sciences of the USA,* No. 30, pp. 10451–10453, 2005.

[180] Mak, K. F., C. Lee, J. Hone, J. Shan, T. F. Heinz, "A new direct-gap semiconductor" *Physics Review Letters,* Vol. 105, No. 13, p. 136805, 2010.

[181] Kuc, A., N. Zibouche, and T. Heine, "Influence of quantum confinement on the electronic structure of transition metal sulfideTS2" *Physical Review B*, Vol. 83, No. 24, p. 245213, 2011.

[182] Liu, Y.-H., S. H. Porter, and J. E. J. Goldberger, " Dimensional reduction of a layered metal chalcogenide into a 1D near-IR direct band gap semiconductor" *American Chemical Society,* Vol. 134, No. 11, pp. 5044–5047, 2012.

[183] Fiori, G., F. Bonaccorso, G. Iannaccone, T. Palacios, D. Neumaier, A. Seabaugh, S. K. Banerjee, and L. Colombo, "Electronics based on 2D materials" *Nature Nanotechnology,* Vol. 9, pp. 768–779, 2014.

[184] Fang, H., S. Chuang, T. C. Chang, K. Takei, T. Takahashi, and A. Javey, "High performance SL WSe2 p-FETSs with chemically doped contacts" *Nano Letters,* Vol. 12, No. 7, pp. 3788–3792, 2012.

[185] Kong, D., W. Dang, J. J. Cha, H. Li, S. Meister, H. Peng, Z. Liu, and Y. Cui, "Few layer nanoplates of BiSe3 and BiTe3 with high tunable chemical potential" *Nano Letters*, Vol. 10, No. 6, pp. 2245–2250, 2010.

[186] Kong, D., J. C. Randel, H. Peng, J. J. Cha, S. Meister, K. Lai, Y. Chen, Z.-X. Shen, H. C. Manoharan, Y. Cui, "Topological insulator nanowires and nanoribbon" *Nano Letters,* Vol. 10, No. 1, pp. 329–333, 2009.

[187] Tenne, R., L. Margulis, M. Genut, and G. Hodes, "PLD grown polycrystalline tungsten disulphide films" *Nature,* Vol. 360, No. 6403, pp. 444–446, 1992.

[188] Feldman, Y., E. Wasserman, D. J. Srolovitz, and R. Tenne, "Highrate, gas-phase growth of MoS2 nested inorganic fullerenes and nanotubes" *Science*, Vol. 267, No. 5195, pp. 222–225, 1995.

[189] Tenne, R., "Inorganic nanotubes and fullerene-like nanoparticles" *Nature Nanotechnology*, Vol. 1, No. 2, pp. 103–111, 2006.

[190] Yin, Z., H. Li, H. Li, L. Jiang, Y. Shi, Y. Sun, G. Lu, Q. Zhang, X. Chen, and H. Zhang, "Single-layer MoS2 phototransistors" *ACS Nano,* Vol. 6, pp. 74–80, 2012.

[191] Tongay, S., J. Suh, C. Ataca, W. Fan, A. Luce, J. S. Kang, J. Liu, C. Ko, R. Raghunathanan, J. Zhou, et al. "Defects activated photoluminescence in two-dimensional semiconductors: Interplay between bound, charged, and free excitons" *Scientific Reports*, Vol. 3, p. 2657, 2013.

[192] Zhang, A., S. You, C. Soci, Y. Liu, D. Wang, and Y.-H. Lo, "Silicon nanowire detectors showing phototransistive gain" *Applied Physics Letters,* Vol. 93, p. 121110, 2008.

[193] Lany, S. and A. Zunger, "Anion vacancies as a source of persistent photoconductivity in II_VI and chalcopyrite semiconductors" *Physical Review B*, Vol. 72, p. 035215, 2005.

[194] Ramesh, A., S-Y. Park, and P. R. Berger, "90nm 32x322 bit tunneling SRAM memory array with 0.5ns write access time, 1 ns read access i time and 0.5 Volt operation" *IEEE Transactions on Circuits and Systems for Video Technology,* Vol. 58, pp. 2432–2445, 2011.

[195] Chynoweth, A. G., W. L. Feldmann, and R. A. Logan, "Excess tunnel current in silicon Esaki junctions" *Physical Review,* Vol. 121, No. 3, p. 684, 1961.

[196] Chang, L. L., L. Esaki, and R. Tsu, "Resonant tunneling in semiconductor double barriers" *Applied Physics Letters,* Vol. 24, pp. 593–595, 1974.

[197] Sokolv, V. N., K. W. Kim, V. A. Kochelap, and D. L. Woolard, "Terahertz generation in submicron GaN diodes within the limited space charge accumulation regime" *Journal of Applied Physics*, Vol. 98, p. 064507, 2005.
[198] Chalk, Jeff D., "Tunneling through a truncated harmonic oscillator potential barrier" *American Journal of Physics,* Vol. 58, No. 2, pp. 147–151, 1990.
[199] Lyo, In-Whan, and Phaedon Avouris, "Negative differential resistance on the atomic scale: implications for atomic scale devices" *Science*, Vol. 245, pp. 1369–1371, 1989.
[200] Chang, J., L. F. Register, and S. K. Banerjee, "Atomistic full-band simulations of monolayer MoS2 transistors" *Applied Physics Letters,* Vol. 103, p. 223509, 2013.

Index